Invitation to Protein Sequence Analysis through Probability and Information

Invitation to Protein Sequence Analysis through Probability and Information

Daniel J. Graham

Department of Chemistry and Biochemistry
Loyola University Chicago

CRC Press
Taylor & Francis Group
Boca Raton London New York

CRC Press is an imprint of the
Taylor & Francis Group, an **informa** business

CRC Press
Taylor & Francis Group
6000 Broken Sound Parkway NW, Suite 300
Boca Raton, FL 33487-2742

© 2019 by Taylor & Francis Group, LLC
CRC Press is an imprint of Taylor & Francis Group, an Informa business

No claim to original U.S. Government works

Printed on acid-free paper

International Standard Book Number-13: 978-0-367-13452-5 (Hardback)

Library of Congress Cataloging-in-Publication Data

Names: Graham, Daniel J. (Daniel Joseph), 1954- author.
Title: Invitation to protein sequence analysis through probability and
information / Daniel J. Graham.
Description: Boca Raton : Taylor & Francis, 2019. | Includes bibliographical
references and index.
Identifiers: LCCN 2018046939 | ISBN 9780367134525 (hbk. : alk. paper)
Subjects: | MESH: Sequence Analysis, Protein—methods | Protein Conformation |
Models, Statistical | Computational Biology | Probability
Classification: LCC QP552.M64 | NLM QU 550.5.S4 | DDC 572/.633—dc23
LC record available at https://lccn.loc.gov/2018046939

Visit the Taylor & Francis Web site at
http://www.taylorandfrancis.com

and the CRC Press Web site at
http://www.crcpress.com

eResource material is available for this title at https://www.crcpress.com/9780367134525.

To Teresa and Jamey

Contents

Preface

In the mid-1980s, I attended an organic chemistry seminar given by a visiting postdoctoral researcher. He worked in a group famous for breakthroughs in natural products synthesis. The audience included professors who had deep-rooted knowledge in the field. I was not one of those professors! I did, however, possess sufficient background to follow the chain of ideas leading to the target molecules. I further appreciated the enormous labor required for the chemistry: work-ups, separations, and structure proofs.

My deep-knowledge colleagues took home a lot more. They marveled—audibly in places—at the ingenuity shown by the researcher in choosing reagents, protecting groups, and reaction steps. They saw not just perseverance at a lab bench, but also elegance as seemingly insurmountable problems were solved. As for the molecules of the hour, my colleagues saw multiple instances of structural beauty, compliments of nature, and the researcher's synthetic prowess. All this appreciation obtained by internal processing skills: my colleagues' ability to comprehend information in the diagrammatic representations of organic compounds.

I am much older now and work at a different university. The seminars feature biochemistry talks on a regular basis. I have thus viewed many presentations on proteins: their catalytic and signaling functions, connections to diseases, and potential as drug targets. The molecules are as organic as any natural products. What I have yet to see, however, is any sharp focus on the amino acid sequences that compose the proteins. To be sure, sequences and their analyses are hot topics in bioinformatics and computer science departments. And sequences lie at the center of research on the "protein folding problem": given a primary structure, what is the tertiary? Ditto for the "function problem": given a primary structure, what is the biochemical function? In the majority of biochemistry talks, however, sequences are presented in passing with few if any expectations of the audience. There is little remarked upon, save for active or mutant sites, or local comparisons of sequence A with B.

There are good reasons for this. For one, the sequences are as remarkable for their inscrutability as for biochemical significance. They certainly do not lend themselves to casual inspection—there is too much to register from a PowerPoint slide and our internal processing skills stop miles short. Another reason is that sequences are easily pushed-aside by the color graphics of folded systems—movies at some talks. These depict the chemically active configurations and are more welcoming to the intuition than disordered letter strings. The visuals engage an audience in discussions of catalytic mechanisms, substrate docking, and more. As for what underpins the graphics, sequences are ignored for the most part or invite similes, for example, "The amino acid side chains can be thought of as words in a long sentence" [1]. Better yet, I have heard allusions at seminars to protein universes and galaxies. The terms are reinforced in the research literature [2].

I have nonetheless found interesting facets in the disordered and pushed-aside, hence the motivation for this monograph. In no way does it crack the inscrutability of amino acid sequences—they read like encrypted text (regardless of alphabet) and present chemical interpretation problems unlike other organic compounds, save for polynucleotides. Rather it is my attempt to study what researchers have pursued over decades, and to explore additional properties with the help of students. The internal processing skills of my yesteryear organic colleagues do not do much for letter string analysis. But the tools of probability and information do, shining light on the lower levels of protein structure. This may not cultivate new lingering on "biological sentences" at weekly seminars. And it will not enable anyone to view an unfamiliar sequence and say "Aha, this describes an oxidoreductase!" But the exposition should enhance the reader's knowledge of how an important class of molecules is designed and put to task in natural systems, and how we can approach class members in hands-on ways. The intended audience is students wishing to learn more about the remarkable information correspondences and probability structures of proteins. Correspondences are pervasive in biochemistry and bioinformatics: proteins share homologies, folding patterns, and mechanisms. Probability structures are just as paramount: folded state graphics reflect Angstrom-scale maps of electron density. In arenas far removed, correspondences and probability structures are centerpieces in the physical chemistry of solutions and phase transitions. They fall under the umbrella of chemical thermodynamics whence applications to proteins can complement the biochemistry and bioinformatics curriculum. In exploring applications, students can acquire firmer footing in mathematical probability and information directed at complex systems. The author is confident that teachers, physical chemists, and biochemists, can assist happily in the journey.

The individuals to acknowledge are numerous. They include the author's research students over the past several years: Eric Andrews, Alexis Antonopoulos, Alinaa Alsaud, Julius Aquas, Jessica Bae, Sam Barlow, Madyson Bondi, James Casey, Bea Chelsea, Diego Cucalon, Jessica Greminger, Shelly Grzetic, Jordan Hauck, Yasir Izhar, Moomal Khan, Marcus Klein, David Korsak, Andrew Oleksijew, Lindsay Oleksijew, Kendrick Reme, Brian Robinson, Maricar Virina, Omar Zahra, and John Zumpf. Thanks also go to classroom students who endured rough drafts of the text with additions and corrections on the fly.

I also cannot appreciate enough my chemistry and biochemistry colleagues: Miguel Ballicora, Jan Florian, Sandra Helquist, Dali Liu, Donald May, Duarte Mota de Freitas, Ken Olsen, Agnes Pecak, and Polina Pine. They have educated me on multiple fronts of protein biochemistry over many years. Thanks are also extended to my yesteryear organic colleagues: Bob Moore, Gabor Fodor, Denis MacDowell, and Kung Wang. They taught me ways of approaching new and fascinating molecular structures. Thanks also to Denise Hall, Carol Grimm, and Sandy Orozco for office support and a welcoming work environment.

I must also acknowledge the writings of Ben-Naim, Bennett, Chaitin, Corey, Dewey, Eigen, Halmos, Jaynes, Karlin, Kaufman, Morawitz, Pande, Randić, and White. Their contributions have enriched and altered my thinking the past several years. On multiple fronts, they have addressed the structural complexity of systems: proteins, organic compounds, and more. These subjects are as pleasing to the intuition as the color graphics at weekly seminars.

Daniel J. Graham
Chicago, Illinois
September, 2018

REFERENCES

1. Appling, D. R., Anthony-Cahill, S. J., Mathews, C. K. 2016. *Biochemistry, Concepts and Connections*, Pearson Education, Hoboken, NJ, p. 125.
2. Holm, L., Sander, C. 1996. Mapping the protein universe, *Science* 273, 595–603.

Introduction

A general chemistry text from the 1970s by George Pimentel and Richard Spratley cites 600,000 compounds having been discovered in nature [1]. At the time, about 900,000 additional molecules had been synthesized in laboratories. Natural and synthetic compounds were being discovered at a rate of 300 per day.

Fast forward to present. A BLAST analysis of an amino acid sequence references ninety million formulae of a *single type* of molecule via either the Swiss-Prot or NCBI websites [2,3]. Clearly times have changed, and not by a little. The millions of primary structures attest to the significance of proteins in the molecular sciences and the advances in genomic and database technology. A 1960s chemistry text by Linus Pauling may have been prophetic: "Proteins may well be considered the most important substances in plants and animals" [4].

The student of proteins has enormous resources at his or her disposal. Online software and databases enable detailed sequence alignments. Evolutionary relationships are established by parsimony analysis and phylogenic tree constructions. Image processing tools enable the quantitative comparison of three-dimensional (3D) configurations. Molecular dynamics and homology software plumb the depths of secondary through quaternary structures. These technologies are staples in biochemistry, bioinformatics, and related education.

Yet there could be—and maybe should be—unease in some circles. In conjunction with aqueous solution environments, the chemical nature of proteins stems from amino acid sequences. Yet the representations for the sequences make little, if any, sense to the chemistry student. Upon completing an introductory course in organic or biochemistry, he or she can detail the formula copied from a textbook or database. For example, KVFE... abbreviates the first few units of the enzymatic protein lysozyme C [5]. The letter string can be expanded as:

Lysine   Valine   Phenylalanine   Glutamic Acid....

However, interpreting the representation beyond the superficial without sophisticated computation is another story. The conventional notions for approaching molecular structure—valence bond orbitals, VESPR theory, resonance structures—fall short when it comes to protein sequences. They find purchase only in a local sense when discussing specific sites of systems, whether folded or denatured.

Matters are all the more bewildering for sets of sequences: viruses typically encode handfuls, for example,

MERIKELRNLMSQSRTREILTKTTVDHMAIIKKYTSGRQEKNPA...

MDVNPTLLFLKVPAQNAISTTFPYTGDPPYSHGTGTGYTMDTV...

MGQEQDTPWILSTGHISTQKRQDGQQTPKLEHRNSTRLMGHC...

while organisms encode thousands [6]. It is easy for the student to think of questions. How is a particular sequence interesting? Where and how should we look at a sequence? Which sequence elements are critical versus incidental? Given the identity of site *i*, what restrictions are imposed at *j*? Given sequence A in a set, what constraints are imposed on B? Are any sequences superfluous to the biological system?

For so many questions, the student looks online to databases and software. Unfortunately, his or her instincts may be devalued in the process. For the results of analyses—alignments, gaps, trees, and folded structure models—obtain rapidly and not with a little opacity. This makes it challenging to spot errors, practice skepticism, and grow a genuine appreciation of sequence and system architecture.

This monograph was motivated by interest in protein sequences (primary structures), both individually and in sets. No one will be reading letter strings KVFER… like a novel anytime soon. However, there are new perspectives to introduce to students which can complement the presentations in biochemistry and bioinformatics textbooks. It is the author's experience that numerous texts and online sources overwhelm the individual with a deluge of information upfront. This leads to confusion in the short term and a tendency to look past primary structures in the long. Hopefully the discussions in the chapters can ameliorate matters to some degree. Among other things, they point to salient characteristics of natural sequences and sets. The characteristics need not be listed in cut and dry terms. Rather they welcome exploration and discovery with the help of probability and information tools. To lay the groundwork, the student can look to (or has already spent considerable time with) books on protein biochemistry and molecular biology [7–10]. For our part, we review only briefly the fundamentals of protein structure and function, and attend more to the complexities. Appreciation of the complexities is critical because it places in perspective the challenges of interpreting sequence information absent sophisticated computation and modeling.

The discussion and exercises of the book are centered on probability and information, just as in present day bioinformatics. The road traveled is different in a few key respects, however. First, we step backward and give renewed attention to "base level" structures. Biochemistry and bioinformatics discussions are initiated typically with primary structures. For lysozyme C in its entirety (the human-encoded version), this is:

KVFERCELARTLKRLGMDGYRGISLANWMCLAKWESGYNTRATNYNAGDRSTDY
GIFQINSRYWCNDGKTPGAVNACHLSCSALLQDNIADAVACAKRVVRDPQGI
RAWVAWRNRCQNRDVRQYVQGCGV

A string such as above is compared exhaustively to database entries, establishing alignments, gaps, and family relations. In contrast, we start conversations at the composition $\leftrightarrow$ base level: the information apart from order in the sequence. For lysozyme C, this is

$$A_{14}V_9L_8I_5P_2F_2W_5M_2G_{11}S_6T_5C_8Y_6N_{10}Q_6D_8E_3K_5R_{14}H_1$$

The primary structure is but one polypeptide isomer consonant with the base, just as benzene is one stable arrangement of C_6H_6. There are multiple structure levels between the base and primary, for example, the set of dimers (di-peptides) KV, VF, FE, … presented by a molecule. We refer to these levels as "base-plus" (base+) which receive systematic attention as well. This is not new-found territory given the ocean of literature, although some technical details are. Composition properties have been central to protein research since the epoch days of Fischer and Sanger [4,11].

Second, a base or above-level structure is explored to construct several types of sets, measures, and vectors. These are objects for comparing the information in complex systems, and the means of drawing out the salient properties. Vectors are nothing more than directed lines in defined coordinate systems not limited to three dimensions. They are abstractions in search of applications such as in mechanics, electromagnetism, and spectroscopy. It should not surprise that vectors find utility in protein sequence analysis. Multiple applications have been advanced by probability researchers over the years. Vectors directed at amino acid strings, bases, and sets should appeal to students (and teachers) because, among other things, they are computationally transparent, requiring only a calculator or spreadsheet program.

Lastly, a protein sequence refers to a real chemical entity. Just as importantly, it charts a synthetic pathway, for the molecules are assembled step-wise at the ribosome sites of cells. Accordingly, there is a pathway structure allied with every sequence which well accommodates analysis by information vectors. Pathways and vectors thus cast additional light on the design of molecules.

The monograph is thermodynamics-inspired. This is in spite of making bare mention of quantities like heat capacity, enthalpy, and temperature. Rather proteins are treated as thermodynamic systems in the most general sense. Each one presents a collection of elements from which multiple sets can be formed and measures assigned. Further, they manifest correspondences that are surprising in places, and yet flow naturally from a few principles. The correspondences help us to make better sense of the representations for molecules and systems at the lower information levels. In classical thermodynamics, the familiar measures include energy, heat capacity, and volume while the tools of the trade are real-variable functions and equations of state. Standard practice directs parametric models such as the van der Waals to draw out the principles of state. This monograph aims at analogous approaches to protein sequences. Only here, the variables—letter strings KVFER... GCGV—are digital. Continuous functions do not apply, but the notions of countable sets, mathematical probability, and information do.

The monograph was adapted from lectures and exercises for a one-semester special topics course on thermodynamics with applications to protein structure analysis. The clientele was advanced undergraduates and beginning graduate students. Only rudimentary knowledge of thermodynamics, organic chemistry, and analytical geometry was expected as a course prerequisite. The same statement holds for readers of this monograph. The writing is (hopefully!) informal and the graphics are generous in number while minimalist in construction. Amino acid sequences have also been presented in full—some repeatedly—using normal-size typeface with accession details found in the appendices. This encourages engaging with primary structures on the printed page or computer screen instead of blowing past them. Protein research of modern day is motivated in large part by biomedical applications. This book does not venture into such vast terrain, although the student may well be motivated by such applications. Perhaps he or she will find the tools and ideas useful down the road.

The chapters of the book are organized as follows.

Chapter One presents the basics of protein composition and assembly, where the molecules come from and what they do. This is review material for readers having completed semesters in organic chemistry and biology. Greater attention is called to the chemical, structural, and biological complexity presented by the molecules. They are atypical systems for reasons which make the chemical inscrutability of sequences not at all surprising. It is all part of the territory.

Chapter Two offers a précis on information and probability. Special attention is given to vector constructions and information applications to protein sequences. The examples make appeal to thermodynamic systems, specifically the liquid and vapor phases of solutions at equilibrium. Their treatment is a short step from applications to protein sequences as discussed in the last section.

Chapter Three focuses on protein structure analysis at the base or composition level. It illustrates signature properties which are less than evident from casual inspection of sequences, secondary structures, and folded state graphics. The properties rise to the surface by experiments involving symbol counting and grouping, and the application of probability ideas. The illustrations center on archetypal molecules like lysozyme, myoglobin, and ribonuclease A and connect with fundamentals of thermodynamics. Included is attention to sets of archetypes to appreciate the variations among base structures. The author has used programming experience to assist with the experiments and illustrations. For the student, the majority of exercises can be completed with the help of spreadsheet and word processing software. Just as helpful, there are numerous internet resources a click away that can detail the compositions of amino acid sequences.

Chapter Four continues the analysis of base structures by focusing on the number of amino acids N that make a protein. As with composition, this property typically receives glancing attention during sequence and 3D configuration inspection. We begin by examining the compressibility of sequences. The discussion segues to the N-distributions of natural protein sets, aka proteomes. Probability and information provide a foundation for discriminating the various distributions and logging correspondences in the process. Some issues of information asymmetry close the chapter.

Chapter Five keeps the spotlight on proteomes with the focus on composition. A number of correspondence properties are illustrated for diverse systems. These reflect the information economy of natural sets. Also addressed are the thermodynamic underpinnings of compositions and correlations with the genetic code. The last part of the chapter is devoted to the topological characteristics of protein sets.

Chapter Six attends to protein structure *between* the base and primary levels. These are referred to as base-plus (base+) levels. The properties for archetypal molecules are illustrated, following symbol counting and grouping and application of probability and information. The base+ properties connect with several facets of protein design and selection, including mass distribution and sequence aperiodicity. The exercises can (yet again) be completed using word processors and spreadsheets. These place a student closer to the internal organization of sequences.

Chapter Seven attends to the auxiliary sequences in proteins. These derive from the disulfide bridges that allow additional covalent paths from the N- to C-terminal groups. The chapter illustrates a diagrammatic and vector approach to path analysis. The angles of the vectors with respect to one another are encapsulated in pair distribution functions. A number of correspondences between systems emerge along the way.

Chapter Eight revisits the core ideas and directs them to sequence writing. The result is a guide to drafting, critiquing, and refining protein sequences. There is endless scope here and room to grow. The designer sequences reflect principles discussed in previous chapters.

Chapter Nine closes the book with a look to horizons for protein structure analysis using probability and information tools.

Several appendices are included for background and context. Appendix One presents essential (and probably familiar) chemical properties of the amino acids. Appendix Two offers a thermodynamic overview of steady state systems—the sources and operating environments of proteins. Appendix Three reviews the Stirling formula for large factorials, which assists probability computations in the text. Appendix Four summarizes the binomial and multinomial distributions which enter several calculations. Appendix Five presents essentials of the genetic code—the underpinnings of proteins. Appendix Six lists the accession details for relevant sequences and sets in the text. Lastly, Appendix Seven lists answers and hints to selected exercises.

Modest training enables a chemistry student to discern and write structures for numerous classes of molecules: aromatics, alkaloids, steroids, etc. Related experience confers skills for discriminating solution types: perfect versus ideal versus non-ideal. The student should want the same capabilities for amino acid sequences and systems. Developing discrimination and writing skills should be a principal objective for readers of this book. The means is not by ponderous theory, memorization, or computer artillery. Rather it is by exploration of much-researched systems and applied probability and information.

The illustrations throughout the book feature archetypes like lysozyme, ribonuclease, and myoglobin. This is more or less the approach taken in organic chemistry textbooks for the various classes of molecules: aromatic, ketones, aldehydes, etc.; ditto for biochemistry books: carbohydrates, fatty acids, etc. [12,13]. Archetypes bear a double-edge, however. They start conversations and help get the major points across. At the same time, they risk the expositions being anecdotal. The risk may be especially acute with proteins given their vast universe, to borrow a seminar metaphor. Every star seems to present a long story unto itself. Hopefully the reader can learn principles of sequence and systems analysis by looking closely at handfuls of the more famous stars. The mathematical tools of the book are accommodating of all proteins in the universe. The reader is encouraged to explore far and wide.

NOTES, SOURCES, AND FURTHER READING

It is impossible to do justice to the protein literature. Accordingly, we make no pretense as to the exhaustiveness of references here and in the remaining chapters. The same statement holds regarding the choice of archetypal systems—every protein seems an archetype of some form. That being said, lysozyme occupies a prominent place in biochemistry and structural biology. It catalyzes the degradation of bacterial cell walls and is a vital component of immune systems. Lysozyme is comparatively easy to crystallize and was one of the first enzymes whose three-dimensional structure was solved at high resolution by X-ray diffraction [14,15]. All told, the protein has been generous with insights about catalytic mechanisms, substrate interactions, and more. In recent years, lysozyme has become a centerpiece in computation-based folding research [16]. Given this foundation, KVFER...GCGV is leveraged throughout this book as a gateway to experiments and takeaways. We direct the reader to pioneers of protein structure analysis for scientific and historical perspective [17–19].

1. Pimentel, G., Spratley, R. D. 1971. *Understanding Chemistry*, chap. 19, Holden-Day, San Francisco, CA.
2. www.UniProt.org
3. www.ncbi.nlm.nih.gov/
4. Pauling, L. 1970. *General Chemistry*, chap. 24, Dover, New York.
5. Muraki, M., Harata, K., Sugita, N., Sato, K. 1996. Origin of carbohydrate recognition specificity of human lysozyme revealed by affinity labeling, *Biochemistry* 35, 13562.
6. Craig, N. L. 2014. *Molecular Biology: Principles of Genome Function*, Oxford University Press, Oxford.
7. Lundblad, R. L. 2006. *The Evolution from Protein Chemistry to Proteomics: Basic Science to Clinical Application*, CRC/Taylor & Francis, Boca Raton, FL.
8. Kyte, J. 1995. *Structure in Protein Chemistry*, Garland Publishers, New York.
9. Hughes, A. B. 2009. *Amino Acids, Peptides, and Proteins in Organic Chemistry*, Wiley-VCH, Weinheim.
10. Hecht, S. M. 1998. *Bioorganic Chemistry: Peptides and Proteins*, Oxford University Press, New York.
11. Sanger, F. 1945. The free amino acid groups of insulin, *Biochem. J.* 39(5), 507–515.
12. le Noble, W. J. 1974. *Highlights of Organic Chemistry*, Dekker, New York.
13. Lehninger, A. L. 1970. *Biochemistry*, Worth Publishers, New York.
14. Crick, F. H. C., Kendrew, J. C. 1957. X-ray analysis and protein structure, in *Advances in Protein Chemistry*, C. B. Anfinsen Jr., M. L. Anson, K. Bailey, J. T. Edsall, eds. Academic Press, New York.
15. Edsall, J. T., Wyman, J. 1958. Problems of protein structure, chap. 3 in *Biophysical Chemistry*, Vol. 1, Academic Press, New York.
16. Radford, S. E., Dobson, C. M., Evans, P. A. 1992. The folding of hen lysozyme involves partially structured intermediates and multiple pathways, *Nature* 358, 302–307.
17. Kendrew, J. C. 1966. *The Thread of Life: An Introduction to Molecular Biology*, Bell Publishers, London.
18. Perutz, M. F. 1992. *Protein Structure: New Approaches to Disease and Therapy*, W. H. Freeman, New York.
19. Tanford, C., Reynolds, J. 2001. *Nature's Robots: A History of Proteins*, Oxford University Press, New York.

Author

Daniel J. Graham was born and raised in San Francisco, California. He received bachelor and doctoral degrees in chemistry from Stanford University and Washington University, St. Louis, respectively. Following postdoctoral research at Boston University, the author joined the chemistry faculty at West Virginia University in 1983. He moved to Loyola University Chicago in 1987 and continues teaching and research in the Department of Chemistry and Biochemistry. The author's teaching focuses on general and physical chemistry while research concentrates on the molecular and thermodynamic applications of information theory. He is the author of *Chemical Thermodynamics and Information Theory with Applications* published in 2011 by CRC Press. He resides in Chicago with his wife.

One

Protein Structure Fundamentals and Complexity

The components, assembly, and sources of proteins are discussed. The levels of structure are illustrated via archetypal globular systems such as lysozyme. Particular attention is given to the complexity so presented—chemical, structural, and biological. This underscores the challenges of protein structure analysis at all levels.

A   WHAT ARE PROTEINS?

Proteins are polymers, but not just any polymers. Groundbreaking advances were made in the polymer chemistry fields beginning in the 1930s and 1940s [1,2]. The applications were materials: nylon, polystyrene, polyvinyl chloride, etc. For these systems, one or two single-molecule units are linked covalently and extensively. The product is a chain-like compound composed of many small molecules. Nylon is a case in point. The so-called Nylon-66 derives from the high-temperature condensation of adipic acid and hexamethyldiamine:

$$HO_2C(CH_2)_4 CO_2H + NH_2(CH_2)_6 NH_2 \xrightarrow{280°C} \left[-CO(CH_2)_4 CO-NH(CH_2)_6 -NH- \right]_n + H_2O$$

The material is structurally complicated owing to the cross-links involving hydrogen bonds. It is famous for durability and applications in the fabric and other materials industries.

And so it goes. The polymer sciences are rich with procedures for linking small molecules to form macroscopic ones. Anhydrides, glycols, phthalates, etc., all prove to be suitable building blocks. Note a critical feature. While the physical structure of a polymer is complicated, the abstract representation is simple. One has only to portray the fundamental unit surrounded by brackets and include a subscript integer n:

$$-[\text{monomeric unit}]_n -$$

This can represented pictorially using organic structure graphs.

Whether textual or graphical, the fundamental unit carries information about *all* parts of the system. It is analogous to the unit cell of a crystal: where an Angstrom- or nanometer-scale piece conveys the structure and organization on a large scale.

Proteins are in a league of their own. They are organic polymers composed of fundamental units, but not just one or two units tell the story. The vast majority host 20 different units in variable amounts. The units are the naturally-occurring amino acids joined by peptide bonds following the exclusion of water.

The simplest amino acids are glycine and alanine:

$$
\text{H}_2\text{N}\!-\!\text{CH}\!-\!\overset{\displaystyle\overset{O}{\|}}{\text{C}}\!-\!\text{OH} \qquad\qquad \text{H}_2\text{N}\!-\!\text{CH}\!-\!\overset{\displaystyle\overset{O}{\|}}{\text{C}}\!-\!\text{OH}
$$
$$
\qquad\quad\underset{\text{H}}{|} \qquad\qquad\qquad\qquad\quad \underset{\text{CH}_3}{|}
$$

Gly ↔ G   Ala ↔ A

In turn, the simplest peptides are represented as:

$$
\text{H}_2\text{N}\!-\!\underset{\text{H}}{\overset{}{\text{CH}}}\!-\!\overset{O}{\overset{\|}{\text{C}}}\!-\!\underset{}{\overset{\text{H}}{\text{N}}}\!-\!\underset{\text{H}}{\overset{}{\text{CH}}}\!-\!\overset{O}{\overset{\|}{\text{C}}}\!-\!\text{OH}
$$

Gly   Gly   Gly   Ala

Ala   Ala   Ala   Gly

Note the rules for joining monomers. The end product requires at least one weak base functional group by way of NH_2—just as in nylon synthesis. It is customary to represent this group on the left. At least one amino acid furnishes a weak acid functionality by way of COOH. It is the customary to place this on the right of the printed page or computer screen. Thus all peptide molecules feature an N-terminal *and* a C-terminal unit. The units that compose a chain are referred to as amino-acid *residues*, that is, what is left of the moieties after condensation and loss of water. The rules of the chemical structure theory are obeyed: *all* the covalent bonds are stable at room temperature. There is free rotation about single bonds but no rotation about the double bonds. There is no crossing of bonds, and so forth. Appendix One presents essential chemical properties of the 20 amino acids.

But the simplest cases telegraph complexity down the line. We have to consider four combinations of glycine and alanine linked by peptide bonds. If 18 more amino acids are added to the palette, we should contemplate the structures of $20^2 = 400$ di-peptides. The numbers increase explosively: a 100-unit polypeptide offers up to $20^{100} \approx 10^{130}$ possibilities. The latter number represents only a lower limit because nature offers about 100 amino acids *beyond* the standard 20 via post-genomic transformations. Proteins present a long and complicated story if only because they allow so many possibilities!

The remaining 18 amino acids are represented as follows.

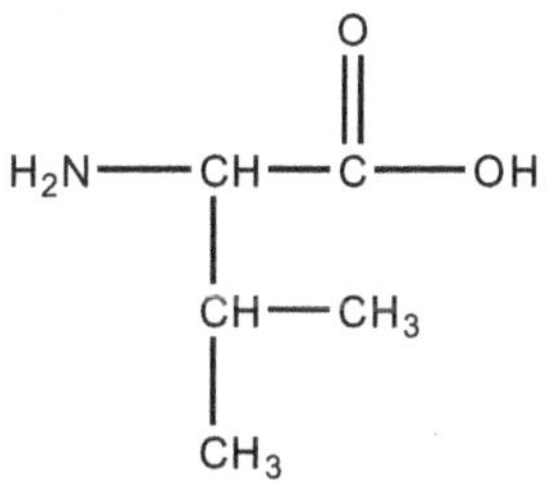

Valine

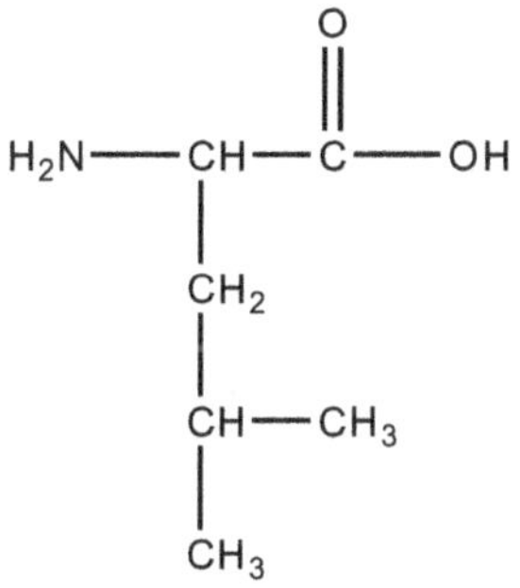

Leucine

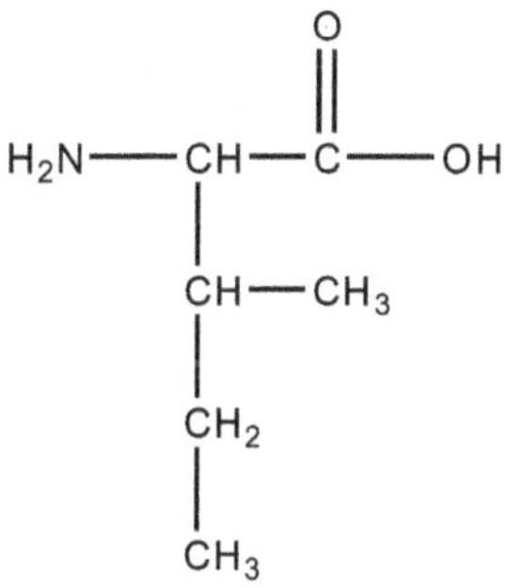

Isoleucine

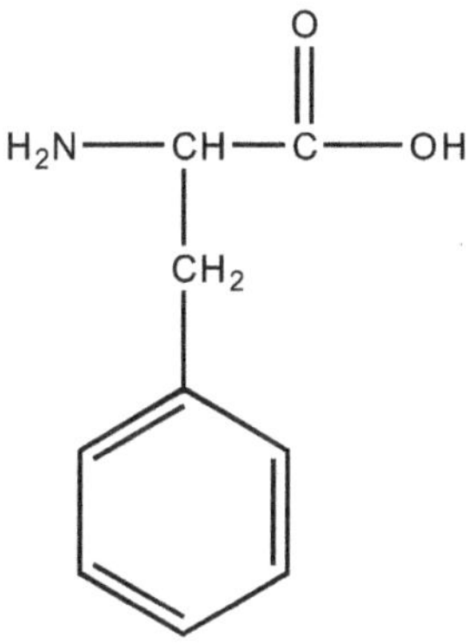

Phenylalanine

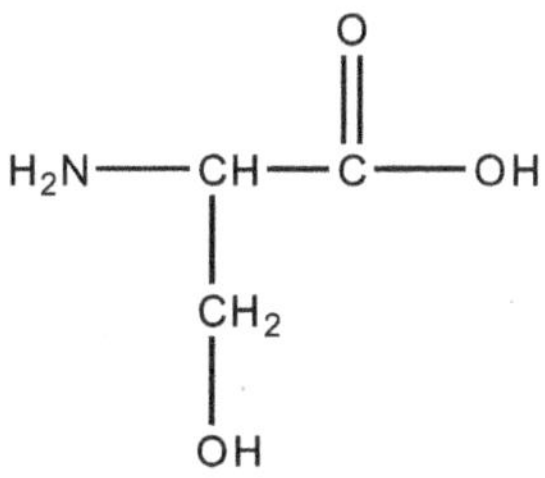

Serine

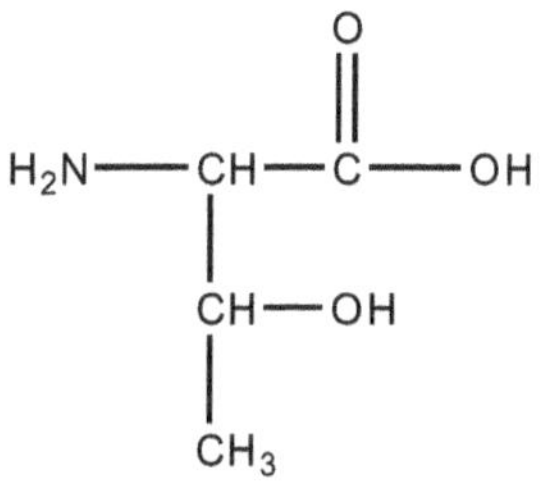

Threonine

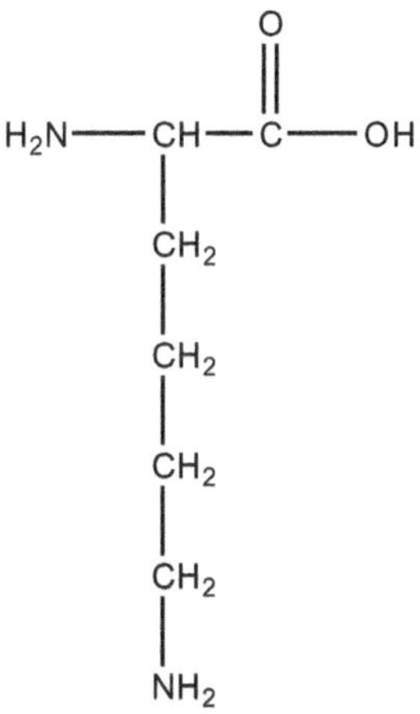

Lysine

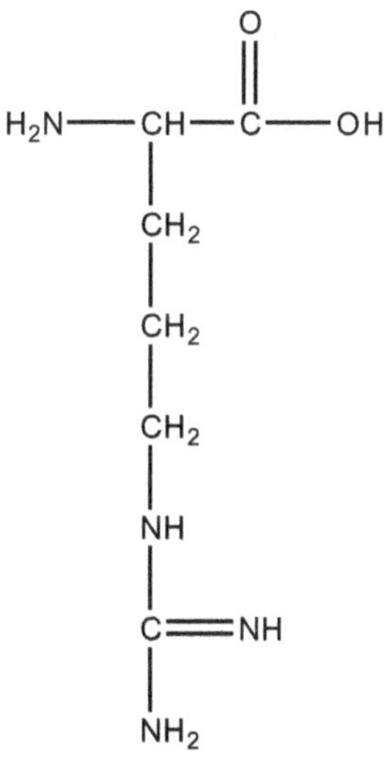

Arginine

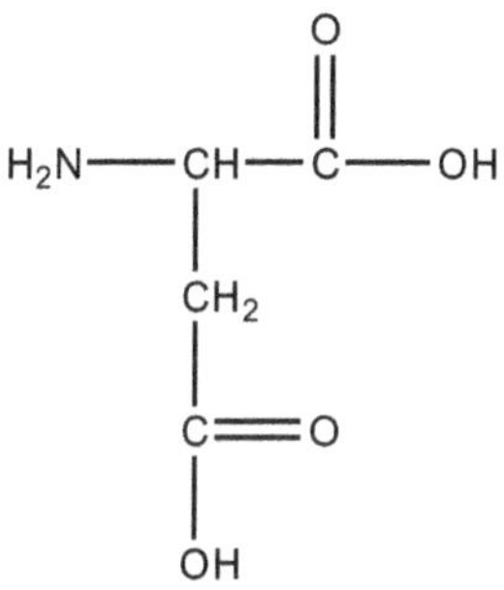

Aspartic Acid

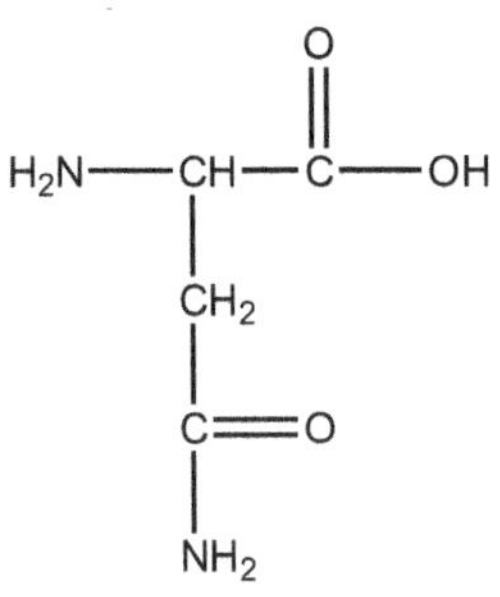

Asparagine

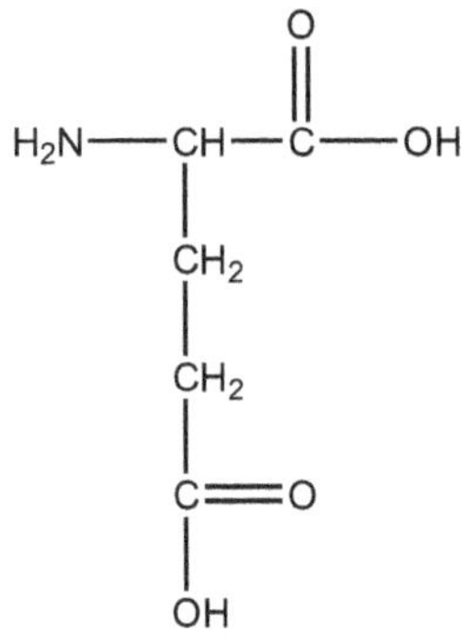

Glutamic Acid

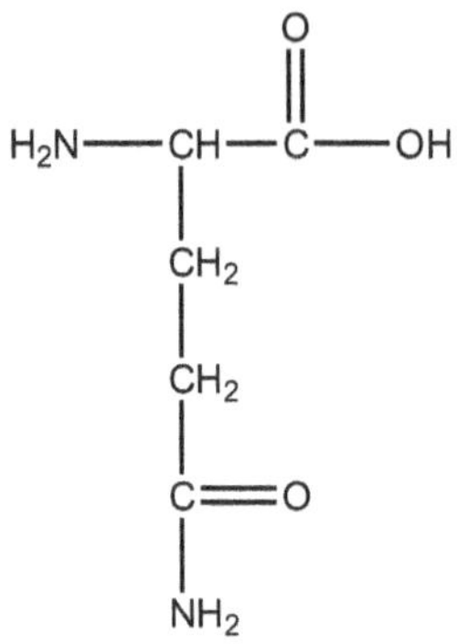

Glutamine

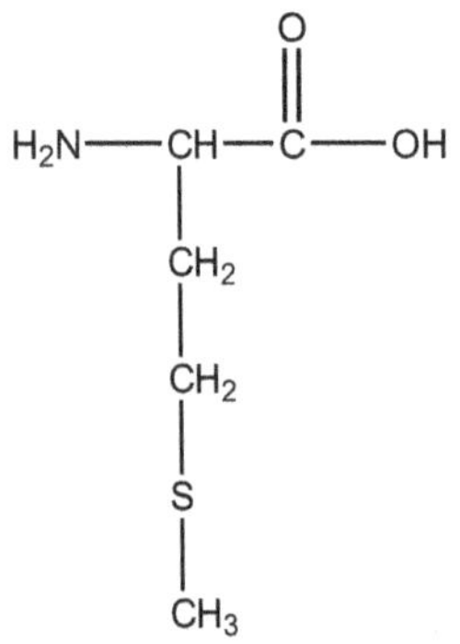

Methionine

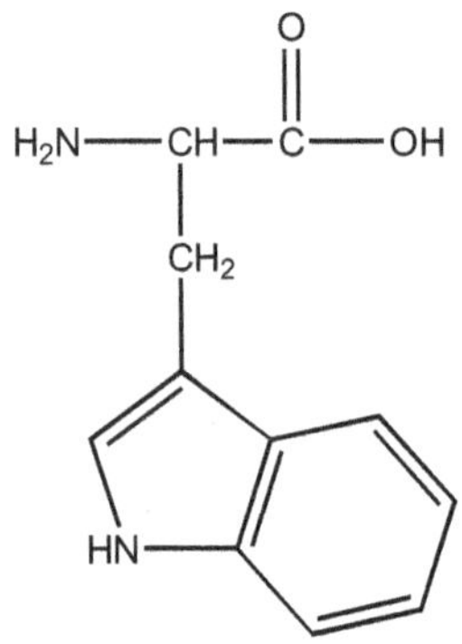

Tryptophan

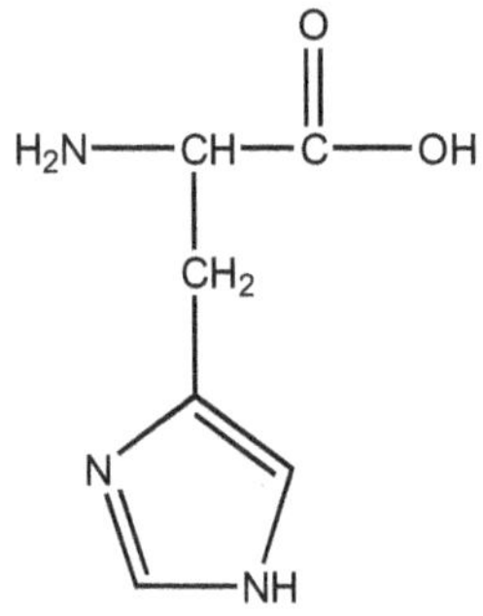

Histidine

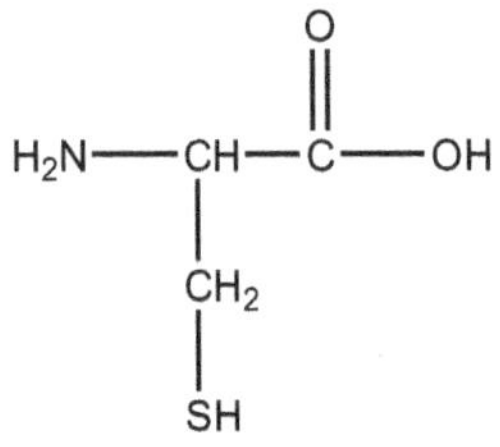

Cysteine

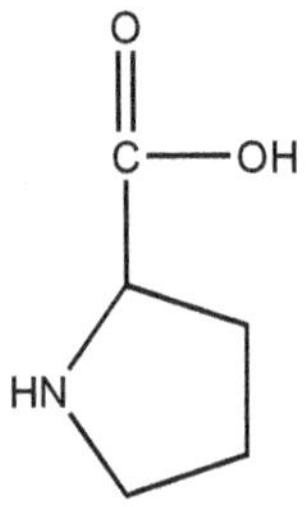

Proline

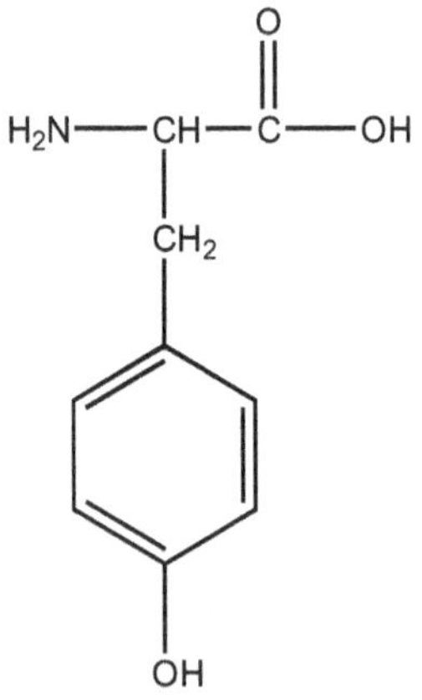

Tyrosine

These are minimalist representations well-suited to blackboards and scratch paper. One must keep in mind, however, that the *real* molecules are three-dimensional (3D) objects, each having a non-trivial fluctuating shape, volume, and surface. Better justice is done using the graphics of computer modeling programs. These capture the flavor of the bond angles and charge clouds—probability structures at the Angstrom scale. Animation of the rotational and vibrational states takes computer representations still closer to reality.

The chemical properties of the amino acids are detailed in classic texts, a few of which are listed in the end-of-chapter references [3–5]. The traits to note are the mix of shared and divergent features. All the amino acids offer a carboxylic acid functional group; all but one carries an amino group. Proline is the exception in that it carries an NH group instead of NH_2; proline is formally termed an *imino* acid. This should not cloud the point that in constructing a protein, an NH is bonded to the carbon atom adjacent to (i.e., α with respect to) the carboxyl group. What renders uniqueness is the substituent that accompanies the amino group. In effect, the chemical character of every amino acid is *tuned* by the α-carbon substituent. The substituent can be as minimalist as a hydrogen atom in glycine, or as elaborate as the rings in tryptophan, tyrosine, and histidine. Note further that save for glycine, every amino acid offers two enantiomeric (mirror-image) forms L and D. The α-carbon is the sole chiral center, the L- or D-handedness dictated by the configuration of four attachments: a hydrogen atom, carboxyl, nitrogen, and substituent groups. Importantly, and perhaps mysteriously, natural proteins are restricted to L-type amino acids [6]. Proteins of the D-variety are obtainable only by artificial methods [7].

Ordinary polymers accommodate shorthand representations, for example, $[-CO(CH_2)_4CO-NH(CH_2)_6-NH]_n$. With proteins, in virtually all cases, the entire sequence of monomers must be communicated. It is cumbersome (and paper-consuming) to draw chemical graphs as above. Fortunately, the sequences can be written in terms of three- or single-letter alphabets. The notation for the standard 20 is mnemonic-friendly [3–5]:

Glycine ↔ Gly ↔ G	Alanine ↔ Ala ↔ A	Valine ↔ Val ↔ V
Leucine ↔ Leu ↔ L	Isoleucine ↔ Ile ↔ I	Phenylalanine ↔ Phe ↔ F
Serine ↔ Ser ↔ S	Threonine ↔ Thr ↔ T	Lysine ↔ Lys ↔ K
Arginine ↔ Arg ↔ R	Aspartic acid ↔ Asp ↔ D	Asparagine ↔ Asn ↔ N
Glutamic acid ↔ Glu ↔ E	Glutamine ↔ Gln ↔ Q	Methionine ↔ Met ↔ M
Tryptophan ↔ Trp ↔ W	Histidine ↔ His ↔ H	Cysteine ↔ Cys ↔ C
Proline ↔ Pro ↔ P	Tyrosine ↔ Tyr ↔ Y	

We note in passing that two additional amino acids are genetically encoded, albeit rarely: seleno-cysteine and pyro-lysine [8]. There are still more amino acids by way of hydroxyl prolines, β-alanine, ornithine, triiodothyronine, and thyroxine [9]. These manifest as a consequence of post-transcription chemistry. However, these are also rare compared with the standard (and should be familiar) 20 (cf. Appendix One). Our sole focus in this book will be on the standard. There is so much to learn about the amino acids either at the lab bench, or from the literature. For now, we sketch brief answers to far-reaching questions: (1) Where do proteins come from? (2) What do they do?

Regarding (1), the molecules are born by the transfer of the weak base sequence information in deoxyribonucleic acid (DNA). Amino-acid strings are thus underpinned by a second type of biopolymer, namely a polynucleotide. The complement of *one* of the chains of a DNA double-helix is synthesized through the action of transcription proteins. The process conserves the chemical information encoded in DNA, re-expressing it in the form of ribonucleic acid (RNA)—yet another biopolymer [9]. The RNA, often by way of multiple intermediates, is subject to extensive cutting and splicing by recognition and other dedicated proteins. The processes ultimately link the *exon* portions of the RNA while the *intron* (intervening) parts are cast aside [10]. The following schematic depicts strands DNA and complement RNA as thin lines. The exon portions are represented as blocks, viz.

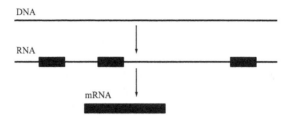

The third-stage molecule is messenger RNA (mRNA), its length matching that of the exon sections combined. The chemical message is mobile in solution as mRNA ferries *the* program that directs protein synthesis. The program is hard-wired in that all sections are composed of electric charges and surfaces. And it can be mass-produced: a single DNA strand is responsible for multiple copies of mRNA.

Each triplet of nucleotides in mRNA codes for a single amino acid. All syntheses require building stock. For proteins, these are the amino acids selectively bound to yet another type of biopolymer, transfer-RNAs (tRNAs) [9]. Prior to being joined by peptide bonds—and excluding water in the process—amino acids are matched with tRNAs; each amino acid is affinity-specific to one or more. The reactions proceed via the coordination of mRNA with an extended population of amino acid-tRNAs: Gly-tRNA$_{Gly}$, Ala-tRNA$_{Ala}$, etc. The coordination site is the ribosome unit of a cell composed of RNAs and additional proteins. The protein is assembled in the direction of the N- to C-terminal end: left-to-right in conventional representations.

Information is at center-stage from start to finish. DNA is the fountainhead of programs for proteins. But the codes are stored most often in mosaic form at low density, especially in higher organisms. Human lysozyme C is composed of 130 amino acids, requiring $3 \times 130 = 390$ nucleotides for synthesis. The nucleotides are contained in four regions of DNA covering 6,000 nucleotides [11]. A form of human myoglobin hosts 154 amino acids, requiring $3 \times 154 = 462$ nucleotides for coding. These occupy three portions of DNA spanning over 10,000 nucleotides [12].

The information is transferred with high fidelity albeit asymmetrically: one cannot backtrack from a protein sequence to determine the program in mRNA, RNA, or DNA. This is because, with the exception of methionine (M) and tryptophan (W), two or more codons (nucleotide triplets) are specific to an amino acid. Lysine (K) is the N-terminal unit of lysozyme; there are *two* possibilities underpinning this "choice" of unit. Valine (V) is the nearest neighbor, having four codon possibilities (cf. Appendix Five). The takeaway is that an mRNA can be translated into a protein sequence, but the converse is not true. Additional information asymmetry arises from the biopolymer lengths: one cannot backtrack from 130-unit KVFER…GCGV to determine the lengths of the source RNA and DNA.

The above is a thumbnail sketch of a *bio*-synthetic process with complete pictures found in the references. The student is directed not only to this literature, but also to classics of protein synthesis in organic chemistry labs [13]. The classics are instructive—and inspiring—because they capture the challenges of obtaining the molecules by artificial means. These include establishing and maintaining the chirality at the α-carbons, avoiding sequence errors, and maximizing the product yield. None of these mountains are easy to scale. The Nobel-prize-winning synthesis of ribonuclease A (124 amino acids) by Merrifield required 369 separate reactions and 11,931 operations [14]. The synthesis of ribonuclease A led by Hirschmann required 22 co-workers [15]. The reader is further encouraged to review the synthetic conundrums presented even by short peptides [16].

Regarding Question (2), the roles of the molecules are famously diverse. They catalyze the chemical reactions that make life possible. Every catalytic protein (aka enzyme) controls the rates of one or more reactions. Proteins further serve as transport vehicles, for example, in dispersing oxygen and disposing of carbon dioxide. The hormonal varieties ferry chemical signals on behalf of the immune and metabolic systems. Tanford and Reynolds have described proteins picturesquely as robots [17]. The robots operate on the nanometer scale and enable organisms to thrive on the macroscopic scale.

Yet the roles of proteins are not restricted to wet chemistry. They form much of the extracellular material of connective tissue and fibers. They are polymers with strength and durability to their name, as with nylon. A fair question is: what tasks are completed by an organism (or virus) independently of proteins? The author cannot think of any and the reader should recall Pauling's statement in the outline section: "Proteins may well be considered the most important substances in plants and animals" [18].

Proteins present an ocean of facts. The reader should enter into Google "protein structure," "protein chemistry," "protein experiments," etc., to appreciate this. We touch upon a few scattered facts to make important points. First, amino acids inspire a cornucopia of synthetics. A number of peptides indeed derive from *non*canonical monomers. Aspartame, aka NutraSweet®, is a household example [19]:

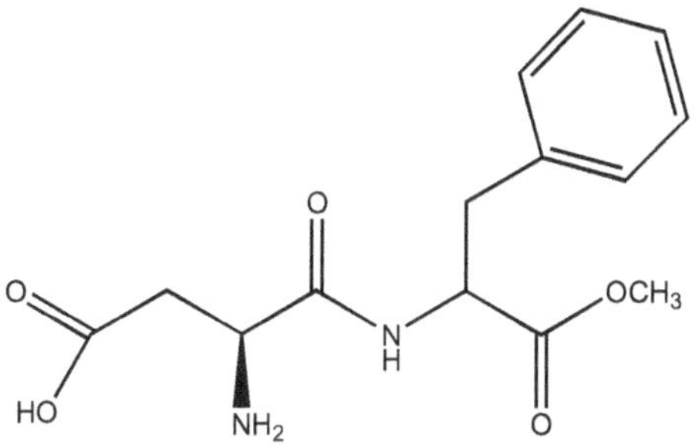

Note how the representation describes a methyl ester of aspartic acid bound to phenylalanine, that is, Asp − Phe = DF. Point being made: aspartame demonstrates that even small peptides are potent carriers of biochemical signals. A molecule need not be a *poly*peptide or even natural to wield biological impact.

Along the same lines, the pharmaceutical compound Lyrica® manufactured by Pfizer is a non-natural *single* amino acid [20]:

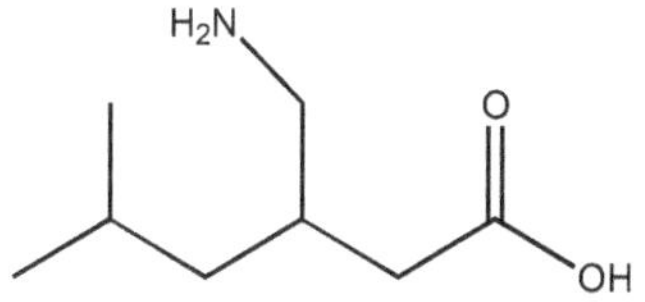

The molecule finds applications toward the control of seizures and nerve pain associated with diabetes. Lyrica® demonstrates that a single, *non-alpha* amino acid can be physiologically potent. Consult the Lyrica® website about the therapeutic uses and side effects. There is an extensive literature surrounding the design, syntheses, and applications of artificial amino acids [21]. The literature is still more extensive for naturally occurring amino acids that are *not* the standard 20.

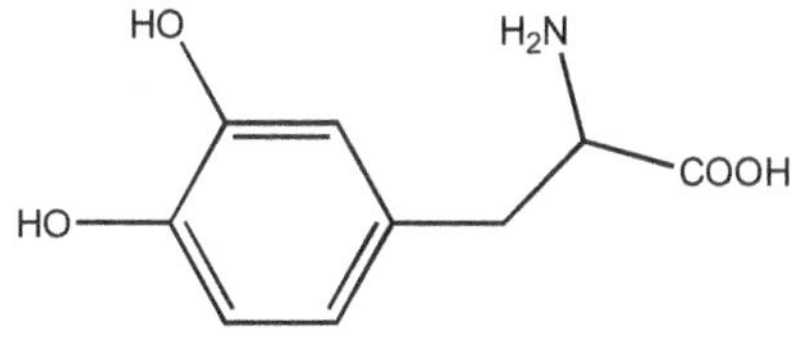

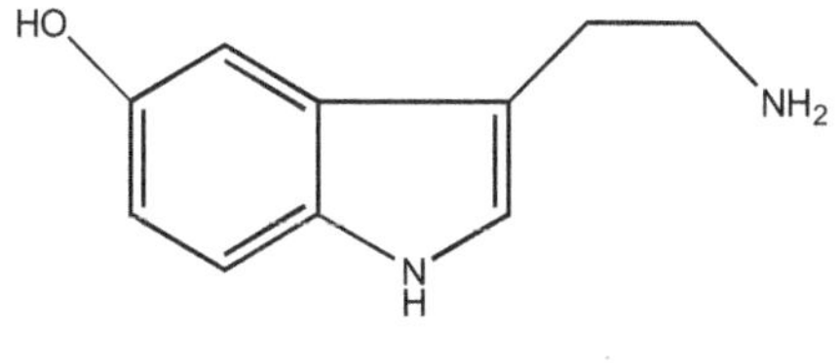

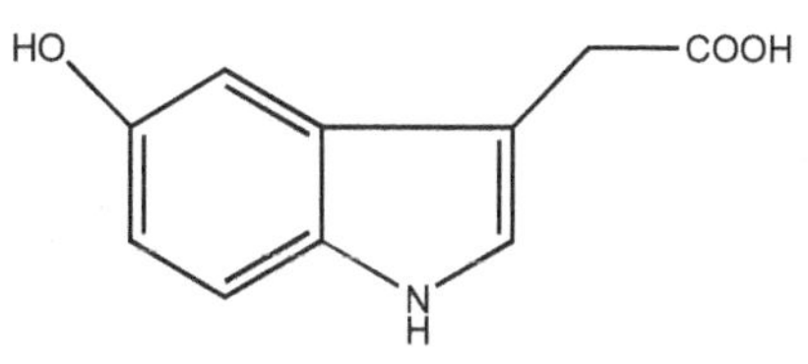

The topmost structure applies to dihydroxyphenylalanine (aka dopa) which is central to the motor operations of vertebrates [22]. The lower left and right feature 5-hydroxytryptamine (aka serotonin) and a metabolic by-product 5-hydroxyindoleacetic acid [9]. The similarities to tryptophan (W) are not coincidental, for the molecules are related by bio-synthetic pathways and figure prominently in the chemistry of emotions. The point is that the standard 20 are not the end-all and be-all in the organism. Among their roles, they serve as precursors for compounds bearing high physiological impact. Molecules which combine the weak acid and base functionalities cover a vast territory of natural products and biological chemistry.

Proteins are robots in the service of organisms. It is then no surprise that many have been designed and synthesized for pharmaceutical purposes. There is a conundrum, however, in that the proteins constructed from naturally occurring amino acids are vulnerable to degradation by proteases—still more proteins in the organism. This has motivated the synthesis of peptoids and polypeptoids [23]. Instead of a substituent group bound to the α-carbon, there is one linked to the amino group. Such modifications render the molecule "invisible" to most proteases. The following illustrate the peptoid counterparts of alanine and asparagines. The structure graphs are followed by the graph for a di-peptoid.

$$
\begin{array}{cc}
\text{HN—CH—C—OH} & \text{HN—CH—C—OH} \\
\quad\;\,| \qquad \| & \quad\;\,| \qquad\qquad \| \\
\text{CH}_3 \qquad \text{O} & \text{CH}_2 \qquad\qquad \text{O} \\
& | \\
& \text{C}=\text{O} \\
& | \\
& \text{NH}_2
\end{array}
$$

$$
\text{HN—CH—C—N—CH—C—OH}
$$

We close this section by noting that when nonscientists are asked for their views on proteins, the answers often focus on dietary issues. Some of the amino acids are said to be essential for humans: histidine (H), lysine (K), tryptophan (W), phenylalanine (F), leucine (L), iso-leucine (I), threonine (T), methionine (M), and valine (V) [24]. The "essential" label acknowledges that humans are able to manufacture the others from scratch, but require the foregoing nine from external sources. "Good" proteins include *all* the essentials while "poor" ones lack one or more. Many an organism is able to synthesize the 20 amino acids necessary for life. We humans fall short on this capacity.

B   LEVELS OF PROTEIN STRUCTURE

Every organic compound presents multiple structure levels. Formulae such as C_3H_6O and $C_5H_{10}N_2O_3$ capture the chemical ground floor properties of acetone and glutamine, respectively. Higher-level structures are represented by graphs such as shown in the previous section. 2D graphs have furnished the international language of chemistry for more than a century.

Still higher-level structures include sets of average bond lengths and angles, and the spreads about averages [25]. The levels do not really terminate because every electronic, vibrational, and rotational state presents its own signature properties—its own set of bond lengths, angles, and deviations.

Proteins present multiple structure levels as well. Given their extraordinariness, however, only select levels are researched intensely. The amino-acid composition provides the most practical information at the base level. For example, the following presents the base or composition of human lysozyme C (hereafter referred to simply as lysozyme):

$$A_{14}V_9L_8I_5P_2F_2W_5M_2G_{11}S_6T_5C_8Y_6N_{10}Q_6D_8E_3K_5R_{14}H_1$$

The order of letters is immaterial as the following carry equivalent information; the bold font indicates the switched components:

$$V_9L_8I_5P_2F_2W_5M_2\mathbf{A_{14}}G_{11}S_6T_5C_8Y_6N_{10}Q_6D_8E_3K_5R_{14}H_1$$
$$\mathbf{H_1}A_{14}V_9L_8I_5P_2F_2W_5M_2G_{11}S_6T_5C_8Y_6N_{10}Q_6D_8E_3K_5R_{14}$$

Base structures are analogous to chemical formulae, for example, benzene $\leftrightarrow C_6H_6$. However, they share much more in common with solution compositions, for example, 2.00 moles of acetone combined with 3.50 moles of chloroform. In turn, protein base structures represent the amino-acid identities and numbers quite apart from order in a sequence and water-exclusion details. As with all molecules, proteins express simplest formulae, for example, lysozyme $\leftrightarrow C_{633}H_{992}N_{200}O_{186}S_{10}$. But aside from conveying atom numbers, these cast only faint light on the chemical functions and biological significance.

The amino-acid sequence provides the next level of information. For lysozyme, this is archived in the Protein Data Bank (PDB) file labeled 1REX using the three-letter abbreviations [26]. An excerpt is listed as follows:

```
SEQRES    1 A  130  LYS VAL PHE GLU ARG CYS GLU LEU ALA ARG THR LEU LYS
SEQRES    2 A  130  ARG LEU GLY MET ASP GLY TYR ARG GLY ILE SER LEU ALA
SEQRES    3 A  130  ASN TRP MET CYS LEU ALA LYS TRP GLU SER GLY TYR ASN
SEQRES    4 A  130  THR ARG ALA THR ASN TYR ASN ALA GLY ASP ARG SER THR
SEQRES    5 A  130  ASP TYR GLY ILE PHE GLN ILE ASN SER ARG TYR TRP CYS
SEQRES    6 A  130  ASN ASP GLY LYS THR PRO GLY ALA VAL ASN ALA CYS HIS
SEQRES    7 A  130  LEU SER CYS SER ALA LEU LEU GLN ASP ASN ILE ALA ASP
SEQRES    8 A  130  ALA VAL ALA CYS ALA LYS ARG VAL VAL ARG ASP PRO GLN
SEQRES    9 A  130  GLY ILE ARG ALA TRP VAL ALA TRP ARG ASN ARG CYS GLN
SEQRES   10 A  130  ASN ARG ASP VAL ARG GLN TYR VAL GLN GLY CYS GLY VAL
```

There are 13 amino acids represented in each row by convention. The above can be condensed in the string of 130 letters:

KVFERCELARTLKRLGMDGYRGISLANWMCLAKWESGYNTRATNYNAGDRS
TDYGIFQINSRYWCNDGKTPGAVNACHLSCSALLQDNIADAVACAKRVVR
DPQGIRAWVAWRNRCQNRDVRQYVQGCGV

Recall the convention: the upper left-most amino-acid *residue*, namely lysine (**K**) in bold, hosts the N-terminal amino group. The lower right-most unit, boldface valine (**V**), hosts the C-terminal carboxylic acid group. Both abbreviation methods communicate what is referred to as the *primary* structure. This arrives experimentally from chemical processes such as Edman degradations, or by physical methods such as laser-induced-desorption, time-of-flight (TOF) mass spectrometry.

Alternatively, the primary structure is determined by translation of the gene (codon sequences) that underpins the protein. Databanks are rich with primary structures, the vast majority originating from gene identification and translation [27].

The primary structure can be written in condensed formats KVFE…GCGV. It can be expanded upon using molecular graphs, viz.

Space on a page or computer screen is rapidly consumed! And the end product provides few clues about what a protein does for (or to!) an organism. The condensed representations are justified in part because their expanded versions are not generous with explanations. The canonical tools for analyzing molecular structure are primarily useful in local applications: the actions of specific sites of folded systems.

Note a few more complications. Numerous proteins present two or more chains of amino acids. The chains can be identical as with many phospholipases: it is as if the message of one molecule is echoed and reinforced by another. They can be different such as for insulin: it is as if one polymer message is combined with and complemented by the other. The point is that a single protein can present multiple primary and base structures.

Given a favorable environment (aqueous solution, temperature, pressure, and pH), a protein adopts higher-level structures, usually spontaneously or with the help of chaperone proteins. These include so-called secondary by way of α-helices and β-sheets. The helices and sheets materialize selectively in parts of the amino-acid chain. For lyzozyme, the placement of helical and sheet structures is noted as follows using bold and underlined italic typeface, respectively.

```
SEQRES    1 A   130   LYS VAL PHE GLU ARG CYS GLU LEU ALA ARG THR LEU LYS
SEQRES    2 A   130   ARG LEU GLY MET ASP GLY TYR ARG GLY ILE SER LEU ALA
SEQRES    3 A   130   ASN TRP MET CYS LEU ALA LYS TRP GLU SER GLY TYR ASN
SEQRES    4 A   130   THR ARG ALA THR ASN TYR ASN ALA GLY ASP ARG SER THR
SEQRES    5 A   130   ASP TYR GLY ILE PHE GLN ILE ASN SER ARG TYR TRP CYS
SEQRES    6 A   130   ASN ASP GLY LYS THR PRO GLY ALA VAL ASN ALA CYS HIS
SEQRES    7 A   130   LEU SER CYS SER ALA LEU LEU GLN ASP ASN ILE ALA ASP
SEQRES    8 A   130   ALA VAL ALA CYS ALA LYS ARG VAL VAL ARG ASP PRO GLN
SEQRES    9 A   130   GLY ILE ARG ALA TRP VAL ALA TRP ARG ASN ARG CYS GLN
SEQRES   10 A   130   ASN ARG ASP VAL ARG GLN TYR VAL GLN GLY CYS GLY VAL
```

Secondary structures are established by diffraction (X-ray and neutron), NMR, and circular dichroism experiments. It is difficult to anticipate such structures *a priori* and a substantial literature is devoted to this. A number of online tools are further available for secondary structure predictions and analysis.

The next level is tertiary. This applies to the folds adopted by a molecule so as to confer the chemical functions. While the primary structure is grounded upon peptide bonds, the tertiary structure springs from longer-range, relatively weak interactions. More than 10^5 tertiary structures have been archived in data banks [28]. The structures determined by X-ray diffraction are typically uninformative about hydrogen atoms due to low scattering cross sections. Tertiary structure files accordingly include the average relative positions of the C, O, N, S atoms of the amino acids, coordinating metals, and the major atoms (i.e., nonhydrogen) of ligands. Another excerpt of the PDB file for lysozyme (1REX) is shown below. The data are presented in the N- to C-terminal amino-acid order of the primary structure. The Cartesian (x, y, z) coordinates for each atom are listed in Angstrom units. The abbreviations are readily interpreted. N, CA, C, and O refer to the respective amino nitrogen, α-carbon, carbonyl carbon, and carbonyl oxygen atoms of the N-terminal lysine. The data place the amino nitrogen atom at average position $(x, y, z) = (1.251, 19.636, 22.152)$ Å.

ATOM	1	N	LYS A	1	1.251	19.636	22.152	1.00	10.85	N
ATOM	2	CA	LYS A	1	1.899	20.606	21.229	1.00	10.55	C
ATOM	3	C	LYS A	1	3.168	20.004	20.624	1.00	10.49	C
ATOM	4	O	LYS A	1	3.910	19.328	21.327	1.00	10.77	O
ATOM	5	CB	LYS A	1	2.282	21.871	22.016	1.00	12.60	C
ATOM	6	CG	LYS A	1	3.243	22.812	21.285	1.00	16.02	C
ATOM	7	CD	LYS A	1	3.634	24.000	22.146	1.00	16.59	C
ATOM	8	CE	LYS A	1	4.454	24.988	21.347	1.00	18.78	C
ATOM	9	NZ	LYS A	1	4.729	26.218	22.143	1.00	22.56	N
ATOM	10	N	VAL A	2	3.384	20.212	19.329	1.00	10.57	N
ATOM	11	CA	VAL A	2	4.597	19.736	18.678	1.00	11.60	C
ATOM	12	C	VAL A	2	5.389	21.024	18.429	1.00	12.41	C
ATOM	13	O	VAL A	2	4.916	21.932	17.730	1.00	11.94	O
ATOM	14	CB	VAL A	2	4.304	19.032	17.338	1.00	13.00	C
ATOM	15	CG1	VAL A	2	5.614	18.569	16.684	1.00	12.39	C
ATOM	16	CG2	VAL A	2	3.366	17.850	17.567	1.00	13.38	C
ATOM	17	N	PHE A	3	6.521	21.159	19.117	1.00	11.56	N
ATOM	18	CA	PHE A	3	7.373	22.338	18.986	1.00	11.69	C
ATOM	19	C	PHE A	3	8.142	22.331	17.689	1.00	11.91	C
ATOM	20	O	PHE A	3	8.417	21.280	17.129	1.00	12.01	O
ATOM	21	CB	PHE A	3	8.433	22.369	20.086	1.00	11.71	C
ATOM	22	CG	PHE A	3	7.953	22.905	21.391	1.00	11.49	C
ATOM	23	CD1	PHE A	3	7.226	22.108	22.263	1.00	12.47	C
ATOM	24	CD2	PHE A	3	8.286	24.186	21.780	1.00	11.68	C
ATOM	25	CE1	PHE A	3	6.833	22.593	23.513	1.00	15.19	C
ATOM	26	CE2	PHE A	3	7.898	24.676	23.022	1.00	12.64	C
ATOM	27	CZ	PHE A	3	7.173	23.879	23.893	1.00	12.84	C
ATOM	28	N	GLU A	4	8.479	23.520	17.209	1.00	12.69	N

Pictures are more appealing than antiseptic text. Thus Figure One illustrates one of many that can be constructed using the data file. As drawn here, the figure presents only the molecular backbone projected in the XY plane. The circles mark the *average* positions of the 130 α-carbons. The large open and filled circles note the N- and C-terminal positions, respectively. Note the representation to be probability-based: the circles mark the positions where α-carbon charge is *expected* to be found, given the crystal diffraction data and best-fit models. Minimalist or elaborate, folded state graphics present probability structures underpinning biochemical functions.

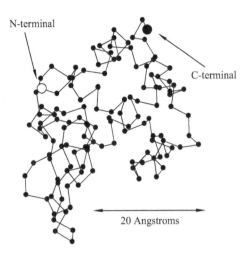

FIGURE ONE Backbone of human lysozyme (PDB 1REX) as projected in the *XY* plane. The coordinates and choice of Cartesian axes derive from X-ray diffraction analysis. The circles mark the positions of α-carbons. Large open and filled circles mark the N- and C-terminal sites, respectively.

It is difficult to anticipate secondary structures with confidence, and considerably more so for tertiary structures. Note how the backbone presents elements which are not easily related to the sequence KVFER….GCGV. Note further that the diffraction data connect only indirectly with the chemical graphs. This is because diffraction methods establish only the average positions of atoms, that is, the first moments of charge distributions. The bond order—single, double, etc., must be inferred.

Attention is then called to another type of covalent bond. A pair of cysteines (C) can support a bond between sulfur atoms, viz.

$$H_2N—CH—C(=O)—OH \quad H_2N—CH—C(=O)—OH$$
$$| \quad\quad | $$
$$CH_2 \quad\quad CH_2$$
$$| \quad\quad |$$
$$S————————S$$

For a folded protein in an oxidative environment, neighboring cysteines typically adopt these bonds. The reader should not be misled by the above diagram, however. The disulfide bond (S–S) is about 2.4 Å in length which compares with ca. 1.5 Å for carbon-carbon single bonds. The C-sites in lysozyme are noted via half-filled circles in Figure Two. They make possible the cross-links which enhance the stability of the folded molecule. This subject is taken up further in Chapter Seven.

Proteins present both as individuals and molecular complexes. All the while, they evince densities typical for organic solids. Consider again the X-ray data that allowed construction of Figures One and Two. The space group of crystalline lysozyme is $P2_12_12_1$ with four molecules per unit cell [26]. The unit cell parameters are shown by experiment to be:

$$a = 57.14 \text{ Å} \quad \alpha = 90.00°$$
$$b = 61.02 \text{ Å} \quad \beta = 90.00°$$
$$c = 33.10 \text{ Å} \quad \gamma = 90.00°$$

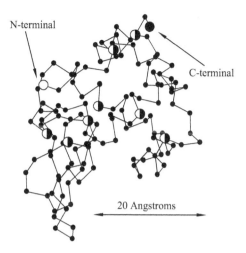

N-terminal

C-terminal

20 Angstroms

FIGURE TWO Backbone of human lysozyme (PDB 1REX) as projected in the *XY* plane. The circles mark the positions of α-carbons. Large open, filled, and half-filled circles mark the respective N- and C-terminal and cysteine sites.

This enables computation of the unit cell volume V using the following formula [18,25]:

$$V = abc \cdot \left[1 + 2\cos\alpha\cos\beta\cos\gamma - \cos^2\alpha - \cos^2\beta - \cos^2\gamma \right]^{1/2}$$

The formula yields $V \approx 1.15 \times 10^5 \text{Å}^3$ whereupon the enzyme density is approximated as follows:

$$\frac{(14{,}720.8 \text{ g/mol}) \times (1.00 \text{ mol}/6.02 \times 10^{23} \text{ molecules}) \times (4 \text{ mol/unit cell})}{(1.15 \times 10^5 \text{ A}^3/\text{unit cell}) \times (10^{-24} \text{ cm}^3/\text{A}^3)} \approx 0.848 \text{ g/cm}^3$$

This compares with the room temperature densities of solid glycine (1.16 g/cm³) and valine (1.23 g/cm³). As a rule, proteins are less dense than their constituent amino acids, but not extraordinarily less: their atoms pack in the manner of a solid phase organic. Further note that the density of a crystalline protein is diminished by the significant free volume and water in the unit cell.

Proteins having two or more N- to C-terminal chains manifest another level of structure, namely quaternary. This refers to the folded configurations of peptide chains with respect to one another. Lysozyme presents a single chain and thus lacks a formal quaternary structure. By contrast, phospholipases typically present chains A and B as illustrated in Figure Three using the same symbol codes as the previous figures. The two chains been distinguished using filled and open circles, respectively. Half-filled circles mark the cysteine sites.

This overview is completed by pointing out two more structure levels plus a critical trait. Regarding levels, a fifth is manifest by proteins which interact significantly and with high specificity. For example, ribonuclease A and ribonuclease inhibitor protein form a van der Waals and chemical-affinity complex [29]. Each party is highly attuned to the exposed sites, shape, and chemical functions of the other. Further, a sixth structure level lies in the networks formed by proteins; the molecules operate in communities, so to speak. As for the critical trait, virtually all molecular interactions, protein and otherwise, operate via electronic surfaces [30]. The folded state graphics have been kept few and minimalist in this book. However, they are taken closer to reality if we imagine wrapping cellophane around the irregular-shape backbones. Protein databanks and molecular modeling programs present dazzling visuals, all based on probabilistic concepts. These drive home that

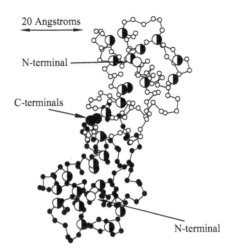

FIGURE THREE   Backbone of phospholipase (PDB 1POB) as projected in the *XY* plane. The circles mark the positions of α-carbons. Large open, filled, and half-filled circles mark the respective N- and C-terminal and cysteine sites. The A and B chains are distinguished using (small) filled and open circles, respectively. Note the C-terminal sites to be in close proximity and to eclipse two of the cysteine sites, as projected in the *XY* plane.

the electronic surfaces are highly specific and exclude the cores from intermolecular recognition and signaling. Sequence and surface relationships are visited in Chapter Six.

C   WHY ARE PROTEINS SO COMPLICATED? CHEMICAL CONSIDERATIONS

The chemical functions of organic compounds have long been communicated using stand-alone graphs.

   When the viewer (e.g., chemistry student) encounters the above, he or she registers definitive internal responses regarding functions, transformations, and activity. Matters are different for protein sequences starting with the facts-and-data deluge. When the viewer encounters

KVFERCELARTLKRLGMDGYRGISLANWMCLAKWESGYNTRATNYNAGDRS
TDYGIFQINSRYWCNDGKTPGAVNACHLSCSALLQDNIADAVACAKRVVR
DPQGIRAWVAWRNRCQNRDVRQYVQGCGV

he or she is inclined to look past the structure, registering little, if anything. If it is elaborated upon using chemical graphs, the deluge is only accentuated. Shorter sequences are no less opaque such as for the ubiquitin:

MALKLIHKEFLELARDPQPHCSAGPVWDDMLHWQATITRPNDSSYLGGVFFL
KFPSDYLFKPPKIKFTNGIYHQR

Longer sequences such as for human-encoded myoglobin seem no more complicated:

MGLSDGEWQLVLNVWGKVEADIPGHGQEVLIRLFKGHPETLEKFDKFKHLK
SEDEMKASEDLKKHGATVLTALGGILKKKGHHEAEIKPLAQSHATKHKIPVKYL
EFISECIIQVLQSKHPGDFGADAQGAMNKALELFRKDMASNYKELGFQG

The letter strings for proteins and polynucleotides drive bioinformatics and companion fields. The objectives include computer hardware and software for comparing and contrasting large quantities of data. A given sequence is probed via programs such as BLAST—basic local alignment sequence tools, in conjunction with international databases [31]. Multiple sequences are compared using MSA—multiple sequence alignment software.

Yet biopolymers of the protein variety present complications significantly beyond letter strings, beginning with the size of the sequence space. If 20 amino acids are programmed by the genetic code, a protein assembled from N number of amino acids poses 20^N possibilities. A student should compute several examples to appreciate this: 20^N where $N = 20, 30, 40, \ldots$. He or she should keep in mind that the earth is composed of "only" 10^{50} atoms [32].

A base structure (composition), by itself, allows multiple variations. For example, the following represent variations on the base theme of lysozyme:

$$\mathbf{A_9}V_{14}L_8I_5P_2F_2W_5M_2G_{11}S_6T_5C_8Y_6N_{10}Q_6D_8E_3K_5R_{14}H_1$$
$$A_{14}\mathbf{V_5}L_8I_5P_2F_2W_5M_2G_{11}S_6\mathbf{T_9}C_8Y_6N_{10}Q_6D_8E_3K_5R_{14}H_1$$
$$A_{14}V_9L_8I_5\mathbf{P_6}F_2W_5M_2G_{11}S_6T_5C_8\mathbf{Y_2}N_{10}Q_6D_8E_3K_5R_{14}H_1$$

Each is obtained by trading the subscripts as noted in bold while conserving the integer set: 14, 9, 8, 5, 2, 2, 5, 2, 11, 6, 5, 8, 6, 10, 6, 8, 3, 5, 14, 1. We count the number of base variants Λ by noting the repeat integers: 14 (twice), 2, 6, 8 (three times each), and 5 (four times). Λ then arrives as:

$$\Lambda = \frac{20!}{3!4!3!3!2!} \approx 2.35 \times 10^{14}$$

But then each variant poses a far greater number of primary structure possibilities, each representing a polymer with identical mass, charge, and covalent bond enthalpy. It is easy to generate the possibilities from either the base structure or sequence. For example, the following describes lysozyme with R10 and V121 interchanged and noted in bold underline:

KVFERCELA**V**TLKRLGMDGYRGISLANWMCLAKWESGYNTRATNYNAGDRS
TDYGIFQINSRYWCNDGKTPGAVNACHLSCSALLQDNIADAVACAKRVVR
DPQGIRAWVAWRNRCQNRD**R**RQYVQGCGV

The interchanges yield a letter string for another protein. To count the total sequence possibilities Ω, we look at the base integers: 14, 9, 8, etc. The integers sum to the total units N:

$$N = 14 + 9 + 8 + \cdots + 5 + 14 + 1 = 130$$

Ω is then follows from:

$$\Omega = \frac{N!}{N_A! N_V! N_P! \ldots N_K! N_R! N_H!}$$

$$= \frac{130!}{14!9!8!\ldots5!14!1!}$$

The factorial expressions get very big very fast. One frequently has to use the Stirling formula (Appendix Three) to estimate the numerator. The simplest version holds that for large integer N:

$$\ln N! \approx N \ln N - N$$

This tells us:

$$N! \approx \exp\left[N \ln N - N\right]$$

Alternatively, if one has access to a programmable calculator or spreadsheet, the following summation is more than doable:

$$\ln N! = \ln N + \ln(N-1) + \ln(N-2) + \cdots + \ln(2) + \ln(1)$$

$$= \sum_{i=1}^{i=N} \ln(i)$$

This enables computation of the logarithm of Ω, both numerator and denominator, via:

$$\ln \Omega = \sum_{i=1}^{i=N_{total}} \ln(i) - \sum_{i=1}^{i=N_{Ala}} \ln(i) - \sum_{i=1}^{i=N_{Val}} \ln(i) - \sum_{i=1}^{i=N_{Pro}} \ln(i) - \ldots$$

The reader should verify that the number of distinct isomers of lysozyme works out to be:

$$\Omega = \frac{130!}{14!9!8!\ldots5!14!1!} \approx 10^{147}$$

Organic compounds offer stable isomers as a rule [33]. For example, consider three of seven possible for C_6H_6:

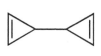

Each represents a bona fide assembly of charges with a unique thermodynamic and chemical personality; the algorithms for obtaining all the stable isomers are nontrivial. As organic compounds, proteins offer exceedingly large Ω, but their isomer constructions that conserve the class identity are simple, involving interchanges within the sequence. Formally speaking, the isomers of a protein form a group with members related by permutations [34]. Primary structure permutations enter discussions in Chapters Six and Eight.

A single protein offers a large number of site variants or mutations as well. For single-site variants, there are 19 alternatives for each amino-acid site. A protein containing N units then offers $v^{(1)} = 19N$ single-site variants.

Double-site variations allow more possibilities and are counted as follows. For each of the two left-most units, there are 19 alternates. The number of two-site variants to the N-terminal and adjacent site is $19 \cdot 19 = 19^2$. But the "second" substitution can be located at any of $N - 1$ sites. The number of two-site variants $v^{(2)}$ begins to look like

$$v^{(2)} = 19 \cdot 19 \cdot 1 \cdot 1 \cdot 1 \cdots + 19 \cdot 1 \cdot 19 \cdot 1 \cdot 1 \cdots + 19 \cdot 1 \cdot 1 \cdot 19 \cdot 1 \cdots$$

$$= 19 \cdot 19 \cdot (N - 1)$$

But the first substitution position need not be the left-most as it has $N - 1$ possibilities. So the *total* number of two-site substitutions takes the form:

$$v^{(2)} = 19 \cdot 19 \cdot (N - 1) + 1 \cdot 19 \cdot 19 \cdot (N - 2) + 1 \cdot 1 \cdot 19 \cdot 19 \cdot (N - 3)$$

$$+ \cdots + 1 \cdot 1 \cdots 1 \cdot 19 \cdot 19 \cdot (N - N + 1)$$

$$= 19 \cdot 19 \times \sum_{i=1}^{i=N-1} (N - i)$$

$$= 3,026,985$$

Suffice to say that even a modest size protein offers enormous room for variations on the chemical themes.

Yet another complexity lies in the wealth of functional groups. Throughout organic chemistry, reaction behavior is comprehended in terms of functional groups and carbon scaffolds. The groups form the centers of transformation while the scaffolds hold parties together. The chemistry of a molecule with one to a few functional groups can be anticipated and controlled to high degree. But proteins host hundreds to thousands of groups—carbonyl, carboxyl, amino, sulfide, and amide. They are chemically active at *every* amino-acid site. As organic compounds go, the reactivity of a protein is difficult to predict (an understatement), much less control.

Complexity stems from the depth and diversity of interactions. Chemical thermodynamics and molecular modeling are adept at anticipating "strong" intra-molecular interactions. These underpin the covalent bonds: C–C, C=O, etc. Weak interactions are another story; for example, involving hydrogen and van der Waals bonds. Covalent bonds derive from nearest-neighbor contacts and are highly directional, whereas hydrogen and van der Waals bonds involve long-distances and multiple neighbors and directions [35]. This is the reason that it is easier to intuit the covalent bond transformations of a molecule compared with condensed phase behavior such as crystallization. How does this apply to proteins? The molecules are complicated because so much of their behavior—folding, substrate recognition, and catalytic activity—hinges on weak interactions. Moreover, multiple forces such as van der Waals are not pair-wise additive and are vexing to model as a consequence [36]. Consider three protein sites A, B, C with van der Waals interaction potentials φ_{AB}, φ_{AC}, φ_{BC}. To compute the total potential, one cannot simply sum $\varphi_{AB} + \varphi_{AC} + \varphi_{BC}$.

Thermodynamics presents still another complexity. All systems are subject to fluctuations. Consider a salt solution represented schematically as:

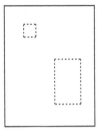

In any region of the sample, the number of ions, solvent, and solute molecules will vary instant to instant. In turn, the density, energy, and concentration will fluctuate about the equilibrium values. But the impact of fluctuations is size-dependent: swings in the energy and concentration carry greater impact in the smaller dotted-rectangle, compared with the larger.

How do fluctuations impact proteins? We have the luxury of $>10^5$ folded structures via data banks. These articulate, however, only the average equilibrium configurations. The fluctuations about the configurations are not incidental. A protein exchanges energy with its environment. At one instant, it holds more than average; at another, less. Statistical thermodynamics provides a handle on the time-average of the internal energy variance [37,38]:

$$\delta U^2 \approx k_{\mathrm{B}} T^2 m C_{\mathrm{V}}$$

U represents the average internal energy of the protein and δU^2 reflects the typical fluctuation. The formula features Boltzmann's constant k_{B}, absolute temperature T, molecular mass m, and the constant volume heat capacity C_{V}. The variance is estimated for lysozyme at human physiological temperature as follows:

$$\delta U^2 \approx \left(1.38 \times 10^{-23} \text{ J/K}\right) \cdot (310 \text{ K})^2 \cdot \left(\frac{14.7 \text{ kg/mol}}{6.02 \times 10^{23} /\text{mol}} \right) \cdot (1.5 \text{ J/g} \cdot \text{K}) \cdot (1000 \text{ g/kg})$$

$$\approx 4.86 \times 10^{-38} \text{ J}^2$$

C_{V} has been approximated by the more accessible C_{p}, the heat capacity at constant pressure equivalent to mass times the specific heat [39]. The standard deviation for the internal energy fluctuations is then estimated as follows:

$$\left(\delta U^2\right)^{1/2} \approx \sqrt{4.86 \times 10^{-38} \text{ J}^2} \approx 2.20 \times 10^{-19} \text{ J}$$

The calculation is placed in perspective by noting hydrogen bond energies to fall in the range 12–20 kJ/mol or 2.00–3.32 × 10^{-20} J per individual bond [40]. Then the typical fluctuation corresponds to the breaking and making of ten or more hydrogen bonds [41]. These bonds could well be crucial to the catalytic activity and other chemical mechanisms of the protein. Representations such as Figures One–Three are static and can do no justice to the fluctuations. However, the significance of fluctuations is further noted by comparing $(\delta U^2)^{1/2}$ with lysozyme's heat of denaturation, namely 280 kJ/mol or 4.65 × 10^{-19} J per molecule [42]. This shows the typical energy fluctuation to approach half of the denaturation enthalpy.

The volume excluded by a protein also changes instant to instant: the polymer trades space back and forth with its solution environment. The volume variance δV^2 is estimated by the thermodynamic formula [37,38]:

$$\delta V^2 \approx k_B T V \beta_T$$

V is the equilibrium volume while k_B and T are as before. The symbol β_T represents the isothermal compressibility. The application to lysozyme yields:

$$\delta V^2 \approx (1.38 \times 10^{-23} \text{ J/K}) \cdot (310 \text{ K}) \cdot (20 \text{ A})^3 \cdot (10^{-10} \text{ m/A})^3 \cdot \left(45 \times 10^{-6} \text{ bar}^{-1}\right) \cdot \left(\frac{1 \text{ bar}}{10^5 \text{ Pa}}\right)$$

$$\approx 1.54 \times 10^{-56} \text{ m}^6$$

Thus the standard deviation is estimated as follows:

$$\left(\delta V^2\right)^{1/2} \approx \sqrt{1.56 \times 10^{-56} \text{ m}^6} \approx 1.24 \times 10^{-28} \text{ m}^3$$

This corresponds to a size fluctuation along a single dimension as large as

$$\sqrt[3]{1.24 \times 10^{-28} \text{ m}^3} \approx 5.0 \times 10^{-10} \text{ m} \approx 5.0 \text{ Å}$$

5.0 Å well exceeds the length of any covalent bond, and is comparable to the distances over which hydrogen and van der Waals bonds operate. Size fluctuations may well be crucial to key interactions between a protein and its substrates, inhibitors, and neighbor proteins. The point is that the biopolymers are complex for multiple reasons, and their average, equilibrium structures provide only parts of stories.

One can go further by examining the third moment of the internal energy, viz.

$$\delta U^3 \approx 2 k_B{}^2 T^3 m C_V + k_B{}^2 T^4 m \frac{\partial C_V}{\partial T}$$

The formula is significant by the first term alone [38]. The second can be ignored as the temperature derivative of C_V is nearly zero at physiological temperatures. For lysozyme we again approximate C_V via C_p giving:

$$\delta U^3 \approx 2 \cdot \left(1.38 \times 10^{-23} \text{ J/K}\right)^2 \cdot (310 \text{ K})^3 \cdot \left(\frac{14.7 \text{ kg/mol}}{6.02 \times 10^{23}/\text{mol}}\right) \cdot (1.5 \text{ J/g} \cdot \text{K}) \cdot (1000 \text{ g/kg})$$

$$\approx 4.2 \times 10^{-58} \text{ J}^3$$

We then look to the cube root:

$$\left(\delta U^3\right)^{1/3} \approx \sqrt[3]{4.2 \times 10^{-58} \text{ J}^3} \approx 7.5 \times 10^{-20} \text{ J}$$

Recall that hydrogen bond energies range from 2.00 to 3.32×10^{-20} J/bond. Hence the cube root of δU^3 equates with two to a few hydrogen bond energies. The lesson is that for systems in which the most probable and average energy are identical, δU^3 is zero. This is not the case with proteins owing to their smallness and nontrivial fluctuations. The structure information in data banks applies to the averages. However, the most probable configurations can present markedly different pictures. This is not to devalue what we learn from data banks. It is only to acknowledge the effects of weak bonds and small-system thermodynamics.

What are other complications? A significant one is the environment of a protein by way of aqueous salt solutions. There are thermodynamically ideal solutions such as hexane/heptane, pentane/isopentane, etc. [43]. And then there are nonideal ones such as chloroform/acetone. Suffice to say that aqueous salt solutions are highly nonideal. Their complexity is marked given that "simple" quantities like molar volume and osmotic pressure cannot be anticipated with high accuracy in advance of experiment.

There is complexity due to the inseparability of structure levels. It is convenient to partition and study levels *as if* they were independent. This is moderately successful for the electronic, vibrational, rotational, and translational states of small molecules at low temperatures. However, the primary, secondary, etc., levels of proteins offer no such luxury in that one level cannot be comprehended apart from the others. The secondary, tertiary, and quaternary levels build upon the base and sequence structures. But bases and sequences are the way they are because they make for biologically active molecules.

Proteins are not singular on this account. Consider the natural product squalene with formula $C_{30}H_{50}$ and representation [44]:

It is a prominent member of the terpene family. Why does squalene present a hydrocarbon backbone as above? The answer is because the molecule is a precursor to steroids such as lanosterol:

Why do steroids express fused rings and chiral centers as above? The answers include that they derive from squalene-type compounds. The reciprocal relationships are apparent if the precursor is represented alternatively:

While the reader could have figured these out by trial-and-error drawings, he or she would be at a loss with proteins. The molecules host many more atoms and functional groups; the relationships are obscured by the myriad configurations.

D WHY ARE PROTEINS SO COMPLICATED? STRUCTURAL CONSIDERATIONS

Consider the meaning of the word "structure." The Merriam–Webster Collegiate Dictionary offers five definitions, plus sub-headings, including "something arranged in a definite pattern of organization." Another definition is "the aggregate of elements of an entity in their relationships to each other" [45]. These phrases seem on the mark. Further note the word "pattern." The same dictionary offers 12 definitions including "a discernible coherent system based on the intended interrelationship of component parts." This definition also seems spot on. Question: What can we gather about the *structure patterns* in proteins?

It should already be apparent that patterns are in short supply in primary structures. The following represent ribonuclease A, insulin precursor, and alpha synuclein, respectively:

KETAAAKFERQHMDSSTSAASSSNYCNQMMKSRNLTKDRCKPVNTFVHESL
ADVQAVCSQKNVACKNGQTNCYQSYSTMSITDCRETGSSKYPNCAYKTTQAN
KHIIVACEGNPYVPVHFDASV

MALWTRLRPLLALLALWPPPPARAFVNQHLCGSHLVEALYLVCGERGFFYTPK
ARREVEGPQVGALELAGGPGAGGLEGPPQKRGIVEQCCASVCSLYQLENYCN

MDVFMKGLSKAKEGVVAAAEKTKQGVAEAAGKTKEGVLYVGSKTKEGVVHG
VATVAEKTKEQVTNVGGAVVTGVTAVAQKTVEGAGSIAAATGFVKKDQ
LGKNEEGAPQEGILEDMPVDPDNEAYEMPSEEGYQDYEPEA

The sequences are interesting if only for their seeming randomness. To borrow a term from Schrödinger, they are *aperiodic* from N- to C-terminal [46]. Among the consequences, if one or a few letters were omitted, it would be most challenging for a viewer to locate their positions, much less infer their identities. If an N-terminal fragment is presented, for example,

SLLEFGKMILEETGK...

it is difficult to register (sans BLAST and databases) where the functional themes are headed. If instead, a C-terminal fragment is presented, for example,

...LYPDFLCKGELKC

it seems impossible to infer units that laid the groundwork. This is in marked contrast to the synthetic pathways of natural products: the sequences make chemical sense on multiple levels to the trained viewer. Total syntheses in labs could never succeed otherwise (cf. Preface).

The aperiodicity is multiplied in not a few cases. Revisit lysozyme with eight cysteines noted in bold.

KVFERCELARTLKRLGMDGYRGISLANWMCLAKWESGYNTRATNYNAGDRS
TDYGIFQINSRYWCNDGKTPGAVNACHLSCSALLQDNIADAVACAKRVVR
DPQGIRAWVAWRNRCQNRDVRQYVQGCGV

These sites allow four disulfide bridges which link parts of the chain otherwise remote from one another [26]. X-ray diffraction data show the following cross-links in the folded state:

$$^{6}C-^{128}C \qquad ^{30}C-^{116}C \qquad ^{65}C-^{81}C \qquad ^{77}C-^{95}C$$

The links diversify the N- to C-terminal sequences to include:

KVFER^{6}C^{128}CGV

KVFERCELARTLKRLGMDGYRGISLANWMCLAKWESGYNTRATNYNAGDRS
TDYGIFQINSRYW^{65}C^{81}CSALLQDNIADAVACAKRVVRDPQGIRAWVAWR
NRCQNRDVRQYVQGCGV

KVFERCELARTLKRLGMDGYRGISLANWM^{30}C^{116}CQNRDVRQYVQGCGV

KVFERCELARTLKRLGMDGYRGISLANWMCLAKWESGYNTRATNYNAGDRS
TDYGIFQINSRYWCNDGKTPGAVNA^{77}C^{95}CAKRVVRDPQGIRAWVAWRN
RCQNRDVRQYVQGCGV

Lysosyme's functions are thus allied with *five* covalently-bonded sequences, one principal and four subsidiaries.

The aperiodicity is largely maintained in the folded state. This is appreciated via plots of the radial distribution function $G(r)$ based on the tertiary structure. $G(r)$ surveys the placement of atoms with respect to different vantage points inside and on the surface of a protein [36]. We imagine sitting on, say, each α-carbon, and noting the distances of other α-carbons in the system. Each α-carbon therefore serves as an origin x, y, z = 0, 0, 0 whereupon $G(r)$ tracks the number of α-carbons at distance $r = (x^2 + y^2 + z^2)^{1/2}$. The function—another type of probability structure—shows which distances are reinforced in the folded state. Where distance reinforcements are lacking, $G(r)$ appears broad and featureless, as opposed to showing sharp peaks.

Figure Four illustrates lysozyme $G(r)$ based on the α-carbon placements; similar plots are obtained using other atom placements, for example, carbonyl carbons and oxygens. The plot manifests a few peaks (distance reinforcements), but otherwise a diffuse background. This communicates a mix of structural order and disorder to the discerning viewer. While a local order is registered about each amino-acid site, only a few distances recur in the range 25–35 Å. In effect, the molecule presents a complex shell structure about each site.

Protein structures are aperiodic on multiple accounts. Even so, disparate molecules show significant correspondences in their radial distributions. $G(r)$ for ribonuclease A appears in Figure Five.

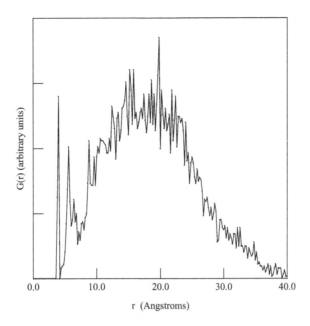

FIGURE FOUR   Radial distribution function of lysozyme based on the positions of α-carbons. The function is assembled using PDB file 1REX. The vertical height is proportional to the number of α-carbons placed (on average) at distance *r* from a reference site.

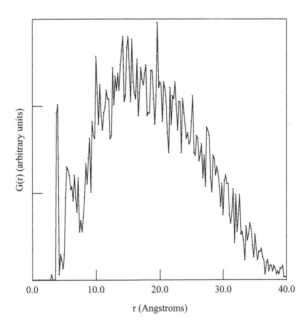

FIGURE FIVE   Radial distribution of α-carbons in ribonuclease A. The distribution is assembled using PDB file 1FS3. The vertical height is proportional to the number of α-carbons placed (on average) at distance *r* from a reference site.

Note how it resembles lysozyme $G(r)$, although this is unapparent from the PDB graphics. The point is accentuated by viewing ribonuclease and lysozyme $G(r)$ together in Figure Six. The peaks are taller for lysozyme (dotted lines) only because the protein contains 130 α-carbons as opposed to 124.

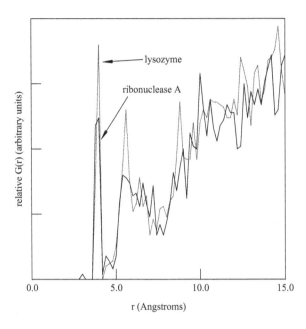

FIGURE SIX Radial distribution functions based on the α-carbons of lysozyme (dotted line) and ribonuclease (solid line). The vertical height is proportional to the number of α-carbons placed (on average) at distance *r* from a reference site.

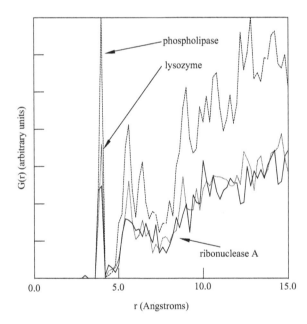

FIGURE SEVEN Radial distribution functions based on the α-carbons of lysozyme (dotted line), ribonuclease (solid line), and phospholipase (dash line).

To close this section, we consider the radial distributions for ribonuclease A, lysozyme, and phospholipase in Figure Seven. The specific functions and disorder carried by the molecules belie their structural similarities. When viewed from an internal and surface perspective, they indeed share multiple nuances of 3D organization.

The takeaway is that folded proteins present *both* structural order and disorder along with corresponding and divergent features. They have plenty of company in this respect. Liquids have densities close to solids, while lacking long-range order. Liquids further present nontrivial shell structures—a few to several depending on the pressure, temperature, solvent, and solutes [47]. Folded proteins are not so unlike this. The subject of shell structures based on pair- and higher-order correlations for protein sets is taken up in Chapter Five.

E WHY ARE PROTEINS SO COMPLICATED? BIOLOGICAL CONSIDERATIONS

Let us consider a principal task of the molecules. An organism maintains itself far from equilibrium over finite time through concentration, thermal, and pressure gradients (cf. Appendix Two). To accomplish this, operations are in place that maintain the critical gradients, but suppress the accumulation of entropy [46,48,49]. As equilibrium states are the maximum entropy ones, organisms strive to avoid these for as long as possible! The flow processes empowered by the gradients generate entropy all the time. Fortunately, the organism has multiple ways to jettison it, for example, by radiating heat to the environment.

But to keep the entropy background tolerable, the organism requires a highly-structured fuel source on a regular basis: it needs to eat. But the organism also has to control the rates of entropy production. Proteins are the molecules in charge of these operations, both directly and indirectly. The simplest organisms such as bacteria require a few thousand proteins to maintain the far-from-equilibrium concentrations of molecules, for example, ATP and ADP. The higher organisms are endowed with tens of thousands of different proteins to get through the day and night.

Then why are proteins so complicated? The question is tantamount to asking why life is complicated. Not life in the philosophical sense, but in the nonequilibrium thermodynamic reality. An organism is a finite-life engine powered by oxidation–reduction reactions having multiple intermediates. How can the polymers which keep the engine running not be complicated?

Proteins have multiple roles. Their enzymatic versions control the rates of chemical reactions in organisms. This includes the reactions whereby DNA is copied, RNAs are synthesized, and mRNAs are the cutting-and-splicing products. The proteins in charge of these operations include polymerases—proteins are thereby responsible for the production of other proteins. Consider further some of the machinery at the production sites [50]:

F_1, F_2, F_3: initiation factors
T, G: elongation factors
R: release factors

All these factors are proteins! Lysozyme, myoglobin, etc., are synthesized through the actions of molecules of their own genre. Clearly proteins are complicated: if they are needed to manufacture one another, their structures and functions intersect with the origins of life [51]. Life requires proteins to get through the day and night. And life requires a highly cooperative collection to manufacture a single protein. Note that systems at the threshold of life rely on proteins just as much. Viruses lack metabolism proteins, but express ones that exploit the manufacturing infrastructure of host organisms. Their agents of invasion, propagation, and virulence are proteins [52].

The complications are multiplicative. If a scientist is interested in a particular protein, he or she cannot help encountering related proteins in the course of study. Ribonuclease A (RNase A) catalyzes the hydrolysis of RNA. But an organism *needs* a mechanism for turning off the catalytic action. It has to guard against its own RNA being hydrolyzed in the wrong places at the wrong times. This requires *another* protein by way of ribonuclease inhibitor (RI). The point is that understanding how RNase A works would seem to require understanding RI-operations [29]. The reverse statement is just as true. Along similar lines are trypsin and its specific inhibitor protein [53]. The functions of the former require the action by the latter in select environments. It does not seem feasible to comprehend the workings of one apart from the other. Protein–protein interactions need not be pair-wise. RNA polymerase complex is composed of five protein subunits: α, β, β', σ, and ω [54]. The operations of one subunit depend on all the others.

The theme of protein cooperation is amplified by observing that inhibitors must not cross boundaries. RI blocks the catalytic functions of RNase A, but must not get in the way of other enzymes. In effect, RI acts as the vast majority of pharmaceutical compounds—drugs are enzyme inhibitors by and large [55]. However, RI chemistry must not impose side effects in the organism. The point is that proteins are complicated as isolated systems *in vitro*. But the molecules are geared to function *in vivo* in clusters and networks. The cooperative phenomena make for structure levels well beyond tertiary and quaternary.

Proteins are further complicated by identity and diversity issues. It is common to speak of ribonuclease A. But such a molecule appears as multiple versions due to the evolution and diversification of species [56]. Each species (humans, horses, whales, etc.) encodes its own form of the enzyme. Even for a given species, there are multiple versions if mutations are taken into account. A famous quote comes from Theodosius Dobzhansky, asserting that nothing in biology makes sense except in light of evolution [57]. Proteins lie at the center of biological structure and function, in tandem with polynucleotides. They are challenging to comprehend *even in* the light of evolution.

The situation is poles apart from nonbiological molecules. Benzene presents various isotopic versions and likewise for proteins. But all benzenes are electronically identical now and centuries from now. In contrast, proteins keep appearing as new versions, or disappearing upon extinction of a species. Each molecule can be thus viewed as an experiment in nature, or perhaps an update in cellular hardware and software. The experiments and updates transpire for better and worse. And their rates of appearance can be very different, depending on the source. For example, the mutations of RNA-encoded viral proteins occur far more frequently than the (DNA-encoded) proteins of organisms [52].

In matters of time, we encounter another source of complexity. A given protein manifests radically different time scales for critical processes. There is the time scale for mutations: years to decades to centuries. The internal units tend to vary slowly so as to preserve the folded structure. The surface residues are subject to higher mutation rates, although the active and recognition sites require conservation. There is the time scale for folding: tens of microseconds to milliseconds. The point is that proteins are complex because they present transformations with time scales that vary by orders of magnitude.

Proteins are complicated because their roles are highly contradictory in places. Ask a person on the street what he or she thinks about the molecules, and the answers might reflect upon principal food groups. Cheese and eggs taste the way they do due to their protein content. Yet people without chemical training would find it surprising that bee and snake venoms are proteins—catalysts that promote the degradation of phospholipids in cellular membranes [58]. This reminds us that while proteins are good for a biological system, they can also be detrimental and even poisonous. There is no shortage of contradictory roles. Chaperone proteins assist the folding of proteins in the organism and are vital to the defense system [59]. Prions effect error-laden folding and imperil the organism. Prions are themselves proteins, playing central roles in neuro-degenerative diseases, among others [60]. Proteins are complex if only for their contradistinctive roles, good and bad.

To close the chapter, we view an enzyme in the most schematic sense. Its job is to accelerate and control the reactions of substrates, so as to yield products at a sufficient rate. One can think of this process in block terms:

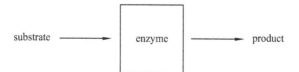

The above portrays protein function as an input/output (I/O) process. But what if the input is wrongly-matched? Then we have:

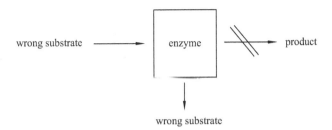

The enzyme discriminates the molecules it encounters to ensure the correct output.

Proteins have just been described in computational terms. Globular proteins are nanometer robots discussed by Tanford and Reynolds [17]. But they can further be viewed as computers responsible for high-fidelity I/O operations. They are not like the everyday computers of laptops and smart phones. Rather enzymes are *Brownian* computers that operate by translational and rotational diffusion in solution environments. This facet is articulated by Bennett in his review of the thermodynamics of computation [61]. Brownian operations occur via thermal collisions with solvent and solute molecules and thus possess an accidental character. It is surprising that chance events can ever be effective—laptops and smart phones could never work this way. For proteins, however, all their actions stem from random collisions and are never absolutely sure things. Yet the molecules succeed in their tasks day-in and day-out. And for every complex facet, there is a gem of fascination and reason to learn more.

The major points of Chapter One are as follows:

1. The building blocks and assembly of proteins are well established and straightforward. However, the number of possibilities for functional group and folded configurations makes for unparalleled chemical diversity and physiological impact. This underpins a complexity rivaled only by polynucleotides—the carriers of the programs directing protein synthesis.
2. Proteins express multiple structure levels that are highly diverse and interwoven. It is convenient to think of levels individually: primary, secondary, etc. However, the details for one level are inextricably linked to the others. Importantly, aspects of structure present correspondences among systems such as illuminated by radial distributions $G(r)$.
3. The experimentally-accessible 3D structures are the equilibrium, time-averaged ones. The effects of thermodynamic fluctuations, however, are significant concerning energy, volume, and other properties. The pictorial representations of proteins are probability-based; they are typically static and should not be interpreted too literally. The real-life objects are dynamic with respect to shape, size, and energy.
4. Proteins are tasked with a spectrum of biological functions. These range from providing connective tissue to operating as Brownian computers in charge of chemical signaling and catalysis. The genome for an organism specifies the protein sequences. However, the functions for the majority of sequences are typically unknown or ill-characterized due to overarching complexity.

EXERCISES

1. Become familiar with the Protein Data Bank (PDB). Review the structure file for lysozyme (1REX). Calculate the distance in Angstroms between the N-terminal nitrogen and C-terminal carbon atoms. Does the distance match that estimated from viewing the folded structure graphics?

2. Write the single- and three-letter alphabet representations of

3. Consider the following sequence fragment:

ARG LEU GLY MET ASP GLY TYR ARG GLY ILE SER LEU ALA

Represent the above using the single-letter alphabet.
Represent the fragment using molecular structure graphs.
Enter the fragment into an online BLAST program. Where does it appear in nature?
Calculate the number of distinguishable polypeptide isomers.
What are the base structures in terms of residues and atoms?
Compute the number of single-site variants. Do likewise for double- and triple-site variants.

4. Consider the fragment MMKSRNLTKDRCKPVNTFVHESLAD. How many arrangements of the protein variety are allowed?

5. Refer to the folded structure of Ribonuclease A (PDB 1FS3). Which cysteines are linked by disulfide bonds? What N- to C-terminal subsequences are formed by the bonds?

6. Consider the sequence:

MKTEWPELVGKSVEEAKKVILQDKPAAQIIVLPVGTIVTYEYRIDRVRLFVDRLD
NIAQVPRVG

Imagine a stockroom that hosts 10^{-15} mol of each isomer related by permutation. How many atoms are in the collection? What is the mass of the collection in grams?

7. Consult a reference table that lists the van der Waals radii of carbon, hydrogen, nitrogen, oxygen, and sulfur atoms. Use these to compute the filling factor for the unit cell of crystalline lysozyme (PDB 1REX). What percent of the unit cell is occupied by material?

8. Several complexities of proteins were described. Write an essay discussing two additional.

9. Proteins are complicated in so many ways. Write an essay discussing ways they are simple.

10. Refer again to the folded structure of Ribonuclease A (PDB 1FS3). Estimate the standard deviation for the internal energy U near physiological temperature. How many hydrogen bonds does this represent?

11. Refer to the folded structure of myoglobin—there are many available via the PDB. Estimate the volume variance near physiological temperature. If the volume fluctuations are concentrated in one or two residues, what distance changes are manifest?

12. Consider a stockroom that hosts a random selection of ten-unit (i.e., $N = 10$) peptides. What are the average molecular weight and the standard deviation? Make a rough sketch of the fraction of peptides versus molecular weight.

13. Write three $N = 30$ sequences at whim. Compare them to Swiss-Prot and NCBI archives using BLAST. Are any close relations identified? Please discuss.

14. Represent the *peptoid* version of DRVRLFVDRLDNIAQVPRVG using molecular structure graphs.

15. Estimate the surface area of folded lysozyme (1REX) in square Angstroms. Do likewise for folded Ribonuclease A (1FS3).

16. Refer to the structure of folded phospholipase (PDB 1POB). Which cysteine sites are joined by disulfide bonds? Count the number of possible disulfide bond configurations.

17. Consider lysozyme in the unfolded state whereby each peptide bond has three possible orientations. Estimate the time required to guess at the correct folded structure, allowing no hints and 10^{-12} s per guess.

18. Consider a protein in the folded state subject to extensive cooling. Predict the effects on structure and then consult the literature for experimental results.

19. Let the previous exercise be directed at the effects of increased pressure applied to folded configurations.

20. The codes for representing protein sequences are used internationally. Other codes would work, however alien and unfriendly they might be mnemonically. What molecule is represented by the following character strings?

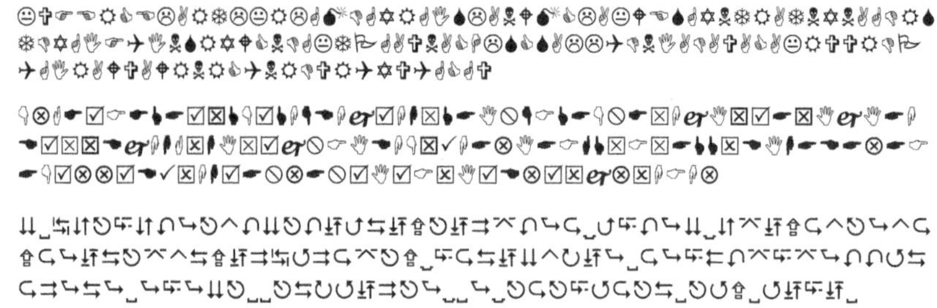

21. Sequences read as encrypted text but admit comic relief. The following represents bona fide stable molecules: SHINEALIGHTINTHEDARK, EATAPPLESDAILY. Can you think of others? And why is the following peptide magic?

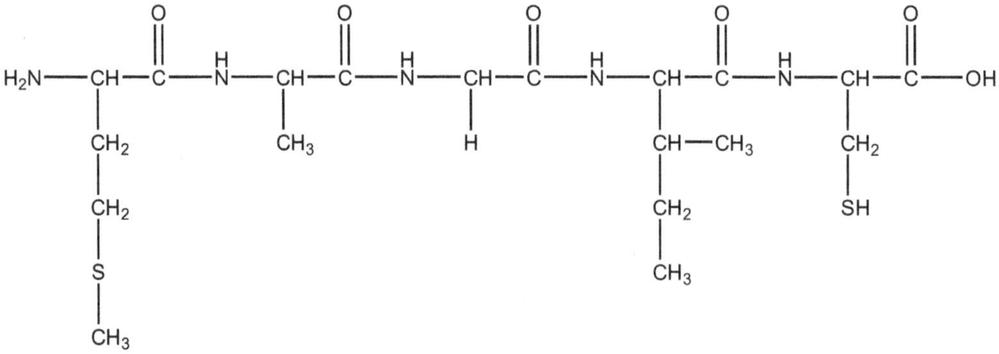

22. Protein sequences readily admit variation by substitution. How many triple-site substitutions are admitted by lysozyme? Can you derive a general formula for m-site substitutions?

NOTES, SOURCES, AND FURTHER READING

The literature is rich with books and articles on the structure and function of proteins; a drop of the ocean is listed below. The same statement holds for the literature surrounding protein and genomic complexity. We call special attention to works by Ben-Naim, Kauffman, and Monod for wide-lens, engaging views [62–66]. Ben-Naim looks to the challenges and controversies of protein thermodynamics while Monod and Kauffman focus on evolutionary issues. We also note the complexities of encoding regions of DNA that underpin protein structure and evolution. The regions are subject to a cornucopia of on–off switching effects as discussed by Alberts [67]. What are the major problems presented by biopolymer sequences, proteins included? Karlin addresses this question succinctly in a review well worth reading [68].

1. Flory, P. J. 1953. *Principles of Polymer Chemistry*, Cornell University Press, Ithaca, NY.
2. Tanford, C. 1961. *Physical Chemistry of Macromolecules*, Wiley, New York.
3. Barrett, G. C. 1998. *Amino Acids and Peptides*, Cambridge University Press, Cambridge, UK.
4. Barrett, G. C. 1985. *Chemistry and Biochemistry of the Amino Acids*, Chapman and Hall, New York.
5. Hughes, A. B. 2009. *Amino Acids, Peptides, and Proteins in Organic Chemistry*, Wiley-VCH, Weinheim.
6. Breslow, R. 2011. The origin of homochirality in amino acids and sugars on prebiotic earth, *Tetrahedron Lett.* 52(32), 4228–4232.

7. Roberts, J. D., Caserio, M. C. 1964. Amino acids, peptides, proteins, and enzymes, in *Basic Principles of Organic Chemistry*, W. A. Benjamin, New York.

8. Appling, D. R., Anthony-Cahill, S. J., Mathews, C. K. 2016. *Biochemistry, Concepts and Connections*, Pearson Education, Hoboken, NJ.

9. White, A., Handler, P., Smith, E. L. 1972. The proteins I, in *Principles of Biochemistry*, McGraw-Hill, New York.

10. Gilbert, W. 1978. Why genes in pieces? *Nature* 271, 501. doi:10.1038/271501a0.

11. Peters, C. W. B., Kruse, U., Pollwein, R., Grzeschik, K.-H., Sippel, A. E. 1989. The human lysozyme gene, *Eur. J. Biochem.* 182, 507–516.

12. Weller, P., Jeffreys, A. J., Wilson, V., Blanchetot, A. 1984. Organization of human myoglobin gene, *EMBO* 3(2), 439–446.

13. Bodansky, M., Klausner, Y. S., Ondetti, M. A. 1976. *Peptide Synthesis*, 2nd ed., Wiley, New York.

14. Merrifield, R. B. 1964. Solid phase peptide synthesis. II. The synthesis of bradykinin. *J. Am. Chem. Soc.* 86, 304.

15. Denkewalter, R. G., Veber, D. F., Holly, F. W., Hirschmann, R. 1969. Total synthesis of an enzyme. I. Objective and strategy. *J. Am. Chem. Soc.* 91, 502. Also see the four papers immediately following.

16. Fleming, I. 1973. Peptide synthesis: Bradykinin, in *Selected Organic Syntheses*, Wiley, New York.

17. Tanford, C., Reynolds, J. 2001. *Nature's Robots: A History of Proteins*, Oxford University Press, New York.

18. Pauling, L. 1970. Biochemistry, in General Chemistry, Dover, New York.

19. Roberts, H. J. 1990. *Aspartame (NutraSweet®): Is it Safe?* Charles Press, Philadelphia, PA.

20. Silverman, R. B. 2008. From basic science to blockbuster drug: The discovery of Lyrica, *Angew. Chem. Int. Ed.* 47(19), 3500–3504.

21. Andrews, M. J. I., Tabor, A. B. 1999. Forming stable helical peptides using natural and artificial amino acids, *Tetrahedron* 55(40), 11711–11743.

22. Romano, F. D. 1979. Effect of Dopamine and Dietary Sodium Deficiency, Thesis (M.S.), Department of Biology, Loyola University Chicago.

23. Bolt, H. L., Cobb, S. I. 2016. A practical method for the synthesis of peptoids containing both lysine-type and arginine-type monomers, *Org. Biomol. Chem.* 14(4), 1211–1215.

24. Otten, J. J., Hellwig, J. P., Meyers, L. D. 2006. DRI, *Dietary Reference Intakes: The Essential Guide of Nutrient Requirements*, National Academies Press, Washington, DC.

25. Wheatley, P. J. 1968. *The Determination of Molecular Structure*, 2nd ed., Dover, New York.

26. Muraki, M., Harata, K., Sugita, N., Sato, K. 1996. Origin of carbohydrate recognition specificity of human lysozyme revealed by affinity labeling, *Biochemistry* 35, 13562.

27. Palzkill, T. 2002. *Proteomics*, Kluwer Academic Publishers, Boston, MA.

28. www.rcsb.org/

29. Chatani, E., Hayashi, R., Moriyama, H., Ueki, T. 2002. Conformational strictness required for maximum activity and stability of ribonuclease A as revealed by crystallographic study of three Phe120 mutants at 1.4 Å resolution, *Protein Sci.* 11, 72–81.

30. Kobe, B., Deisenhofer, J. 1995. A structural basis of the iterations between leucine-rich repeats and protein ligands, *Nature* 374, 183–186.

31. www.expasy.org/blast/

32. Hand, D. J. 2014. *The Improbability Principle: Why Coincidences, Miracles, and Rare Events Happen Every Day*, Scientific American/Farrar, Straus, and Giroux, New York.

33. le Noble, W. J. 1974. *Highlights of Organic Chemistry*, Dekker, New York.

34. Passman, D. S. 1968. *Permutation Groups*, W. A. Benjamin, New York.

35. Scheiner, S. 1997. *Molecular Interactions: From van der Waals to Strongly Bound Complexes*, Wiley, New York.

36. Goodstein, D. L. 1985. *States of Matter*, Dover, New York.

37. Hill, T. L. 1986. *An Introduction to Statistical Thermodynamics*, Dover, New York.

38. Cooper, A. 1976. Thermodynamic fluctuations in protein molecules, *Proc. Natl. Acad. Sci. U. S. A.* 73(8), 2740–2741.

39. Wang, P. H., Rupley, J. A. 1979. Protein-water interactions. Heat capacity of the lysozyme-water system, *Biochemistry* 18(12), 2654–2661.

40. Pauling, L. 1960. *The Nature of the Chemical Bond*, 3rd ed., Cornell University Press, Ithaca, NY.

41. Glaser, R. 2005. *Biophysics*, Springer, New York.
42. Blumlein, A., McManus, J. J. 2013. Reversible and non-reversible thermal denaturation of lysozyme with varying pH at low ionic strength, *Biochim. Biophys. Acta* 1834(10), 2064–2070.
43. Prigogine, I. 1957. *The Molecular Theory of Solutions*, North Holland Publishing Company, Amsterdam.
44. Johns, W. F. 1976. *Steroids*, Butterworths, London.
45. Merriam-Webster's Collegiate Dictionary, 10th Edition, 1999, Merriam-Webster, Inc., Springfield, MA.
46. Schrödinger, E. 1948. *What is Life?* Cambridge University Press, Cambridge, UK.
47. Rice, S. A., Gray, P. 1965. *The Statistical Mechanics of Simple Liquids*, Interscience, New York.
48. Yourgrau, W., van der Merwe, A., Raw, G. 1982. *Treatise on Irreversible and Statistical Thermophysics*, Dover, New York.
49. Küppers, B.-O. 1990. *Information and the Origin of Life*, MIT Press, Cambridge, MA.
50. Lehninger, A. L. 1970. *Biochemistry*, Worth Publishers, New York.
51. Dyson, F. 1999. *Origins of Life*, Cambridge University Press, Cambridge, UK.
52. Kuo, L. C., Olsen, D. B., Carroll, S. S. 1996. *Viral Polymerases and Related Proteins*, Academic Press, San Diego, CA.
53. Ye, S., Loll, B., Berger, A. A., Muelow, U., Alings, C., Wahl, M. C., Koksch, B. 2015. Fluorine teams up with water to restore inhibitor activity to mutant BPTI. *Chem. Sci.* 6, 5246–5254.
54. McKnight, S. L., Yamamoto, K. R. 1992. *Transcriptional Regulation*, Cold Spring Harbor Laboratory Press, Plainview, NY.
55. Copeland, R. A. 2005. *Evaluation of Enzyme Inhibitors in Drug Discovery*, Wiley, Hoboken, NJ.
56. Raines, R. T. 1998. Ribonuclease A, *Chem. Rev.* 98(3), 1045–1066.
57. Dobzhansky, T. G. 1973. Nothing in biology makes sense except in the light of evolution, *Am. Biol. Teacher* 35(3), 125–129; JSTOR 4444260.
58. Moseley, W. 1987. *The Phospholipases*, Plenum Press, New York.
59. Gierasch, L. M., Horwich, A., Slingsby, C., Wickner, S., Agard, D. 2016. *Structure and Action of Molecular Chaperones: Machines that Assist Protein Folding in the Cell*, World Scientific, Hackensack, NJ.
60. Wetzel, R., Kheterpal, I. 1999. *Amyloid, Prions, and Other Protein Aggregates, Methods in Enzymology*, Academic Press, San Diego, CA.
61. Bennett, C. H. 1982. Thermodynamics of computation—A review, *Int. J. Theor. Phys.* 21, 905.
62. Ben-Naim, A. 2013. *The Protein Folding Problem and Its Solutions*, World Scientific, Hackensack, NJ.
63. Ben-Naim, A. 2016. *Myths and Verities in Protein Folding Theories*, World Scientific, Hackensack, NJ.
64. Kauffman, S. A. 1993. *The Origins of Order: Self-Organization and Selection in Evolution*, Oxford University Press, New York.
65. Kauffman, S. A. 1995. *At Home in the Universe: The Search for Laws of Self-Organization and Complexity*, Oxford University Press, New York.
66. Monod, J. 1971. *Chance and Necessity: An Essay on the Natural Philosophy of Modern Biology*, Knopf, New York.
67. Alberts, B. 2008. *Molecular Biology of the Cell*, Garland Science, New York.
68. Karlin, S., Buche, P., Brendel, V., Altschul, S. F. 1991. Statistical methods and insights for protein and DNA sequences, *Annu. Rev. Biophys. Biophys. Chem.* 20, 175–203.

Two Essentials of Information

Information is central to protein structure analysis. This chapter offers a précis based on systems, states, and mathematical probability. Attention is given to the different types of information and the distance measures between states. Liquid and vapor solutions offer familiar systems amenable to characterization by information vectors. They prepare the stage for exploring the lower-level structures of proteins.

A PRELIMINARIES

The preceding chapter focused on the structure and complexity of proteins. The molecules are organic polymers, but not just any polymers. Why so? The answers include that they express extraordinary amounts of information. The biology canon teaches that an amino-acid sequence, in conjunction with the solution environment, holds information for directing the folding process [1–6]. A folded configuration, solvent, and counter-ions carry information directing the chemical functions. Revisit the sequence for lysozyme:

KVFERCELARTLKRLGMDGYRGISLANWMCLAKWESGYNTRATNYNAGDRS
TDYGIFQINSRYWCNDGKTPGAVNACHLSCSALLQDNIADAVACAKRVVR
DPQGIRAWVAWRNRCQNRDVRQYVQGCGV

This can be viewed as a high-level representation of *part* of the program that steers the polymer to an ensemble of folds, *and* empowers recognition and catalytic functions. The program is robust and rapidly executed given favorable thermal and aqueous conditions. If lysozyme is substituted in a few places, the polymer and solution are still able to "find" ensemble members in a few milliseconds. Recall that lysozyme manifests more than one sequence internally. Disulfide bridges, hydrogen, and van der Waals bonds allow multiple paths from the N- to C-terminal sites [7]. The primary structure delineates *one* path, its information *only part* of the picture. All this begs the question: what is information?

Check the nearest dictionary for definitions. Then consider two ideas. First is that information is ultimately material, the substance of which controls how work is performed [8]. CDs and DVDs are round objects with special coatings. Their light-passage properties direct the output of playing devices; their material and fabrication constitute information. The same principle applies to proteins: compositions, folds, and solvent are the means to control biochemical operations. Information is matter and details. All molecules and environments hold information of one form or another.

Yet information presents an abstract side as well, which is the second idea. It can be regarded as a measure of "how far" one's knowledge moves as a system transits from one state to another [9]. "How far" does not refer to a Euclidean distance, but rather a metric grounded on observation and expectation. If *our* knowledge changes by watching a system change from an initial to final state, then we, *another* system so interfaced, acquire information. If a recording device alters in response to a system, then *it* gains information which can be transferred elsewhere. Information depends as much on observers, devices, and circumstances as it does on the systems of interest. And with certain assumptions in place, the amount can be quantified. This is rooted in three propositions [10–12].

The first is that information is additive for independent events. How a dime lands post-flipping provides information if we choose to look at it. But the event—flipping, coming to rest, and our observation—has nothing to do with, say, a bridge hand we are dealt. We receive information from two sources, the total being the sum of two independent quantities.

The second proposition is that highly improbable events have the potential for granting more information than probable ones. If we observe a dime land heads twenty times in a row, the process taking order into account yields more information than eleven heads interspersed with nine tails.

The third proposition is an extension of the second. Sure-to-occur events offer zero information. The sun rose this morning and we knew it was going to do so. Our state of knowledge did not change when we saw the morning light.

The mathematical function which engages the three propositions is as follows:

$$I = K \cdot \log_b prob$$

where I represents the information measure, b is a base selected for computing logarithms (e.g., 2, e, 10), and $prob$ is the event probability. K is a constant usually defined as -1. It is traditional, although not exclusively so, to work in base-two ($b = 2$), whence the unit attached to I is "bit," short for binary digit.

We see why the function does justice to events and observations. Independent events A and B have *joint* probability equal to the product of individual probabilities [13]:

$$prob_{AB} = prob_A \cdot prob_B$$

The information is additive as our intuition tells us. Taking $K = -1$, we have

$$I_{AB} = -\log_2\left(prob_{AB}\right) = -\log_2\left(prob_A \cdot prob_B\right)$$

$$= -\log_2\left(prob_A\right) - \log_2\left(prob_B\right)$$

Improbable events carry high information whereby *our* state of knowledge moves a great deal when we see them occur. Twenty dimes flipped and all landing heads express probability:

$$prob_{20\,\text{heads}} = \left(\frac{1}{2}\right)^{20} \approx \frac{1}{1.05 \times 10^6}$$

The information gained by the observation is as follows:

$$I = -\log_2\left(\frac{1}{2}\right)^{20} \approx -\log_2\left(\frac{1}{1.05 \times 10^6}\right)$$

$$\approx \frac{-1}{\ln(2)} \cdot \ln\left(\frac{1}{1.05 \times 10^6}\right)$$

$$\approx 20 \text{ bits}$$

By contrast, the event of eleven heads mixed with nine tails is not so surprising: our knowledge state alters only a little. The probability can be shown to be:

$$prob_{11\,\text{heads},9\,\text{tails}} = \frac{20!}{11!9!} \times \left(\frac{1}{2}\right)^{11} \times \left(\frac{1}{2}\right)^{9}$$

$$\approx 0.160$$

The information gain is

$$I = -\log_2(0.160)$$

$$\approx \frac{-1}{\ln(2)} \cdot \ln(0.160)$$

$$\approx 2.64 \text{ bits}$$

Why is information in the abstract form important? The reasons are the same as for probability. Both notions reflect upon our sense of possibility for a system and conditions. Where an event probability is regarded as low, the information to be gained upon observation is potentially high. Furthermore, high-gain circumstances motivate fresh digging into systems and processes. Finding a rabbit in a hat has exceedingly low probability, or so an audience is encouraged to believe, but a magician is able to do so in every performance with probability 1. The high information gained by the audience at the rabbit's appearance sparks not just entertainment, but also questions about magic tricks. The magician acquires *zero* information upon seeing the rabbit. Its appearance was never in doubt!

B   QUERIES, INFORMATION, AND PROBABILITY

Information is intertwined with systems, states, and communication [14]. For its measure to be established, the following apparatus must be in place one way or another:

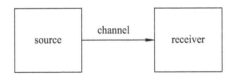

The source is a system of interest, while the channel and receiver are part of the experimental configuration. The receiver includes the scientist and recording media such as computers, notebooks, and flash drives. Valid experiments are those which yield results unknown in advance. The simplest experiments (in principle) seek answers to yes/no questions, for example,

Does the sample contain magnesium?
Does lysozyme hydrolyze cellulose?
Does ribonuclease A also hydrolyze DNA?

The results are thereby restricted to two outcomes.
Let the number of possible outcomes be represented by Ω. Then for an experiment aimed at one of the above questions,

$$\Omega = 2 = 2^1 = 2^{\text{number Y/N Qs}}$$

The number of Y/N questions equates with the exponent which, in turn, establishes the information in bits. Information is material and configuration details, but the quantification proceeds with the questions addressed by observation. Alternatively, we can think of bits as equivalent to the number of symbols needed to record an observation. Different results call for different symbols such as 0 and 1. When a question is answered, one symbol is applied while the other is placed aside. Experiments which address multiple questions require the most code for record keeping. Such experiments move the scientist's state of knowledge the farthest. By the same token, if the scientist *knows* the outcome of an experiment because a colleague has already performed it and shared the results, then $\Omega = 1$ and $I = \log_2\Omega = 0$ bits. His or her state of knowledge does not move anywhere.

But then things turn complicated. While an outcome may be unknown in advance, some are more plausible than others. It is important to consider the amount of information *expected* from an experiment.

Consider a procedure aimed at a yes/no question, with "yes" held to be far more likely than "no." The degree of plausibility is gauged by probability having a universal measure between 0 and 1. Suppose the scientist anticipates "yes" and "no" with probabilities 0.999 and 1–0.999, respectively. If he or she performs the experiment, and finds "yes," the exercise provides little information. In contrast, if the outcome proved "no," that would be quite surprising, and the information gain would be high. But such a gain is improbable, or so the scientist thinks.

We introduce S_i, the surprisal—measure of surprise—of an outcome labeled by index *i*:

$$S_i = -\log_2(prob \text{ of } i\text{th event})$$

S_i equates with the information posed by *the i*th event, as opposed to some other event labeled by *j*. When *all* possible events are taken into account, the *expected* value (i.e., average) is as follows:

$$I = \langle S_i \rangle = -\sum_i prob(i) \cdot \log_2(prob(i))$$

This is the formula for the Shannon information [10–12,14]. It is closely related to the entropy of mixing in solutions and the entropy function in statistical thermodynamics [15–17]. The expectation of *I* is indeed often referred to as the Shannon entropy. Information is material and configuration details, but its abstraction is rooted in thermodynamics and statistical physics.

There is a subtle point being made, namely that if information pertains to states of knowledge, then so does probability. Furthermore, information is a by-product of any event that alters probability. When a scientist gains information, the probabilities of states change across the board. When and where probability holds steady, the scientist gains zero information. There are countless examples which demonstrate this. The following one is discussed by Hamming and, from a somewhat different perspective, by Parzen [18,19].

Imagine a deck of shuffled cards and inquire: what is the probability of drawing the ace of spades on the second card? The answer depends not only on the system ↔ cards, but also *our* state of knowledge.

Consider two scenarios. In the first, we draw the top card and look at it. Then we draw the second card and view it. In the second scenario, we draw the top card *without* looking at it. Then we draw and look at the second card.

In the first scenario, the probability of finding the ace of spades to be the top card is 1/52. If the card is indeed the ace of spades, then the probability of the second card being the ace of spades is 0/51. If the top card is not the ace of spades, then the second-card probability is 1/51. There is nothing counter-intuitive going on.

But then we consider the second scenario. We did not view the top card. It could very well be the ace of spades—or not. As for the second card, we compute the probability of it being the ace of spades as follows:

$$prob = \frac{51}{52} \times \frac{1}{51} = \frac{1}{52}$$

Because the cards are placed independently *and* randomly in the deck, the probabilities of the possible draw events multiply. 51/52 is the probability of the top card *not* being the ace of spades. 1/51 is the probability that the second card *is* the ace of spades. The overall probability is 1/52—the *same* as the top card in scenario one *being* the ace of spades! The probability of the second card being the ace of spades is not 0/51 or 1/51, because such values would require *our* state of knowledge to have changed somehow. The probability is 1/52 because, by not viewing the first or any other card, our knowledge of the deck remained unaltered.

Examples such as cards provide a minimalist and highly portable definition of probability [9]:

$$prob \text{ of event } A = \frac{\text{number of possibilities } \textit{favorable to } A}{\text{total number of } \textit{equally likely} \text{ possibilities in } X}$$

This can be represented pictorially as follows:

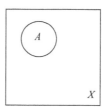

The circle encloses *A*, a set of points within the total sample space *X*. The probability of observing an event located *somewhere* in *A* via some random process can be imagined as the ratio of the "sizes" (aka *measures*) of *A* and *X*. Let the sizes be denoted by $|A|$ and $|X|$. Then

$$prob(A) = \frac{|A|}{|X|}$$

Note that in defining the probability of an *A*-event, we automatically arrive at another: that of an event *not* in *A*. Such events make up the complement of *A* denoted by A^c, viz.

$$prob(A^c) = 1 - \frac{|A|}{|X|}$$

This shows probabilities to present not as singles, but rather in sets containing two or more elements: they constitute *set* functions as opposed to *point* functions. Note further how the minimalist definition connects with our intuitive sense of information. If we were to learn that an event landed the system in *A*, we would know *more* about the location in the sample space, compared with an A^c-landing—the boundaries on the *A*-points are more confining.

Ratios and pictorials are classical and by no means the final word on probability and information. Nor are they appropriate for every situation—see Hamming, Parzen, Fine, and di Finetti for enlightenment [18–21]. However, the definitions, pictures, and one's intuition suffice where the states of a system are in full view. Fortunately, the accommodating venues include the letter strings that represent protein sequences. Probability, like information, is multi-faceted because, in so many situations, a quantity depends on the observer's state of knowledge [9,21]. Moreover, the notions are fluid because the measures change with every observation. Indeed, a single observation alters an entire set of probabilities and the information so underpinned. We provide more details as follows.

Experimental outcomes come in multiples while symbols and ordered sets are the means for representing them. An experiment may have a set of outcomes {*a*, *b*, *c*, *d*, *e*} with corresponding probabilities {0.10, 0.20, 0.050, 0. 55, 0.10}. The probabilities (fractions) sum to 1 by the convention. A set is not presented as {0.25, 0.00, 0.30, 0.00, 0.10, 0.35} for this would imply an impossible outcome and infinite information upon its occurrence. We instinctively minimize the number of set elements to consider [22]. The sample space becomes the *smallest possible* for the events of interest. Revisit the matter of coin tossing.

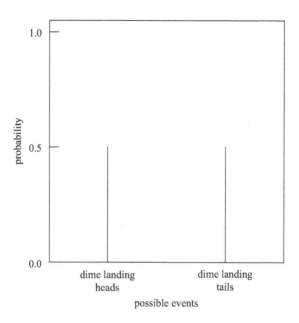

FIGURE ONE   The probability distribution that applies to a fair coin *before* it is tossed.

A coin adopts one of two states following a toss. The outcomes and probability sets are {heads, tails} and {0.50, 0.50}. Both sets reflect our knowledge of the system ↔ coin and experiment ↔ tossing. If we have no reason to suspect something fishy, we view the event probabilities as equal—one landing scenario seems as likely as the other. We practice the principle of maximum indifference, lacking supplementary data or mystical powers [9,23]. In picture form, the event and probability sets look like Figure One.

A single toss, landing, and observation yield information I in amount:

$$I = -\frac{1}{2}\log_2\left(\frac{1}{2}\right) - \frac{1}{2}\log_2\left(\frac{1}{2}\right)$$

$$= -\log_2\left(\frac{1}{2}\right)$$

$$= +\log_2\left(2^1\right)$$

$$= 1 \text{ bit}$$

But now imagine that the dime lands heads. *Now* the picture looks like Figure Two.

By acquiring 1.00 bit of information, we arrive at new, much-reduced sets {heads} and {1}. There are no other possible states that can be accessed by the system at rest. If we were to leave the room and return, we would know exactly what to expect for the state of the coin. To be sure, the sets can always be restored to {H, T} and {0.50, 0.50}. But this requires work on our part and a new experiment with a to-be-determined outcome. The point is that acquiring information alters not just a single probability, but rather a set.

A dictionary lists multiple definitions of information. Unsurprisingly, the abstract form manifests more than one way. The Shannon information reflects the amount *we* expect from an experiment. A second manifestation acknowledges the fallibility of our knowledge and inclination to make assumptions. Consider two probability sets A and B and the following weighted sum:

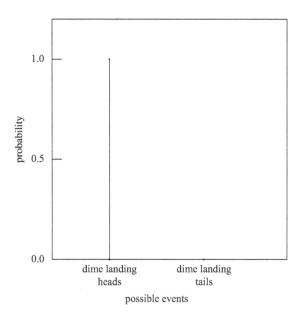

FIGURE TWO   The probability distribution that applies to a fair coin *after* it has been tossed and observed to land heads.

$$K_I^{(A,B)} = + \sum_i prob(A_i) \cdot \log_2 \left[\frac{prob(A_i)}{prob(B_i)} \right]$$

The logarithm argument is a ratio of two probabilities with the summation leading to the Kullback–Leibler information [24]. This can be viewed as the penalty that *would* be assessed if we assumed—erroneously or not—that the *A*-events were paired with *B*-probabilities [25]. If our assumption is sound, there is zero penalty: $prob(A_i)/prob(B_i) = 1$ whence $\log_2(1) = 0$ for all i. Our knowledge does not alter by the experiment (although our assumptions probably become further cemented!). If the assumption is incorrect, there are bits to pay depending on the "gap" separating *A* and *B*. It can be shown that in all cases, $K_I \geq 0$.

There are two principles at work. The first is that information operates as a commodity and even type of currency. This makes sense: it costs work, time, and storage media to trap and preserve it. K_I teaches that information grounded upon incorrect assumptions warrants a penalty or forfeiture of some kind. We have to spend resources to bring our knowledge to where it should be.

The second idea is the *A, B* separation. For example, let *A* and *B* be {3/4, 1/4} and {1/3, 2/3}, respectively. Then we have:

$$K_I = \frac{3}{4} \log_2 \left[\frac{\frac{3}{4}}{\frac{1}{3}} \right] + \frac{1}{4} \log_2 \left[\frac{\frac{1}{4}}{\frac{2}{3}} \right]$$

$$= \frac{3}{4} \log_2 \left[\frac{9}{4} \right] + \frac{1}{4} \log_2 \left[\frac{3}{8} \right]$$

$$= \frac{\frac{3}{4} \ln \left[\frac{9}{4} \right] + \frac{1}{4} \ln \left[\frac{3}{8} \right]}{\ln(2)}$$

$$\approx 0.524 \text{ bits}$$

This can be viewed as a non-Euclidean distance separating sets A and B But note the asymmetry. We customarily think of distances between objects as:

It is as far from A to B above as B to A, right? This is not the case with probability sets and comparisons via K_I. Assume (erroneously) that the B events with *true* probabilities $\{1/3, 2/3\}$ were allied with $A \leftrightarrow \{3/4, 1/4\}$. Then we have:

$$K_I^{(B,A)} = \frac{1}{3}\log_2\left[\frac{\frac{1}{3}}{\frac{3}{4}}\right] + \frac{2}{3}\log_2\left[\frac{\frac{2}{3}}{\frac{1}{4}}\right]$$

$$= \frac{\frac{1}{3}\ln\left[\frac{4}{9}\right] + \frac{2}{3}\ln\left[\frac{8}{3}\right]}{\ln(2)}$$

$$\approx 0.553 \text{ bits}$$

The distance is close to, but not the same as previously! It is just as noteworthy that K_I values do not always satisfy the triangle inequality. Consider three sets A, B, C: $\{0.75, 0.25\}$, $\{0.50, 0.50\}$, $\{0.25, 0.75\}$. It can be shown that the following inequality does not hold across the boards:

$$K_I^{(A,C)} \leq K_I^{(A,B)} + K_I^{(B,C)}$$

Suffice to say that K_I is a bridge connecting new experiments with old assumptions, the latter often in error. But then another type of information connects experiments with hard facts. This is *conditional* information, which is really just a refinement of ideas already presented. All probability is conditional in one or more ways. When playing bridge, we have to know the number of cards, suits, and colors to assign probability values. The values are conditioned by our knowledge and assumptions of a righteous deck.

But now imagine a friend who draws a card, inspects it, and then asks *you* to guess the identity. You rightly figure the probability of the card being the ace of spades (or something else) as 1/52. But let your friend volunteer the color of the card. Clearly this increases your knowledge. If the card is black, your view of the ace of spades probability increases from 1/52 to 1/26. If he then shows the card, your information gain is $-\log_2(1/26)$ bits. This is less than $-\log_2(1/52)$ because you had already acquired $-\log_2(1/2)$ bits via the color. If the friend stated that the color was red, your measure of the ace of spades probability drops to zero. Regardless of color, your information upon learning the card identity is capped at $-\log_2(1/26)$ bits. It is because the friend supplies a hard fact about the system.

Conditional information impacts how much is gained from further inquiry. The amount hinges on cognition of one or more properties. Notation-wise, we represent conditional probability via *prob(A|B)*: the probability of observing A *given* knowledge that B, which may or may not relate to A, is the case. The expression is written equivalently as $prob_B(A)$. Conditional probability is the basis for conditional information whereby the surprisal becomes:

$$S_{(A|B)} = -\log_2 prob(A \mid B)$$

Note that conditional probability need not exceed unconditioned, and conditional information need not be less than unconditioned. What if B has nothing to do with A? This would be the case if the friend described minutiae of the reverse side of the card. From your perspective, the probability of the card being the ace of spades would remain at 1/52, and the information gained from learning the card would be $-\log_2(1/52)$ bits. In other words,

$$S_{(A|B)} = -\log_2 prob(A \mid B) = -\log_2 prob(A)$$

if A and B are independent. Conditional probability and information connect with our common sense when dealing with uncertainty. It is easy to cite examples with everyday systems such as cards and board games. To travel beyond the superficial, however, we appeal again to pictorials.

Let us imagine a system of interest with many possible states. Let the states be categorized (grouped) as A, B_1, B_2, B_3, and B_4 based on measurable properties. An experiment enables the system to express *one* of the possible states. Let a state correspond to a cluster of points in a category region, the location unknown in advance of the experiment. Furthermore, let the probability of "landing" within a category be proportional to the size or span of the category. These stipulations revisit our classical definition of probability. The sample or total state space may then be visualized by Figure Three. The A- and B-regions are specified by the boundaries. The number of possible states in each category is proportional to the area for that category.

But note the special feature. The B-regions are exclusive of one another: their territories do not overlap, although they do to various degrees with A. An experimental result corresponds to a landing *somewhere* in the total space. The landing zone—cluster of points—is *sure* to lie in one of the B-regions while it may or may not lie in A. Sure-to-occur events have probability 1. Thus we have:

$$prob(B_1) + prob(B_2) + prob(B_3) + prob(B_4) = 1$$

Furthermore, the probability of landing the point in any particular region, say A, equates with an area ratio:

$$prob(A) = \frac{|A|}{|B_1| + |B_2| + |B_3| + |B_4|}$$

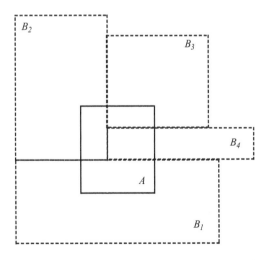

FIGURE THREE Schematic for the spaces allotted to five states, four of them being mutually exclusive. A is bounded by the solid line while the B_i regions are bounded by dotted lines.

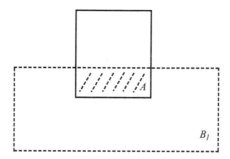

FIGURE FOUR   Schematic for overlapping state spaces. The cross-hatched area is proportional to the probability that, given an experimental result lies in B_1, the result also lies in A.

We are taken immediately to conditional probability. If tests show that the state of the system lies in, say, B_1, what is the probability that the state *also* lies in A? The answer is given pictorially by Figure Four. The cross-hatches mark the intersection of A and B_1. The A probability is *conditioned* by B_1 and given by

$$prob(A \mid B_1) = \frac{|A \cap B_1|}{|B_1|}$$

But the areas are proportional to their total enclosed points: they are *proportional* to probability measures. This tells us that:

$$prob(A \mid B_1) = \frac{prob(A \cap B_1)}{prob(B_1)}$$

whereupon

$$prob(A \mid B_1) \cdot prob(B_1) = prob(A \cap B_1)$$

This is a multiplication formula that arises prominently in probability and information analysis. Note that if there were no intersection, then

$$prob(A \mid B_1) = \frac{|A \cap B_1|}{|B_1|} = \frac{|\varnothing|}{|B_1|} = 0$$

This symbol $\varnothing$ represents the empty or vacuum set having zero measure. The above ratio applies if an experimental result corresponding to B_1 shows an A co-assignment to be impossible. This is obviously not the case in Figures Three and Four.

An experimental result corresponds to points in the state space. What is the probability of the points falling within A? Yet again, the answer is represented by a ratio. But now think of it in terms of Figure Five. The figure shows $|A|$ to be the sum of four areas of intersection:

$$prob(A) = \frac{|A \cap B_1| + |A \cap B_2| + |A \cap B_3| + |A \cap B_4|}{|B_1| + |B_2| + |B_3| + |B_4|}$$

$$= \frac{prob(A \cap B_1) + prob(A \cap B_2) + prob(A \cap B_3) + prob(A \cap B_4)}{prob(B_1) + prob(B_2) + prob(B_3) + prob(B_4)}$$

But the denominator sums to 1 while each numerator term can be represented as a product. This directs us to

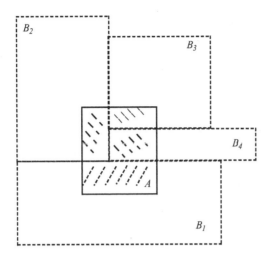

FIGURE FIVE   Revisited view of the state space. Each of the B_i-categories "contributes" to the probability of landing in A.

$$prob(A) = prob(A \mid B_1) \cdot prob(B_1) + prob(A \mid B_2) \cdot prob(B_2) + prob(A \mid B_3) \cdot prob(B_3)$$

$$+ \, prob(A \mid B_4) \cdot prob(B_4)$$

This can be condensed as

$$prob(A) = \sum_{i=1}^{4} prob(A \mid B_i) \cdot prob(B_i)$$

The above asserts what is called the law of *total* probability [26]. It describes how the likelihood of event A is determined by *all* the mutually exclusive events that lead to it. Alternatively the above equations often referred to as the *partition* formula: the B_i regions do not overlap, but their union spans the total sample space. Partitions will be a critical concept in the analysis of protein structure information in Chapters Three and beyond.

But now consider matters in reverse. If tests show that an experiment landed on points in A, what is the probability that the points *also* lie in, say, B_1? The answer refers again to the cross-hatches marking the A, B_1 intersection in Figure Four. This A-*conditioned* probability is as follows:

$$prob(B_1 \mid A) = \frac{|A \cap B_1|}{|A|} = \frac{prob(A \cap B_1)}{prob(A)}$$

But the law of total probability and the multiplication formula enable us to re-write the above as follows:

$$prob(B_1 \mid A) = \frac{prob(A \mid B_1) \cdot prob(B_1)}{\displaystyle\sum_{i=1}^{4} prob(A \mid B_i) \cdot prob(B_i)}$$

This is an illustration of the rule of Bayes and Laplace [9,26]. It relates an event A to its possible "causes." The rule is a cornerstone of mathematical probability and information, although its application proceeds often with controversy. There is nothing untoward about the rule. However, there is typically uncertainty in one's knowledge of causes, not to mention the assumptions of mutual exclusivity and completeness.

Probability and information are tools that apply to countless systems. These would include the symbol strings KVFER...GCGV that represent protein sequences. This does not mean that the sequences have anything to do with games of chance—although evolution involves Brownian events over millennia. It is rather that *our* knowledge of the sequences is incompletely drawn—it is daunting to impossible to infer missing pieces or spot erroneous ones in unfamiliar strings. This hearkens back to the complexities in Chapter One. Probability and information, however, provide a framework for growing knowledge so as to shine light on new questions. This is demonstrated in Section D and is a recurring theme in subsequent chapters. All applications entail counting and grouping the symbols in sequences, and quantifying probability and information measures. The information in sets of sequences can further be quantified, bearing in mind the asymmetries of K_I. But before demonstrating applications, there is one more tool to present, one that quantifies the distances between probability sets in a symmetric way.

C   PROBABILITY SETS AND INFORMATION VECTORS

We become familiar with vectors from analytic geometry, mechanics, electricity and magnetism, and thermodynamics [27,28]. While mathematically abstract, they are easily imagined via arrows. Vectors in relation to probability and information provide another application, one useful to the analysis of sequences and thermodynamic systems in general. Things work as follows.

Probability sets A and B offer information measures I and K_I. The sets further accommodate vector representations, viz.

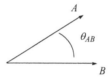

Each element of A and B is paired with an allowed state $\leftrightarrow$ experimental outcome. The number of states defines the dimension of the vector space. The probability of each state then establishes the vector strength along a coordinate. Together, the probabilities determine precisely how a vector points.

Different probability sets confer different vector orientations as above. The distance between two vectors arises from the angle θ_{AB} having the following properties:

$$\theta_{AB} \geq 0$$

$$\theta_{AB} = 0 \text{ for } A = B$$

$$\theta_{AB} = \theta_{BA}$$

Furthermore, given three sets A, B, C, the triangle inequality holds:

$$\theta_{AC} \leq \theta_{AB} + \theta_{BC}$$

The symmetry relation $\theta_{AB} = \theta_{BA}$ is particularly important. It ensures that it is as far from the A- to B-vectors as it is from the B- to A-vectors. This property is deficient when comparing sets via K_I. The triangle relation is just as critical, viz.

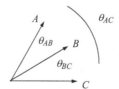

We are assured the distance from A- to C-vectors is never greater than from A- to B-*plus* from B- to C-vectors. In short, the angles present as *true* distances that conform to our sense of space.

The expressions for vectors are straightforward:

$$\vec{V}^{(A)} = \sum_{i=1}^{N} prob_i^{(A)} \cdot \hat{x}_i$$

$$\vec{V}^{(B)} = \sum_{i=1}^{N} prob_i^{(B)} \cdot \hat{x}_i$$

The $prob_i$ are the elements of the probability sets which multiply unit vectors $\hat{x}_i$. The notation is conventional: the arrow over a V indicates a quantity having both direction and magnitude while the summations are taken over all N dimensions. The unit vectors can be imagined as equal-length arrows pointing at right angles to each other, that is,

$$\theta_{ij} = \frac{\pi}{2}, i \neq j$$

In so doing, they meet the following conditions:

$$\hat{x}_i \cdot \hat{x}_j = 1, i = j$$

$$\hat{x}_i \cdot \hat{x}_j = 0, i \neq j$$

As an example, let A and B, as in Section B, be {3/4, 1/4} and {1/3, 2/3}, respectively. Then we have:

$$\vec{V}^{(A)} = \sum_{i=1}^{N=2} prob_i^{(A)} \cdot \hat{x}_i = \frac{3}{4} \cdot \hat{x}_1 + \frac{1}{4} \cdot \hat{x}_2$$

$$\vec{V}^{(B)} = \sum_{i=1}^{N=2} prob_i^{(B)} \cdot \hat{x}_i = \frac{1}{3} \cdot \hat{x}_1 + \frac{2}{3} \cdot \hat{x}_2$$

All should be clear so far. To measure the angle between vectors, however, the strengths need to be re-scaled so that they, like unit vectors, have lengths equal to 1. As things stand, the vectors have nonuniform lengths less than 1, for example,

$$\left| \vec{V}^{(A)} \right| = \sqrt{\left(\frac{3}{4} \cdot \hat{x}_1 + \frac{1}{4} \cdot \hat{x}_2 \right)^2} = \sqrt{\left(\frac{3}{4} \cdot \hat{x}_1 \cdot \frac{3}{4} \cdot \hat{x}_1 \right) + \left(\frac{1}{4} \cdot \hat{x}_2 \cdot \frac{1}{4} \cdot \hat{x}_2 \right)}$$

$$= \sqrt{\left(\frac{3}{4} \right)^2 + \left(\frac{1}{4} \right)^2} \approx 0.791$$

The rescaling is completed by computing square roots. The components become $\left(prob_i \right)^{1/2}$ whereupon

$$\left| \vec{V}_1 \right| = \sqrt{\sum_{i=1}^{i=N} \left(prob_i \right)^{1/2} \cdot \left(prob_i \right)^{1/2}} = 1$$

The 1-subscript signifies a unity length. Dot products then quantify the projection of two vectors in the standard way:

$$\vec{V}_1^{(A)} \cdot \vec{V}_1^{(B)} = \left| V_1^{(A)} \right| \left| V_1^{(B)} \right| \cos \theta_{AB}$$

$$= 1 \cdot 1 \cdot \cos \theta_{AB}$$

The angle arrives by the inverse cosine function:

$$\theta_{AB} = \cos^{-1} \left\{ \sum_{i=1}^{i=N} \left(prob_i^{(A)} \right)^{1/2} \cdot \left(prob_i^{(B)} \right)^{1/2} \right\}$$

For A and B above, we have:

$$\theta_{AB} = \cos^{-1} \left\{ \sqrt{\frac{3}{4}} \cdot \sqrt{\frac{1}{3}} + \sqrt{\frac{1}{4}} \cdot \sqrt{\frac{2}{3}} \right\} \approx \cos^{-1}(0.9082)$$

$$\approx 0.432 \text{ radians} \approx 24.7°$$

θ_{AB} measures the *gap* between vectors *and* the probability sets upon which they are based. Large gaps reflect sets and systems that are markedly disparate in their information expression. Small gaps reflect near-equal expression. This method of gauging distance is intuitive and finds applications in statistics, thermodynamics, and quantum mechanics, to name a few subjects [29–31]. It is just as applicable to protein sequences as shown in Section D and beyond.

To demonstrate the vector approach, and to set the stage for sequence analysis, we visit briefly the thermodynamics of solutions. These are everyday lab samples formed by mixing components in the liquid phase: ethanol and water, hexane and cyclohexane, benzene and toluene, etc. [32]. Each combination offers infinite possibilities and need not be binary; for example, hexane, benzene, and toluene form a ternary solution. Importantly, a preparation yields *not* just one solution, but rather two or more, depending on the components and conditions. The second solution is typically a vapor that, if allowed, establishes equilibrium with the liquid. We demonstrate how a vector can be constructed for each phase based on the mole fractions. The simplest examples involve ideal solutions, although the vector approach applies just as well to nonideal solutions.

Ideal solutions feature organic molecules which are structurally very similar, for example, 2-methyl-pentane (2MP, below left) and 3-methyl-pentane (3MP, right):

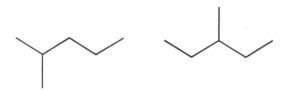

The *CRC Handbook of Chemistry and Physics* offers extensive thermodynamic data (vapor pressure, heats of vaporization, etc.) for organic compounds such as above [33]. Ideal solutions have signature properties which include adherence to Raoult's Law [15]. This holds that the equilibrium partial pressure (p) of a vapor phase component is proportional to the component mole fraction (X) in the liquid. The proportionality constant is the vapor pressure $p°(T)$ of the *pure* liquid component, dependent on temperature T. Thus for a 2MP, 3MP binary system at equilibrium:

$$p_{2MP}^{(vap)} = p_{2MP}°(T) \cdot X_{2MP}^{(liq)}$$

$$p_{3MP}^{(vap)} = p_{3MP}^{\circ}(T) \cdot X_{3MP}^{(liq)}$$

The liquid mole fractions are determined by the preparation details while the vapor fractions are subsidiary, equating with the pressure fractions:

$$X_{2MP}^{(vap)} = \frac{p_{2MP}^{(vap)}}{p_{2MP}^{(vap)} + p_{3MP}^{(vap)}}$$

$$X_{3MP}^{(vap)} = \frac{p_{3MP}^{(vap)}}{p_{2MP}^{(vap)} + p_{3MP}^{(vap)}}$$

For added simplicity, we are assuming ideal gas behavior by the vapor phase.

Solutions are a short step from probabilistic thinking. Imagine selecting a molecule at random from, say, the liquid portion of a 2MP, 3MP solution. The probability of the molecule being 2MP equates with the mole fraction $X_{2MP}^{(liq)}$. Corresponding statements hold for 3MP in the liquid phase and molecules in the vapor phase. Thus, the mole fractions do double duty: they provide mixing details *and* probability measures for experiments of the real or contemplative variety. The proviso is that we assume macroscopic samples at equilibrium temperature and pressure, and that fluctuations and surface effects can be ignored.

Ideal solutions do not form azeotropes wherein the liquid and vapor mole fractions are identical [15]. Thus for virtually all preparations, the vectors representative of equilibrium phase compositions point in different directions, viz.

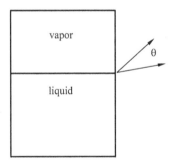

That being said, the more ideal the solution, the *closer* the vectors align under equilibrium conditions, viz.

vapor phase

liquid phase

In effect, the composition of one phase carries significant information about another. This will be a recurring theme in our examination of protein and proteome compositions. For now, it is worthwhile to examine the angles for a variety of solution conditions. Similar computations apply to individual protein sequences approached by vector methods, for primary structures are like solutions prepared at the nanometer scale.

To begin, the vapor pressure of an organic compound, and its dependence on temperature, is encapsulated (cf. *CRC Handbook*) in the following (or equivalent) formula:

$$\log_{10} p = \frac{-0.2185\alpha}{T} + \beta$$

Pressure p and temperature T are measured in torr and degrees Kelvin, respectively. The symbols α and β represent constants specific to a component. A few examples are:

	α	β
2-methyl-pentane	7,676.6	7.944630
3-methyl-pentane	7,743.9	7.947042
Cyclohexane	7,830.9	7.662126
n-hexane	7,627.2	7.717119

Imagine a 2MP, 3MP system at $T = 294$ K with liquid mole fractions 0.400 and 0.600, respectively. The partial pressures in the vapor follow from Raoult's Law and the *Handbook* data:

$$p_{2\mathrm{MP}}^{(\mathrm{vap})} = p_{2\mathrm{MP}}^{\circ}(T) \cdot X_{2\mathrm{MP}}^{(\mathrm{liq})}$$

$$= 10^{\frac{-0.2185\alpha}{T} + \beta} \cdot 0.400 \approx 69.4 \text{ torr}$$

$$p_{3\mathrm{MP}}^{(\mathrm{vap})} = p_{3\mathrm{MP}}^{\circ}(T) \cdot X_{3\mathrm{MP}}^{(\mathrm{liq})}$$

$$= 10^{\frac{-0.2185\alpha}{T} + \beta} \cdot 0.600 \approx 93.3 \text{ torr}$$

The ideal vapor mole fractions become:

$$X_{2\mathrm{MP}}^{(\mathrm{vap})} = \frac{p_{2\mathrm{MP}}^{(\mathrm{vap})}}{p_{2\mathrm{MP}}^{(\mathrm{vap})} + p_{3\mathrm{MP}}^{(\mathrm{vap})}} \approx \frac{69.4}{69.4 + 93.3} \approx 0.426$$

$$X_{3\mathrm{MP}}^{(\mathrm{vap})} = \frac{p_{3\mathrm{MP}}^{(\mathrm{vap})}}{p_{2\mathrm{MP}}^{(\mathrm{vap})} + p_{3\mathrm{MP}}^{(\mathrm{vap})}} \approx \frac{93.3}{69.4 + 93.3} \approx 0.574$$

We can now construct vectors for the two phases in equilibrium:

$$\vec{V}^{(\mathrm{liq})} = \sum_{i=1}^{N=2} X_i^{(\mathrm{liq})} \cdot \hat{x}_i = 0.400 \cdot \hat{x}_{2\mathrm{MP}} + 0.600 \cdot \hat{x}_{3\mathrm{MP}}$$

$$\vec{V}^{(\mathrm{vap})} = \sum_{i=1}^{N=2} X_i^{(\mathrm{vap})} \cdot \hat{x}_i = 0.426 \cdot \hat{x}_{2\mathrm{MP}} + 0.574 \cdot \hat{x}_{3\mathrm{MP}}$$

The angle between the vectors, given rescaling is:

$$\theta_{\mathrm{liq,vap}} = \cos^{-1}\left\{ \sqrt{0.400} \cdot \sqrt{0.426} + \sqrt{0.600} \cdot \sqrt{0.574} \right\} \approx \cos^{-1}(0.9997)$$

$$\approx 0.0264 \text{ radians} \approx 1.53°$$

The phase vectors are nearly aligned because the mole fraction sets are so close together. The solutions are nearly perfect!

The same computations apply in reverse, although are a little more challenging. Imagine a 2MP, 3MP sample at $T = 294$ K with *vapor* mole fractions 0.400 and 0.600, respectively. Then we have:

$$X_{2\mathrm{MP}}^{(\mathrm{vap})} = 0.400 = \frac{p_{2\mathrm{MP}}^{(\mathrm{vap})}}{p_{2\mathrm{MP}}^{(\mathrm{vap})} + p_{3\mathrm{MP}}^{(\mathrm{vap})}} = \frac{p_{2\mathrm{MP}}^{\circ} \cdot X_{2\mathrm{MP}}^{(\mathrm{liq})}}{p_{2\mathrm{MP}}^{\circ} \cdot X_{2\mathrm{MP}}^{(\mathrm{liq})} + p_{3\mathrm{MP}}^{\circ} \cdot X_{3\mathrm{MP}}^{(\mathrm{liq})}}$$

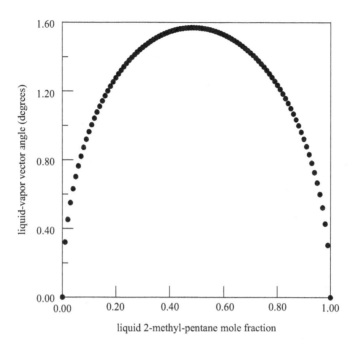

FIGURE SIX   The plot shows the angle between the liquid-vapor phase vectors for solutions of 2- and 3-methyl pentane as a function of the 2-methyl pentane liquid mole fraction. The temperature used for the computations was 294 K.

$$X_{3MP}^{(vap)} = 0.600 = \frac{p_{3MP}^{(vap)}}{p_{2MP}^{(vap)} + p_{3MP}^{(vap)}} = \frac{p_{3MP}^{\circ} \cdot X_{3MP}^{(liq)}}{p_{2MP}^{\circ} \cdot X_{2MP}^{(liq)} + p_{3MP}^{\circ} \cdot X_{3MP}^{(liq)}}$$

The liquid mole fractions are obtained after some algebra:

$$X_{2MP}^{(liq)} \approx 0.374 \quad X_{3MP}^{(liq)} \approx 0.626$$

The angle between the vectors, after rescaling is:

$$\theta_{liq,vap} = \cos^{-1}\left\{\sqrt{0.374} \cdot \sqrt{0.400} + \sqrt{0.626} \cdot \sqrt{0.600}\right\} \approx \cos^{-1}(0.9996)$$

$$\approx 0.0270 \text{ radians} \approx 1.53°$$

The vectors are closely aligned here as well!

The method of vector analysis applies to all mole fraction combinations, and to *all* solutions, ideal and otherwise. It applies even when the system is *not* a solution per se. A pure sample is where the vapor and liquid have identical compositions. For pure 3MP, the mole fraction set is {1, 0} for *both* {$3MP_{liq}$, $2MP_{liq}$} and {$3MP_{vap}$, $2MP_{vap}$}. The liquid and vapor phase vectors can be imagined as exactly collinear. The addition of 2MP to form a solution rotates the vectors with respect to each other.

Solutions are infinitely tunable whereby the angles over the span of mole fractions are readily computed. An example is plotted in Figure Six. The vectors never stray far apart given the similarities of the molecular components.

D   INFORMATION AND PROTEIN SEQUENCES

It may seem a stretch to see how the foregoing ideas connect with amino-acid sequences. On the one hand, they *look* random and evocative of multi-component solutions. On the other hand, it is not clear how events and probabilities make contact. A letter string KVFER…GCGV seems just that and nothing more.

The contact is made by exploring sequences much more than casually. For lysozyme and ribonuclease A, we have, respectively [7,34]:

KVFERCELARTLKRLGMDGYRGISLANWMCLAKWESGYNTRATNYNAGDRS
TDYGIFQINSRYWCNDGKTPGAVNACHLSCSALLQDNIADAVACAKRVVR
DPQGIRAWVAWRNRCQNRDVRQYVQGCGV

KETAAAKFERQHMDSSTSAASSSNYCNQMMKSRNLTKDRCKPVNTFVHESL
ADVQAVCSQKNVACKNGQTNCYQSYSTMSITDCRETGSSKYPNCAYKTTQAN
KHIIEGNPYVPVHFDASV

The above constitutes knowledge about real physical entities. The bridge to information obtains by the questions we pose and experiments we conduct. The latter do not require lab benches and glassware, but only the means to discriminate and count symbols. Our efforts *are* experiments in that we never know precisely the outcomes in advance, our chemical training notwithstanding. In turn, letter strings yield *new* information and *move* our knowledge farther than it would otherwise be. There is really no limit—a sequence does not pose a single Shannon, Kullback–Leibler, or surprisal value. Rather the bits hinge on what we are curious about, and our capacity to phrase questions. For example, let us imagine that a site is chosen by some random process in lysozyme. We inquire:

1. How surprised would we be upon learning the identity?
2. How many bits would be assessed if we anticipated the identity based on knowledge of ribonuclease?

The questions are elementary, and yet there are multiple answers to each to enrich our perspective. This is because the information depends not just on the protein, but also on what *we* elect to call a site. For example, we could consider a site as having two possibilities: alanine ↔ A and not-alanine ↔ L, I, P, …, K, R, H. Or a site could pose twenty possibilities: A, V, L, …, K, R, H. A site need not be monomeric, for it could be a dimer (e.g., AV), trimer (e.g., AVL), or cluster (e.g., KVFERCEL). Site selection need not be so abstract. We can imagine the process as a thermal collision between the protein and, say, a solvent molecule over the course of Brownian motion. Information will always depend on the protein structure *and* our practice of curiosity. The sequences are simply the raw materials for transforming our knowledge and curiosity. In thermodynamic terms, the sequences are *the* systems of interest, the low-level states describable by counting variables. Or in probability language, sequences are the event spaces for which multiple sets can be formed and measures assigned. Critically, not all questions are within bounds. What is our surprise upon learning that a site is part of a β-sheet? An α-helix? Does a particular site bind ATP? Is the protein impacted by a specific inhibitor? Does the protein fold this way or that way? Does it fold at all? While these may be genuine curiosities, we can only address such questions by looking beyond the letter strings toward wet chemistry and diffraction (or NMR) data. Justifiable questions throughout this book are restricted to the sequences—and nothing more. They are more akin to: is the following true for integer n [35]?

$$\sqrt{n}^{\sqrt{n+1}} < \sqrt{n+1}^{\sqrt{n}}$$

The answer is within our grasp by expending work and we acquire as much as 1.00 bit of information in the process. It is the same for exploring protein sequences: the answers to questions are always in full view and require only modest work grouping and counting symbols. As with mathematical expressions, efforts can be assisted by computation, in particular, using elementary programming, spreadsheets, and word processors. We illustrate a few examples while further practice is encouraged in the end-of-chapter exercises.

Let us start in the simplest way and regard sites as monomeric and of two types, alanine and not-alanine. We regard every site as equally accessible in a random selection process. It is quick work using the find function of a word processor to count letters. We identify 14A in $N = 130$ lysozyme and 12A in $N = 124$ ribonuclease A where N marks the total letters in a sequence. The sample spaces are $X_{lysozyme}$ and $X_{ribonuclease\,A}$ with 130 and 124 respective possibilities. An event makes contact with a site and connects with our minimalist definition of probability:

$$prob \text{ of } A = \frac{\text{number of possibilities favorable to A}}{\text{total number of equally likely possibilities in } X}$$

$$prob \text{ of not-A} = \frac{\text{number of possibilities unfavorable to A}}{\text{total number of equally likely possibilities in } X}$$

Then if A were *the* site selected in lysozyme, our surprise would be:

$$S_A = -\log_2\left(\frac{14}{130}\right) = \frac{-\ln\left(\frac{14}{130}\right)}{\ln(2)} \approx 3.22 \text{ bits}$$

But our expectation is considerably lower and given by a weighted average:

$$\langle S \rangle = -\left(\frac{14}{130}\right)\log_2\left(\frac{14}{130}\right) - \left(\frac{130-14}{130}\right)\log_2\left(\frac{130-14}{130}\right) \approx 0.492 \text{ bits}$$

The reason is that the A-sites form the minority party. We expect *not* to encounter one of them by random event.

The numbers are different if the sites are labeled by the full 20-symbol alphabet. We count for lysozyme:

14A	2F	5T	8D
9V	5W	8C	3E
8L	2M	6Y	5K
5I	11G	10N	14R
2P	6S	6Q	1H

The numbers reflect the base structures introduced in Chapter One. Our surprise at selecting an A-site remains at 3.22 bits. But our expectation involves a more extended sum:

$$\langle S \rangle = -\sum_i^{i=20} f_i \log_2 f_i$$

The i subscripts refer to each type of amino acid, viz.

$$\langle S \rangle = -f_A \log_2 f_A - f_V \log_2 f_V - f_L \log_2 f_L - \cdots$$

$$= -\left(\frac{14}{130}\right)\log_2\left(\frac{14}{130}\right) - \left(\frac{9}{130}\right)\log_2\left(\frac{9}{130}\right) - \left(\frac{8}{130}\right)\log_2\left(\frac{8}{130}\right) - \left(\frac{5}{130}\right)\log_2\left(\frac{5}{130}\right) - \cdots$$

$$\approx 4.08 \text{ bits}$$

Question (2) concerned penalties based on outright false assumptions. We compare the alanine fraction with that of ribonuclease A. With only two types of sites, we have:

$$K_I = + \sum_i^{i=2} f_i \log_2 \left(\frac{f_i}{q_i} \right)$$

$$= \left(\frac{14}{130} \right) \log_2 \left(\frac{\frac{14}{130}}{\frac{12}{124}} \right) + \left(\frac{130-14}{130} \right) \log_2 \left(\frac{\frac{130-14}{130}}{\frac{124-12}{124}} \right) \approx 9.52 \times 10^{-4} \text{ bits}$$

The penalty is miniscule (perhaps surprisingly so) because the two enzyme sequences are highly similar concerning their A-fractions. This is not so obvious when the primary, secondary, and tertiary structures are viewed on computer screens or PowerPoint slides. As such, it motivates digging further into what the sequences and base structures are all about. The penalty levied on the basis of 20-letter assumptions is left as an exercise. Suffice to say that each calculation enables a comparison of systems of interest. It is akin to discriminating the mole fractions of different phases as in Section C.

This brings us to the last tool of the chapter and a question: what is the angle in radians between the information vectors for lysozyme and ribonuclease A? Alternatively, how close are low-level structure elements of the two proteins?

Just as a thermodynamic solution, a protein sequence can be allied with one or more information vectors. The component number is determined by how we assign the states. Our binary case (A, not-A) presents two-dimensional (2D) vectors while the case of 20-states calls for 20D vectors. We are unable to imagine things visually in 20D, but fortunately, we never have to. Regardless of dimension, the vectors so assembled always have unity magnitude and it is their dot products that we care about. It is straightforward to construct 2D and 20D types following the guidelines of Section C. In 2D for A, not-A states we have:

$$\vec{V}_{\text{lysozyme}}^{(2D)} = \sqrt{\frac{14}{130}} \hat{x}_A + \sqrt{\frac{130-14}{130}} \hat{x}_{\text{not A}}$$

$$\vec{V}_{\text{ribonuclease A}}^{(2D)} = \sqrt{\frac{12}{124}} \hat{x}_A + \sqrt{\frac{124-12}{124}} \hat{x}_{\text{not A}}$$

Vector dot products lead us to:

$$\theta_{\text{lys,rib A}} = \cos^{-1} \left\{ \sqrt{\frac{14}{130}} \cdot \sqrt{\frac{12}{124}} + \sqrt{\frac{130-14}{130}} \cdot \sqrt{\frac{124-12}{124}} \right\} \approx \cos^{-1}(0.99983)$$

$$\approx 1.81 \times 10^{-2} \text{ radians}$$

In 20D, things look like:

$$\vec{V}_{\text{lysozyme}}^{(20D)} = \sqrt{\frac{14}{130}} \hat{x}_A + \sqrt{\frac{9}{130}} \hat{x}_V + \sqrt{\frac{8}{130}} \hat{x}_L + \sqrt{\frac{5}{130}} \hat{x}_I + \sqrt{\frac{2}{130}} \hat{x}_P + \cdots$$

$$\vec{V}_{\text{ribonuclease}}^{(20D)} = \sqrt{\frac{12}{124}} \hat{x}_A + \sqrt{\frac{9}{124}} \hat{x}_V + \sqrt{\frac{2}{124}} \hat{x}_L + \sqrt{\frac{3}{124}} \hat{x}_I + \sqrt{\frac{4}{124}} \hat{x}_P + \cdots$$

It should be clear how to complete the calculation to obtain the angle between 20D vectors. The angle is not inconsequential given the component diversity of the two proteins. Yet other ways of viewing them, for example, alanine, not-alanine, illuminate closely aligned vectors. This teaches something important about the biopolymers, especially the enzymatic ones, and will be elaborated upon in subsequent chapters.

The major points of Chapter Two are:

1. Information is material by way of the composition and configuration for a system. By expressing information, a system harbors a capacity for controlling how work is performed, either within itself, or in connection with another system. All systems present information of multiple types and the letter strings used to represent protein sequences are no exception.
2. Information and its abstract form are rooted in probability and the states of a system. The acquisition of information is concomitant with experiments which alter probability values and our knowledge and sense of the system. The results of an experiment do not change a single probability value, but rather an entire set.
3. Information and probability are time-honored analytical tools. Their extensions include the rule of Bayes and Laplace, the Kullback–Leibler information, and vectors for measuring the distance between probability sets.
4. Protein sequences present facts and data, just as the recipes for multi-component solutions. Information obtains by the questions we submit and is the return on modest work invested. Addressing questions is tantamount to experiments, the procedures carried out by symbol counting and grouping.

EXERCISES

1. Compute the information in bits provided by three cards drawn in succession from a complete and well-shuffled deck. In scenario one, the cards are viewed individually as they are drawn. In scenario two, only the third card is viewed.
2. Consider ideal solutions of cyclohexane and n-hexane. Construct a diagram analogous to Figure Six appropriate to room temperature conditions. How does the diagram change if the temperature is raised by 10°C?
3. Review the major ideas of Appendix Four. Then compute the probability of observing eleven heads and nine tails in twenty coin flips irrespective of order. How does the value change if we take order into account, for example, by observing eleven heads in a row followed by nine tails in a row?
4. Consider a room temperature solution of ethanol and water at the azeotropic point. Refer to thermodynamic tables for details. Construct the information vectors for the liquid and vapor phases. Compute the angle between the vectors.
5. Show that the Kullback–Leibler information is positive definite: that in all cases, $K_I \geq 0$.
6. Consider three coins A, B, and C with disparate landing probabilities. Is the following statement always true: $K_I^{(A,C)} \leq K_I^{(A,B)} + K_I^{(B,C)}$?
7. Consider the *possible* disulfide bridge configurations in lysozyme (cf. Chapter One). How many bits of information are acquired when a scientist learns the correct configuration? What assumptions are being made for this calculation?
8. Consider an experiment which selects a single amino-acid site of lysozyme at random such as by a thermal collision. How many bits of information are obtained if *we* were to learn the identity? What assumptions are made in arriving at the answer?
9. This is a two-party exercise. Let one individual learn the primary structure of a protein of interest by reading the literature. The second individual must then ask yes/no questions (and receive honest answers) in order to identify the first ten amino acids. What is the total number of questions asked? Does the number seem reasonable based on one's knowledge about proteins and information quantification?

10. Consider 1.00 micromole of thoroughly-hydrolyzed lysozyme molecules in weak acid solution. Imagine drawing single amino acids from solution at random. What is the number of alanines *expected* in 1.30 million draws? What is the *most likely* number drawn?

11. Complete the discussion of information vectors in Section D. Construct the 20D vectors for lysozyme and ribonuclease A and compute the angle between them.

12. Calculate the penalty in bits for assuming that the amino-acid fractions in lysozyme are the same as for ribonuclease A.

13. A single amino-acid site is chosen at random in lysozyme in pH-neutral aqueous solution and found to be charged. What is the probability that the site has a basic substituent attached to the α-carbon? What is the probability that the substituent has an acidic functionality?

14. Probability theory is the formal study of triples $(S, \mathbb{S}, P)$. S denotes a sample set of events while $\mathbb{S}$ represents a class of sets. P is a function which ascribes fractional real numbers greater than or equal to zero to the subsets of S. Discuss how these ideas apply to a protein sequence.

15. Revisit the sequence for lysozyme with the amino-acid sites numbered by index k: $1 \leq k \leq 130$. We count left-to-right starting from the N- to C-terminal site. Define sets a, b, and c by the amino acids represented in the following sections:

$$a: 1 \leq k \leq 10 \qquad b: 11 \leq k \leq 20 \qquad c: 21 \leq k \leq 30$$

    a.   Show that $a \cap (b \cup c) = (a \cap b) \cup (a \cap c)$.

    b.   Show that $a \cup (b \cap c) = (a \cup b) \cap (a \cup c)$.

16. Revisit the sequence for lysozyme with the amino-acid sites numbered. Consider the molecular weights of the first twenty amino acids: KVFER... $\leftrightarrow$ 146.2, 117.2, 165.2, 147.2, 174.2... g/mol. They present as random variables—a viewer unfamiliar with lysozyme would not know what is coming at, say, $k = 21$. Use the weight series to illustrate Markov's inequality:

$$prob(mwt > \varepsilon) \leq \frac{\langle mwt \rangle}{\varepsilon}.$$

The brackets denote the average molecular weight in the series. Are there limits placed on ε in grams per mole?

17. Consider the same weight series as in the previous exercise. Illustrate Chebyschev's inequality by considering:

$$prob\left(\left|mwt - \langle mwt \rangle\right| > \delta\right) \leq \frac{\sigma_{mwt}^2}{\delta^2}$$

σ_{mwt}^2 denotes the variance of the molecular weight. Are there limits placed on δ?

18. A set of states aligns with a set of probabilities. What probability set corresponds to the $\{B_i\}$ represented in Figure Three? To answer this question, consider the relative areas for each state. Construct a prior-to-experiment distribution as in Figure One. Do likewise for a post-experiment distribution.

19. Revisit the minimalist definition of probability in pictorial terms, viz.

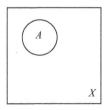

Is $prob(A)$ independent of $prob(A^c)$? Discuss this in light of the notion of conditional probability.

20. Liquid-vapor systems in equilibrium seldom conform exactly to Raoult's Law. Consider the case of positive deviations for a binary solution of components A and B, that is,

$$p_A^{(vap)} > p_A^\circ(T) \cdot X_A^{(liq)}$$

How do the deviations affect the angle between information vectors for the liquid and vapor phases? Do negative deviations have the same effect?

NOTES, SOURCES, AND FURTHER READING

Chemistry and biochemistry students gain experience with probability and information beginning in the first year. Laboratories are the venues for significance testing of data toward hypothesis acceptance or rejection. Quantum chemistry, molecular modeling, and crystallography courses feature spatial probabilities through charge density maps. Statistical thermodynamics addresses the state probabilities of equilibrium systems. Bioinformatics courses teach extensively about the statistical significance of sequence alignments. The list goes on; it is impossible for students *not* to cross paths with the probability structures of interesting systems.

The topics in this chapter deal primarily with mathematical probability. This subject underpins the applications in science, but is centered on notions of sets, measures, and Boolean algebras. The references offer a sampling of textbooks in these areas. The reader is encouraged to consult the writings of Jaynes, Halmos, and Kolmogorov [9,12,36]. The probability facets of proteins fall into two basic camps: statistical thermodynamic and sequence analysis. The reader is directed to Hill and Ben-Naim for expositions on the former [17,37]. Works by Karlin and White and texts by Yockey and Vidyasagar supply indispensible reading for the latter [38–41]. A text by the author illustrates connections between chemical thermodynamics and information theory [10]. Pande and co-workers have addressed the lack of randomness in natural protein sequences [42]. Dewey has considered the information and complexity of protein sequences [43,44]. Chaitin has established the fundamentals of algorithmic complexity with applications to general sequences [45].

1. Lehninger, A. L. 1975. *Biochemistry*, 2nd ed., Worth, New York.
2. Stryer, L. 1975. *Biochemistry*, W. H. Freeman, San Francisco, CA.
3. Appling, D. R., Anthony-Cahill, S. J., Mathews, C. K. 2016. *Biochemistry, Concepts and Connections*, Pearson Education, Hoboken, NJ.
4. White, A., Handler, P., Smith, E. L. 1972. The proteins I, in *Principles of Biochemistry*, McGraw-Hill, New York.
5. Tanford, C., Reynolds, J. 2001. *Nature's Robots: A History of Proteins*, Oxford University Press, New York.
6. Anfinson, C. B. 1973. Principles that govern the folding of protein chains, *Science* 181, 223–230.
7. Muraki, M., Harata, K., Sugita, N., Sato, K. 1996. Origin of carbohydrate recognition specificity of human lysozyme revealed by affinity labeling, *Biochemistry* 35, 13562.
8. Landauer, R. 1993. Information is Physical, Proc. Workshop of Physics and Computation, p. 1, IEEE Com. Sci. Press, Los Alamos, NM.
9. Jaynes, E. T. 1979. Where Do We Stand on Maximum Entropy? in *The Maximum Entropy Formalism*, M. Tribus, R. D. Levine, eds., MIT Press, Cambridge, MA.
10. Graham, D. J. 2011. *Chemical Thermodynamics and Information Theory with Applications*, Taylor and Francis Groups, CRC Press, Boca Raton, FL.
11. Applebaum, D. 1996. *Probability and Information: An Integrated Approach*, Cambridge University Press, Cambridge, UK.
12. Halmos, P. R. 1944. The foundations of probability theory, *Am. Math. Monthly* L1(9), 493–510.
13. Itô, K. 1984. *Introduction to Probability Theory*, Cambridge University Press, New York.
14. Shannon, C. E. 1948. A mathematical theory of communication, *Bell Sys. Tech. J.* 27, 379.

15. Denbigh, K. 1971. *The Principles of Chemical Equilibrium*, Cambridge University Press, Cambridge, UK.
16. Ben-Naim, A. 2006. The entropy of mixing and assimilation: An information-theoretical perspective, *Am. J. Phys.* 74(12), 1126–1135.
17. Hill, T. L. 1986. *An Introduction to Statistical Thermodynamics*, Dover, New York.
18. Hamming, R. W. 1991. *The Art of Probability for Scientists and Engineers*, Addison-Wesley, Redwood City, CA.
19. Parzen, E. 1960. *Modern Probability Theory and Its Applications*, Wiley, New York.
20. Fine, T. L. 1973. *Theories of Probability: An Examination of Foundations*, Academic Press, New York.
21. De Finetti, B. 1975. *Theory of Probability: A Critical Introductory Treatment*, Wiley, New York.
22. Yeh, R. Z. 1973. *Modern Probability Theory*, Harper & Row, New York.
23. Tribus, M., Levine, R. D., eds. 1979. *The Maximum Entropy Formalism*, MIT Press, Cambridge, MA.
24. Kullback, S. 1997. *Information Theory and Statistics*, Dover, New York.
25. Brillouin, L. 1956. *Science and Information Theory*, Academic Press, New York.
26. Reza, I., Chacon, P., Kac, M. 1988. *Basic Stochastic Processes: The Mark Kac Lectures*, Macmillan, London.
27. MacDuffee, C. C. 1949. *Vectors and Matrices*, Mathematical Association of America, Ithaca, NY.
28. Weinhold, F. 1976. Geometric representation of equilibrium thermodynamics, *Acc. Chem. Res.* 9, 232.
29. Wootters, W. K. 1981. Statistical distance and Hilbert space, *Phys. Rev. D* 23, 357–362.
30. Braustein, S. L., Caves, C. M. 1994. Statistical distance and the geometry of quantum states, *Phys. Rev. Letts.* 72, 3439–3443.
31. Kowalski, A. M., Martin, M. T., Plastino, A., Rosso, O. A., Casas, M. 2011. Disturbances in probability space and the statistical complexity setup, *Entropy* 13, 1055–1075. doi:10.3390/e13061055.
32. Ricci, J. E. 1951. *The Phase Rule and Heterogeneous Equilibrium*, Van Nostrand, New York.
33. 1972. *Handbook of Chemistry and Physics*. 52nd ed., Chemical Rubber Company, R. C. Wheast, ed., Cleveland, OH.
34. Raines, R. T. 1998. Ribonuclease A, *Chem. Rev.* 98(3), 1045–1066.
35. Halmos, P. R. 1983. *Selecta*, Springer-Verlag, New York, p. 212.
36. Kolmogorov, A. N. 1956. *Foundations of the Theory of Probability*, Chelsea Publishing Company, New York.
37. Ben-Naim, A. 1992. *Statistical Thermodynamics for Chemists and Biochemists*, Plenum Press, New York.
38. Karlin, S., Buche, P., Brendel, V., Altschul, S. F. 1991. Statistical methods and insights for protein and DNA sequences, *Annu. Rev. Biophys. Biophys. Chem.* 23, 407–439.
39. White, S. H. 1994. Global statistics of protein sequences: Implications for the origin, evolution, and prediction of structure, *Annu. Rev. Biophys. Biomolec. Struct.* 23, 407–439.
40. Yockey, H. P. 1992. *Information Theory and Molecular Biology*, Cambridge University Press, Cambridge, UK.
41. Vidyasagar, M. 2014. *Hidden Markov Processes: Theory and Applications to Biology*, Princeton University Press, Princeton, NJ.
42. Pande, V. S., Grosberg, A. Y., Tanaka, T. 1994. Nonrandomness in protein sequences: Evidence for a physically driven stage of evolution, *PNAS USA* 91, 12972–12975.
43. Strait, B. J., Dewey, T. G. 1996. The Shannon information entropy of protein sequences, *Biophys. J.* 71, 148.
44. Dewey, T. G. 1997. The algorithmic complexity of a protein, *Phys. Rev. E* 56, 4545.
45. Chaitin, G. J. 1987. *Algorithmic Information Theory*, Cambridge University Press, Cambridge, UK.

Three Protein Structure Analysis at the Base Level

Proteins offer information in abundance and their structures motivate endless questions. This chapter illustrates properties of the base level which are not so apparent from casual inspection of sequences, secondary structures, and folded-state graphics. The properties are investigated using tools of the previous chapter in conjunction with symbol counting and grouping. The illustrations center on archetypal globular proteins such as lysozyme, ribonuclease A, and myoglobin. Experiments direct us to a framework for reading and discriminating base structures. Further insights arrive by exploring the variations among bases. The chapter is closed by attention to the reciprocal relations between base and primary structures.

A PRELIMINARIES

Chapter One presented the fundamentals of protein components, assembly, and structure levels. Complexities arose at the outset due to the length, diversity, and aperiodicity of amino-acid sequences. Collectively these properties make the typical sequence appear cryptic and random to a viewer. It seems impossible for a viewer—even expert—to spot errors or supply inferences for unfamiliar sequences with confidence. It appears just as impossible to project how a given sequence relates to another; each presents its own puzzle and special case. Revisit the representations for human lysozyme C, bovine ribonuclease A, and human myoglobin [1–3]:

KVFERCELARTLKRLGMDGYR**GI**SLANWMCLAKWESGYNTRATNYNAGDRS
TDY**GI**FQINSRYWCNDGKTPGAVNACHLSCSALLQDNIADAVACAKRVVR
DPQ**GI**RAWVAWRNRCQNRDVRQYVQGCGV

KETAAAKFERQHMDSSTSAASSSNYCNQMMKSRNLTKDRCKPVNTFVHESL
ADVQAVCSQKNVACKNGQTNCYQSYSTMSITDCRETGSSKYPNCAYKTTQAN
KHIIVACEGNPYVPVHFDASV

MGLSDGEWQLVLNVWGKVEADIPGHGQEVLIRLFKGHPETLEKFDKFKHLK
SEDEMKASEDLKKHGATVLTALG**GI**LKKKGHHEAEIKPLAQSHATKHKIPVKY
LEFISECIIQVLQSKHPGDFGADAQGAMNKALELFRKDMASNYKELGFQG

Multiple abbreviations are in place (i.e., single letters instead of molecular graphs) and the absence of patterns precludes further ways to shorten the strings, much less relate them to each other. This may not seem the case at first and second glance. For example, **GI** in bold underline, appears in three places of lysozyme. To condense the formula, **GI** could be replaced by, say, the emoji ☺. Then the primary structure could be represented more economically (and amusingly) using $130 - 6/2 = 127$ symbols:

KVFERCELARTLKRLGMDGYR☺SLANWMCLAKWESGYNTRATNYNAGDR
STDY☺FQINSRYWCNDGKTPGAVNACHLSCSALLQDNIADAVACAKRVVRDPQ☺
RAWVAWRNRCQNRDVRQYVQGCGV

The information savings are illusory, however. The viewer would have a 21 letter alphabet with which to contend and the novelty symbol would be useless in base and higher-level analysis. Besides, **GI** appears only once in myoglobin, and nowhere in ribonuclease A and countless other proteins. Matters would be different if sequences presented as follows:

$$MVVVGGGVVVVDETTTT....$$
$$MHHPPPPPTTTSSSVSSSC...$$

Then they could be condensed as $MV_3G_3V_4DET_4...$, $MH_2P_5T_3S_3VS_3C...$, etc., with correlations more easily rising to the surface—they would not burden us with so much information. The reality is otherwise whence diversity and randomness are signature traits.

Except when they are not. Evolution *does* confer patterns in places. Consider the keratin sequence:

ITYFAPMCNIITYFAPMCNIITYFAPMCNIITYFAPMCNIITYFAPMCNIITYFAPMCNIITYFAPM
CNIITYFAPMCNIITYFAPMCNIITYFAPMCNIITYFAPMCNIITYFAPMCNIITYFAPM
CNIITYFAPMCNIITYFAPMCNIITYFAPMCNIITYFAPMCNIITYFAPMCNIITYFAPMCL

ITYFAPMCNI is a repeating unit whence the above can be written as follows:

$$(ITYFAPMCNI)_{16}ITYFAPMCL$$

But keratins lack the chemical sophistication of enzymes, signal carriers, and more. Periodic sequences are generally restricted to fibrous molecules and connective tissue [4–6]. The matter of sequence information compression is taken up in Chapter Four. For now we concentrate on the composition or "parts list" properties.

Biochemistry texts survey protein compositions across the spectrum and turn quickly to folds, functions, and physiological roles [7]. As we are traveling a different road, we will step backward and examine more closely information at the base level. This is expressed in the number and identity of amino-acid units, irrespective of order. The attention is warranted because the base presents collectively in advance of gene transcription and protein synthesis. Moreover, all the structure levels (helices, β-sheets, etc.) and functions connect with the base. The thermo-chemical properties of solutions take root in component identities and mole amounts [8]. The same should be true for proteins, their components conferred by the coding regions of genomes.

The base structures of lysozyme, ribonuclease A, and myoglobin express, respectively, as follows:

$$A_{14}V_9L_8I_5P_2F_2W_5M_2G_{11}S_6T_5C_8Y_6N_{10}Q_6D_8E_3K_5R_{14}H_1$$
$$A_{12}V_9L_2I_3P_4F_3W_0M_4G_3S_{15}T_{10}C_8Y_6N_{10}Q_7D_5E_5K_{10}R_4H_4$$
$$A_{12}V_7L_{17}I_8P_5F_7W_2M_4G_{15}S_7T_4C_1Y_2N_3Q_7D_8E_{14}K_{20}R_2H_9$$

The order of the components is immaterial. Important rather is how the information—much reduced from the sequences—can be explored for lessons.

To set the stage, we observe that base structures are highly irregular and idiosyncratic. This is no revelation given the aperiodicity of the sequences. Yet the irregularities and idiosyncrasies manifest in spite of size, mass, and functional similarities. Lysozyme and ribonuclease A express comparable unit numbers ($N = 130, 124$), weights and folded configuration radii, and both are hydrolases. Yet the contrasts are several: lysozyme contains 5W while ribonuclease A lacks W. Lysozyme has 14R while ribonuclease A has 4R. There are features in common: both proteins contain 6Y, 8C, 9V, and nearly equal Q. The list can be made tediously long—longer if comparisons are extended to myoglobin, keratins, etc.

But the key information is and should be about "design." Proteins are prepared from amino-acid collections within cells. What "choices" and "strategies" are registered at the ground floor?

We know that the genomes underpinning proteins evolve over millennia—hence, the quotation marks in places. The marks acknowledge that evolution does not design this or chose that; evolution just happens. Yet the lengths and randomness of sequences obscure the thermo-chemical picture. It is difficult to say what makes a favorable or unfavorable base.

Matters are clear at the atomic level for small organics. The relationship between composition and structure is reciprocal. Thus, a chemistry student has no problem interpreting:

$$C_7H_{10}O \quad C_5H_{10} \quad C_{10}H_5 \quad CH_7O_{10}$$

With a little thought, he or she recognizes that stable isomers are allowed by the two groups on the left.

$C_7H_{10}O$ and C_5H_{10} are accordingly favorable base structures in atomic terms. $C_{10}H_5$ and CH_7O_{10}, by contrast, offer no stable arrangements as portrayed by Lewis diagrams. Models for atomic valence and electron sharing provide the means for judgment [9].

Since the polypeptides represented by virtually all sequences are electronically and thermodynamically stable, the same models are no help discriminating bases. Thus we need to view them alternatively and at a supra-atomic level through probability and information. Clearly not just any base will do. We need to assess what is preferred through natural selection and what seems avoided. Yet probability and information depend on *our* state of knowledge—our take on a system. And since our knowledge is fluid (or should be), we need to experiment with bases from more than one perspective.

Our first will start from scratch: we question whether *any* strategies are at work. Or could a base arrive just as easily (and ironically) by chance? This is a fair question given all the complexities. Proteins manifest biochemical functions subject to evolution. But do their parts lists reflect processes that are indiscriminately random? We doubt this judging from juxtapositions such as 14R and 1H in lysozyme, not to mention folding and solubility demands. Were selection forces absent, the component numbers should be more evenly matched. All the same, it is valuable to explore this perspective, for it offers first lessons. And rather than just list facts and figures, we illustrate experiments with lysozyme which can serve as a gateway for studying other proteins.

For our first effort, we imagine drawing $N = 130$ amino acids (as in lysozyme) from a large reservoir. Let each component A, V, L, ... have equal representation, ignoring all issues of solvent, temperature, pressure, and molecular interactions. The reservoir might be pictured as follows:

```
VVHVNVYVVLTVVYLVMGADRTCWKGIWEWPSHIVCNEMNILGSVNEQDV
PRVNGHRCSTCCIFAIGPAPGYGRHGYFKCIAKECYALPHYSSWAFMQDA
QEGPEFKENYGWDYLYLDAKVPWHYAWQMGISAGGKCLTDKERNIFTCMV
WRTAARDPVHNTQQFPFFEIQNHGQCCTKSFDYVNLEIKIMKEFWWLELQ
SMPQYAIPFAAEELCLLAMDTGIDWTNLQPMCTESAHFFNRKNNKEFKWQ
NWEYNGREPSLPCRVDHWYIHDYPDIGRSFQYFMCFDGLLVLRIGECGEE
ENYPMHEYDNDFCWLHSFERMVYPNQHLSPSMEVRTCKMPYIKLVITFSL
RNFMCGDWMWMCEFGNSPTHNCRPYIRRAPIIMRPFQEHAQDCTIDDSMI
```

```
TNTSQPKKATTNTFKMWSYFINNMYMFEWDIGWHHATFSHHLASMFRAWQ
LRVYCWKEPCKTTCGVERRMRECYKQSHVQMYYRAENRKKVNPHEVRFWP
YAEKHLKWWFWDKYFSSACLHHVVDWHCLVIDGESCDFNPQWKMCQKWLE
KAWAGKKVKVSNCARTRARDTWEVSTRKTAVGFVRKPEWAPCWRHHHMMS
VHGFGNMILIKKGYQWYPFTQMTNKHYMLTMRYWFPQDLQSWYDPIMHPI
SSYILDGVHSFMQASGMPIRQIDWCASNFWGMWDSDICRSCCDGGNVTFY
KGNFIQCAAMIARQKVADQGYISNYFGHECPACYHDPSPADGSKSLPVCT
LPESTYMLQEMKMCGAHSMQLSCIILHIMQREFDQNYDVAPEKWLKRYNC
ESQTLCFMAGDQRPTTSGIRKTDVQERQLMGQNSQNLCTHAHWSRTNYPV
VFRYNNDQFHRPHHIDFIATRTHDEYPTCDCKTKLQDRAVPNGNVWAGQN
RKGPMEAKWVNIWIWNICVGLTSTVTMIWYHLRMQWMQLYLQFLHHYFFG
DPLIQEFGTYAEPCQDQHESDVNHYWMCWPKGASTTKIHDTNDILHRLWM
```

If we practice unbiased selection, some $N = 130$ bases would look like:

$$A_7V_2L_4I_{10}P_8F_{11}W_5M_7G_3S_6T_6C_{10}Y_8N_{12}Q_7D_6E_0K_6R_5H_7$$
$$A_9V_3L_6I_9P_4F_5W_4M_{14}G_7S_6T_7C_8Y_6N_6Q_9D_5E_4K_3R_9H_6$$
$$A_8V_4L_3I_8P_7F_6W_2M_7G_{10}S_9T_6C_6Y_5N_4Q_7D_7E_8K_{10}R_4H_9$$

The base for lysozyme is one of *many* possible. It would follow from collecting sites such as in bold:

```
XXXXXXXXXXXXXXXXXXYVXXXXXXXXXXXVXXXXXXXXXXXWXXXXXXX
XXQEXXXXXXXXXRXXFXXGXXXXNIXKXXXXXXXXXMXXXXXXXXXXXXX
XXXXXXXNGXXXXKXXXXXXXXXXXXXXXXXXXXIXXCXXXARXXXXXXX
XLXXXXXXXXXLXXXXXXXXXXXXXXXXXAXXXXXXPSDXXXXXXXXXXXX
XDXXXXXXXXXXXXXXXGXXXXXXXDLRXXXFXXXXXXXXSXXXXXXXXXX
XXHXXXRXXXXXXXXXQXXXAXXVXXXRXVXXXXXXXXXXXXXXXXAXXX
XXXXSXXXXXXXXXXXRXXXXXXXXXXXXXXXXXXXXXXXXXXXXXXXY
XXXXXQXXXXXXXXXVXVXXXXXXXXXXXXXXTXXXXXXXAXXXXXXXXA
XXXXXXXXXXAAXXXXXXXXXXXXXXXXXXXIXXXXXXXXXXXXYLXXXXA
XXXXXXNXXXXCXXXXXXXXXXNXGXXXXCXXXXXXAXGXXYXXXXXXXXX
XXXXXXXXLCXXXXXXXXYXXXXXXXXXXXVXXXGXXXNXXXXXXXXQXXG
XNXXRXXXXXXXXXXXXXXXXXXXXXXXXDXWXXXXXXXXXXXXXXXXXX
XXXXXXXXXGXEXXEXXXXXXXXXXXXXXXXSXXXRXXXXTXLXXXXXXX
XXXXSXXXXXXXXXXXXXXXXXXXXXXXXXTXPXXXXXXXGXXXAXGXXMX
XXXVXXXXGAXXXXCXXXXXXXXXXXXXXXXCXXQXAXXDXXXXXXWXXXX
XXCXXXXXXXXXXXXXXXXXXXXXXXXXXXXXXXXXIXXXXXXXNXXXX
XXXXXXXXXXXXXXXXXXAXXXXXXXXXXXXXXXKXXXXXXVXXXXXXXXXI
XXXXXXXXXKXXXXXXXXXXXXXXDXXRXXXXXXXXXXXXXXXXXXTXQX
XXXXXXYXXWRXXXXXRXLXXXLXXXXRXXXXXXTRXXXXXDXXXXXXS
XXXNXXXXXKXXXXXXXDXXWXXXXXXXXXNXCXXXXXXXXXXXXRXXNXXX
```

The others would be bypassed, so marked by X. The boldface sites form a set of major significance, for they hold information matching the lysozyme base:

```
WYRSCQPKVGDDKTANQASN
NARGADVNDWCHVRGTCSDI
FYAYANQRVLLRLALYGAKV
LGCRFGEMRGITRMRACYGV
CQTNQVGKERSCKIDTSDWG
NWIRALLQNCAAAIVALRGN
RWSEYPRNDV
```

If synthesis of the enzyme had *only* this set available, the molecule could still obtain. How plausible is the set by chance?

As a warm-up, we consider a single component, namely alanine (A). The special set hosts 14A. The binomial distribution (Appendix Four) establishes the probability of obtaining k "successes" in N independent random draws [10]:

$$prob(k) = C(N,k) \cdot p^k \cdot q^{N-k}$$

It applies because the order is irrelevant to a base structure. We call drawing alanine from solution a "success" and anything else a "failure." $C(N, k)$ is needed to go further:

$$C(N,k) = \frac{N!}{k!(N-k)!}$$

For lysozyme with $k = 14$, $N = 130$, we have:

$$prob(k = 14, N = 130) = \frac{130!}{14!(130-14)!} \times p^k \cdot q^{N-k}$$

But what are p and q? For unbiased draws, alanine is one of twenty equally likely components. The probability p then equates with the mole fraction (we ignore solvent and ion contributions):

$$p = X_{Ala} = \frac{1}{20} = 0.050$$

In turn,

$$q = 1 - X_{Ala} = 1 - \frac{1}{20} = \frac{19}{20} = 0.95$$

We will assume the draw probabilities hold constant by taking the reservoir to be infinitely large. It follows that:

$$prob(k = 14, N = 130) = \frac{130!}{14!(130-14)!} \times \left(\frac{1}{20}\right)^{14} \cdot \left(\frac{19}{20}\right)^{130-14}$$

The Stirling formula (cf. Chapter One and Appendix Three) can handle the larger factorials while an inexpensive calculator takes care of the smaller ones [11]. We have:

$$\ln 130! \approx 130 \ln 130 - 130$$

$$\ln(130-14)! \approx 116 \ln 116 - 116$$

$$14! \approx 8.72 \times 10^{10}$$

$$\left(\frac{1}{20}\right)^{14} \approx 6.10 \times 10^{-19}$$

$$\left(\frac{19}{20}\right)^{130-14} \approx 2.61 \times 10^{-3}$$

The pieces connect to give:

$$prob(k = 14) \approx \frac{e^{130\ln 130 - 130}\left(6.10 \times 10^{-19}\right)\left(2.61 \times 10^{-3}\right)}{\left(8.72 \times 10^{10}\right) \cdot e^{116\ln 116 - 116}} \approx 3.28 \times 10^{-3}$$

This compares favorably with $prob(k = 14) \approx 3.48 \times 10^{-3}$ without appeal to Stirling. Then how likely is 14A in 130 draws? The answer corresponds to about 1 chance in 300. Information-wise, the event of obtaining the special set, viewed in A, not-A $\leftrightarrow$ A, A^c terms underpins surprisal:

$$S_{14A, N=130} = -\log_2(3.48 \times 10^{-3}) \approx 8.17 \text{ bits}$$

Apparently the base of lysozyme is improbable on alanine grounds alone, absent selection bias. Upon inspecting the base, our surprise is equivalent to observing eight flipped dimes all landing heads. The base includes 14R. Our level of surprise applies to *two* components of the enzyme.

We took the draw probability to be $p = 1/20$ and viewed the special set as one entity. However, we can invert the optics. We can accept that p is unknown prior to experiment, and imagine some fraction f of mechanisms favorable to alanine; $1 - f$ mechanisms are unfavorable. If our experiment finds k successes in N draws, then the probability that f lies *somewhere* in the interval

$$p < f < p + \Delta p$$

is estimated by the beta distribution [12]:

$$\frac{(N+1)!}{k!(N-k)!} \cdot p^k \cdot (1-p)^{N-k} \Delta p$$

By exploring combinations of p and Δp, we arrive at terms consistent with observation. Given $k = 14$A in $N = 130$ draws, taking $p = 0.100$ and $\Delta p = 0.07$, we find:

$$\frac{(130+1)!}{14!(130-14)!} \cdot 0.100^{14} \cdot (1-0.100)^{130-14} \cdot 0.07 \approx 0.924$$

This shows our maximum indifference assignment of $p = 1/20$ to be considerably off the mark. Instead we can be 90% sure that the base draw probability for alanine (arginine, too!) is close to 1/10. The lesson does not alter: without selection bias, the lysozyme base is surprising by its alanine and arginine content.

The experiment can be extended one-by-one to all the components of lysozyme, keeping in mind expectation $\langle k \rangle$ and variance σ^2. For $N = 130$, $p = 1/20$, these are for the binomial distribution [10]:

$$\langle k \rangle = Np = 130 \cdot \frac{1}{20} = 6.50$$

$$\sigma^2 = Npq = 130 \cdot \frac{1}{20} \cdot \frac{19}{20} \approx 6.18$$

The standard deviation σ becomes:

$$\sigma \approx \sqrt{6.18} \approx 2.49$$

The expectations frame a window $\langle k \rangle \pm \sigma \approx 6.50 \pm 2.49$. Seven components besides alanine and arginine appear in amounts *outside* this window. This demonstrates the lysozyme base surprising on multiple accounts.

Yet a base need not be inspected one component at a time. The information allied with A *and* V (or other pair) can be explored by the *trinomial* distribution [13]:

$$prob(k,l,N) = \frac{N!}{k!l!(N-k-l)!} p^k \cdot \theta^l \cdot (1-p-\theta)^{N-k-l}$$

This applies to N draws with k "successes," l "failures," and $(N-k-l)$ "neutrals." For lysozyme with 14A and 9V, with A and V each assigned (constant draw) probability 1/20, we have:

$$prob(k=14,l=9,N=130) = \frac{130!}{14!9!(130-14-9)!} \times \left(\frac{1}{20}\right)^{14} \times \left(\frac{1}{20}\right)^{9} \times \left(\frac{18}{20}\right)^{130-14-9}$$

The reader should verify that $prob(k = 14, l = 9, N = 130) \approx 2.52 \times 10^{-4}$ corresponding to one chance in 3,960 and surprisal 11.9 bits.

The information expressed by the total base can be weighed using the multinomial distribution [14]. This is a general extension of the binomial and trinomial distributions. If independent, random draws apply, we have for lysozyme:

$$prob(k=14,l=9,m=8,\ldots) = \frac{130!}{14!9!8!5!2!2!5!2!11!6!5!8!6!10!6!8!3!5!14!1!}$$

$$\times \left(\frac{1}{20}\right)^{14} \times \left(\frac{1}{20}\right)^{9} \times \left(\frac{1}{20}\right)^{8} \times \ldots$$

$$= \frac{130!}{14!9!8!5!2!2!5!2!11!6!5!8!6!10!6!8!3!5!14!1!} \times \left(\frac{1}{20}\right)^{14+9+8+\ldots}$$

The reader should verify that

$$prob \approx 4.91 \times 10^{-24}$$

The probability underpins surprisal:

$$S \approx -\log_2(prob) = -\frac{\ln(prob)}{\ln(2)} \approx 77.4 \text{ bits}$$

The calculations show the lysozyme base to have *vanishing* probability, absent selection bias. But we need to practice caution on three accounts, the first being that *any* base is improbable by chance events. This makes sense. The number of *possible* structures for $N = 130$ is ponderously large as discussed in Chapter One. The bases restricted to the integer set 14, 9, 8, 5, ... alone allow

$\Lambda = \dfrac{20!}{3!4!3!3!2!} \approx 2.35 \times 10^{14}$ possibilities. And Λ is greater still if other integer combinations are considered for $N = 130$. Then if one component combination is as likely as another, the probability of observing any particular combination is nearly zero. In picture terms, a base such as for lysozyme corresponds to a drop in a vast sea of possibilities:

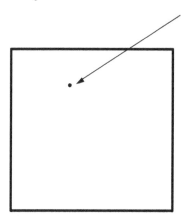

A random process which lands on the drop has low probability indeed. To better appreciate this, consider other $N = 130$ combinations such as:

$$A_7V_2L_4I_{10}P_8F_{11}W_5M_7G_3S_5T_6C_{10}Y_8N_{12}Q_7D_6E_1K_6R_5H_7$$

Then estimate the time required to obtain the set from random, independent draws at 10^{-12} seconds per structure assembly. The time is long indeed as can be explored. Then to view a base in the proper light, we need to survey the sea of possibilities. How large is *its* drop compared to others with equivalent N?

We conduct surveys by simulating $N = 130$ random draws, and computing the probability and surprisal of each collection that obtains. By sampling several thousand bases, we can approximate the distribution across the sea. This is the lesson of Figure One. Each point corresponds to a base obtained by 130 draws. Each component of a base is *expected* in amount $Np = 6.50$; thus we anticipate the most probable integers for A, V, L, etc., to be 6 or 7. The vertical axis marks the sum of squares of deviations from expected:

$$\sigma^2 = \sum_{j=1}^{20}\left(k_j - \langle k \rangle\right)^2$$

The k_j equate with the integers obtained by random draws: 4, 7, 9, etc. The horizontal axis marks the surprisal in bits obtained via the multinomial probability. The point for lysozyme is noted via the large filled circle.

We surmise that, absent selection forces, the base for lysozyme is highly improbable—even among the highly improbable. If things were otherwise, the large filled circle would place more in the crowd or, better yet, close to the origin. This assessment is firmed by the information measures. Structures in which the k_j stay close to maximum indifference expectations demonstrate surprisals near 50 bits and minimal σ^2. In contrast, lysozyme is an outlier in far right field: its base is tens of bits *more* surprising than the crowd bases.

For comparison, experiments and analysis directed at the base for ribonuclease A lead to:

$$-\log_2\left(prob_{\text{ribonuclease A}}\right) \approx 78.6 \text{ bits}$$

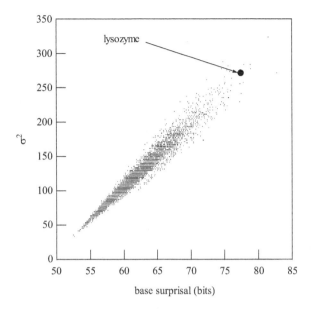

FIGURE ONE Distribution of surprisals and deviations from expected for $N = 130$ independent, random assemblies. The probability of drawing each amino acid is taken to be 1/20, irrespective of identity. The point for lysozyme is marked using the large filled circle.

This is also about 30 bits more surprising than the most probable base structures from unbiased selection.

Experiments with myoglobin and other proteins are left as exercises. The results for lysozyme, ribonuclease A, and myoglobin are emblematic, especially for globular proteins. Absent selection, archetypal bases strike us as highly improbable and surprising, even among the improbable and surprising.

The second reason for caution is that the multinomial distribution does not distinguish the component labels, only total probabilities. Since we are taking individual draw probabilities to be uniformly 1/20, base structures such as:

$$\mathbf{A_5}V_9L_8I_5P_2F_2\mathbf{W_{14}}M_2G_{11}S_6T_5C_8Y_6N_{10}Q_6D_8E_3K_5R_{14}H_1$$
$$A_{14}V_9L_8I_5P_2F_2W_5M_2G_{11}\mathbf{S_1}T_5C_8Y_6N_{10}Q_6D_8E_3K_5R_{14}\mathbf{H_6}$$
$$A_{14}\mathbf{V_{11}}L_8I_5P_2F_2W_5M_2\mathbf{G_9}S_6T_5C_8Y_6N_{10}Q_6D_8E_3K_5R_{14}H_1$$

warrant the same level of surprise as lysozyme—their drops in the sea have identical cross sections. The subscripts have been interchanged in places, as marked in bold, keeping the integer set as a whole intact. Hence the special point in Figure One applies not just to lysozyme, but rather to an extended class of base structures. A base is specified by integer and component sets. Every set underpins a large population of variants.

The third (and final) reason for caution may have been apparent from the start. Base structures are more than surprising by chance. This could be evidence that the component fractions in the source solutions are skewed. The draw probabilities for A and R appear close to 0.10. This could signal that indiscriminate selection applies, but in environments where A and R account for 20% of the building stock. Another interpretation is that the draws from solution are not independent of each other. The selection of a particular unit A, L, R, etc., impacts others down the line. The lesson stands that proteins reflect discrimination and selection processes, even at the ground floor. But then what mechanisms would favor some components over others? A suggestion comes from a source of complexity.

Proteins derive from Brownian processing of DNA, RNA, and mRNA by helicases and polymerases [15]. Their syntheses are advanced by the Brownian actions of tRNAs [7]. Proteins fold and encounter substrates through translational and rotational Brownian motion (cf. Chapter One). These considerations point to a mechanism: if proteins are Brownian devices and the products of Brownian processing, perhaps their compositions reflect as much by adopting Brownian signatures. Maybe base structures attest to the favoring of genomes toward some amino acids over others. The genetic code dedicates 61 of 64 codons to the standard 20: four codons are specific to alanine, four to valine, six to leucine, one for tryptophan, and so forth (cf. Appendix Five) [16]. This motivates another experiment.

We imagine the stock solution as before where the amino acids have equal representation. But we practice deliberate biases during draws. Let our bias directed toward an amino acid be weighted by the number of codons specific to it, for example, let retrieving leucine (L) be six times more likely than tryptophan (W). We repeat the warm-up exercise such that the binomial distribution still applies. Thus we call alanine retrieval a "success," and not-alanine a "failure." Let the draw probability be p be not 1/20 but rather 4/61 due to the genetic-code details. Then the probability of *not* drawing alanine is $q = 1 - 4/61$. The special set with information matching the lysozyme base is the same as before:

```
WYRSCQPKVGDDKTANQASN
NARGADVNDWCHVRGTCSDI
FYAYANQRVLLRLALYGAKV
LGCRFGEMRGITRMRACYGV
CQTNQVGKERSCKIDTSDWG
NWIRALLQNCAAAIVALRGN
RWSEYPRNDV
```

How likely is *the* set from events weighted by the genetic code?

We have:

$$prob(k = 14, N = 130) = \frac{130!}{14!(130-14)!} \times \left(\frac{4}{61}\right)^{14} \cdot \left(\frac{61-4}{61}\right)^{130-14}$$

This takes us to

$$prob(k = 14, N = 130) \approx 0.0228$$

The result corresponds to 2 chances in 100 and surprisal 5.46 bits. The 14A were surprising from unbiased ($p = 1/20$) draws. The same statement holds if the genetic-code weights are applied.

Moving forward, the multinomial distribution addresses the total base:

$$prob(k = 14, l = 9, m = 8, ...) = \frac{130!}{14!9!8!5!2!2!5!2!1!1!6!5!8!6!10!6!8!3!5!14!1!}$$

$$\times \left(\frac{4}{61}\right)^{14}_A \times \left(\frac{4}{61}\right)^{9}_V \times \left(\frac{6}{61}\right)^{8}_L \times \left(\frac{1}{61}\right)^{5}_I \times \cdots$$

The specific draw probabilities reflect the genetic code:

$$p_{Ala} = \frac{4}{61}, p_{Val} = \frac{4}{61}, p_{Leu} = \frac{6}{61}, ...$$

The reader should verify that the analysis leads to

$$prob \approx 2.19 \times 10^{-23}$$

This corresponds to surprisal:

$$-\log_2(prob) \approx 75.3 \text{ bits}$$

As in our first experiments, we find the total probability vanishing and surprise level high. But as before, every base structure is improbable and surprising. We still have to examine where lysozyme places in the sea of possibilities. The results of assembling several thousand genetic-code weighted structures appear in Figure Two. The axes are the same as in Figure One and the placement for lysozyme is noted.

We learn that with the genetic-code weights applied, the base for lysozyme is *still* highly improbable, even among the highly improbable. As before, the *expected* structures place near the origin in the lower left. Note that unlike the first experiment, the Np_i are diverse, for example,

$$N \cdot p_{\text{Ala}} = 130 \times \frac{4}{61}$$

$$N \cdot p_{\text{Trp}} = 130 \times \frac{1}{61}$$

The base for lysozyme places in the far outfield, given selection weighted by the genetic code. The selection details disperse the crowd as the individual draw probabilities are spread out considerably.

The takeaway is that genetic-code biases do not move us closer to selection criteria much less a strategy. This is driven home by Figure Three which shows the component numbers versus number of codons specific to components. The plot shows the base and genetic-code particulars to be only

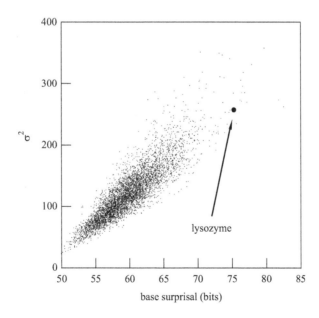

FIGURE TWO Distribution of base surprisals and deviations from expected for $N = 130$ independent, random assemblies. The probability of selecting an amino acid is proportional to the number of codons specific to it in the genetic code. The point for lysozyme is placed by the large filled circle.

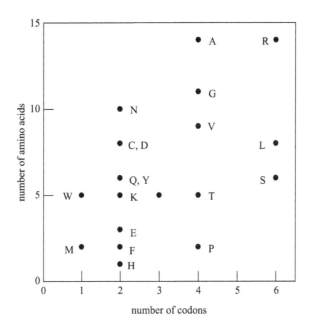

FIGURE THREE   Number of amino acids in lysozyme versus the number of codons specific to the amino acids, according to the genetic code.

weakly correlated: $R^2 \approx 0.232$. Note especially the points for H, F, E, K, Q, Y, C, D, and N. These are allied with identical codon numbers. However, their contributions to lysozyme span the range of low to middle to high.

We have undertaken experiments with a famous archetype and have learned about unlikely-to-apply scenarios. Other inquiries require only graphing. For example, the molecular weights for the amino acids are well established along with solubility, heats of formation, and ionization constants (cf. Appendix One and protein literature). Perhaps a base structure reflects a tendency toward certain physical characteristics. We consider weights briefly while other properties can be explored as special projects for students.

Chapter One showed the 20 amino acids to be mass-diverse. There are the light ones like glycine (G) and alanine (A) and heavies such as tryptophan (W). Organisms need to show restraint about material and energy consumption, not to mention entropy production [17]. Every atom contributes roughly k_B (Boltzmann's constant, cf. Chapter One) of entropy to a system [18]. Perhaps the base structures of proteins like lysozyme are tilted toward the lighter components. But there could be tradeoffs: perhaps the heavies dampen energy and volume fluctuations and enhance the stability of the folded state.

Figure Four plots the number of amino-acid components in lysozyme versus molecular weight. A left-to-right downward trend is weakly apparent: the linear correlation coefficient $R^2 \approx 0.121$—less than in Figure Three! This suggests that material economy is practiced to a small and inconsistent degree. The inconsistencies stand out in several places: arginine (R) is a heavy, and yet a major component; proline (P) is a lightweight and contributes but little to the base.

Is there a more illuminating avenue to explore? It may appear that the way forward lies in structure analyses *above* the base—looking at the primary, secondary, and folded configurations. But then recall a complexity from Chapter One, namely that there is no single version of a protein. Lysozyme is not a fixed object like benzene but rather a work in progress with eons of history. Different versions operate across time, species, and environments. Databases bear this out: the formulae for hundreds of lysozyme and related proteins are available from the NCBI and Swiss-Prot websites. Then perhaps we should not be concentrating on a single drop in the sea. This viewpoint

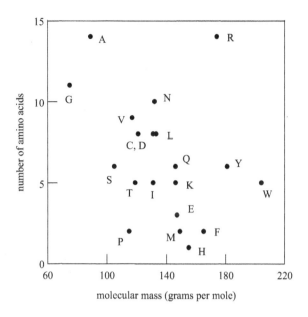

FIGURE FOUR Number of amino-acid components in lysozyme versus component molecular mass.

sets too narrow an aperture and overlooks a critical facet of proteins. Better that we consider neighborhoods in which the drop is one resident among many, viz.

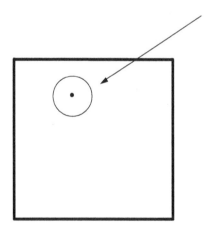

A neighborhood such as enclosed by the circle is home to a large population of closely related structures.

The way forward takes variations into account. Human lysozyme C (KVFER...GCGV) is a certain way now, but subject to change at a later date. Further, the base and sequence need not have been the fixtures of yesteryear. Chapter One presented protein variations as hardware and software updates for better or worse. But updates cannot sink ships by extinguishing critical functions or introducing deleterious ones. And they cannot jeopardize the plasticity—a protein has to accommodate changes down the line. This encourages variations to be surgical and incremental, and to keep doors open. For example, via gene mutations or transcription errors, alanine might appear at a site occupied by valine giving base structure

$$A_{14+1}V_{9-1}L_8I_5P_2F_2W_5M_2G_{11}S_6T_5C_8Y_6N_{10}Q_6D_8E_3K_5R_{14}H_1$$

Two fewer methyl groups should not matter much, right? Or a mutation could insert aspartic acid in place of glutamic acid, giving

$$A_{14}V_9L_8I_5P_2F_2W_5M_2G_{11}S_6T_5C_8Y_6N_{10}Q_6\mathbf{D_{8+1}E_{3-1}}K_5R_{14}H_1$$

Then the base and sequence would contain one less methylene group. These are conservative modifications and should not rock the boat too much.

Chemical fungibility is a trait throughout nature, especially when it comes to proteins. The ramifications include that a folded state can obtain through *many* amino-acid sequences: the primary (and base) information specify the folds given favorable solution environments, but the converse is not true. The situation presents another instance of information asymmetry.

Yet chemical fungibility does not commence with biological systems. Organic molecular behavior is governed throughout by functional and skeletal groups, and the effects of substitutions [9]. Consider acetone (C_3H_6O):

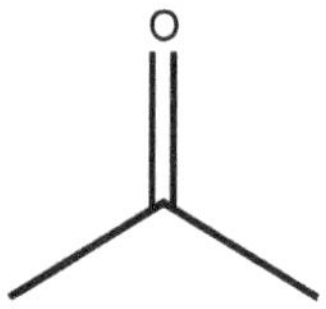

Its chemistry and solution behavior is not altered significantly if skeletal attachments are added.

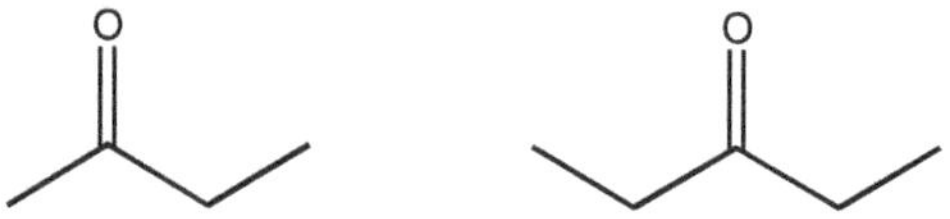

Consider the chlorinated pentene (C_5H_9Cl):

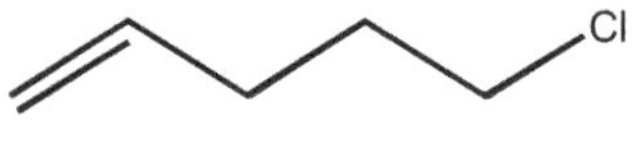

The physical and chemical changes are minor with the following modifications:

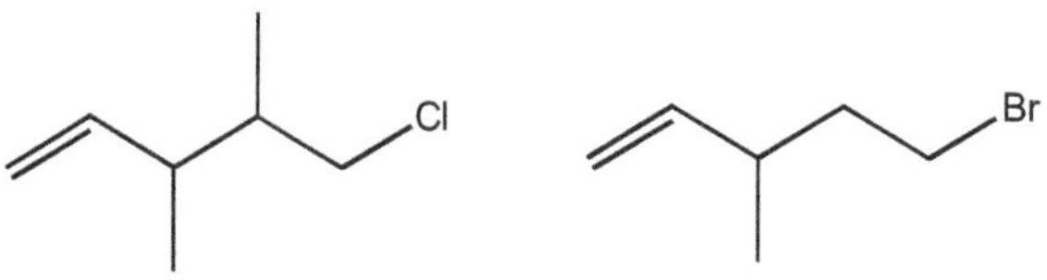

Conservative substitutions impact a system quantitatively, for example, concerning solubility, excluded volume, and reaction rates. However, they maintain the overall behavior qualitatively.

To go further with experiments, we need a lens for viewing amino acids less as individuals and more as family representatives. One representative can serve in place of another should biological need, chance, or expedience arise—although some representatives undoubtedly bring more effective chemistry to the task. But which families should we focus on and what should be the procedural details? Since the answers are critical to base structure analysis, we need to digress and consider what precisely makes a family. Then we can return to the business of experiments.

B A DIGRESSION ON SETS AND PARTITIONS

It should not surprise that the amino acids supported by the genetic code pose multiple family arrangements. Three are as follows:

Amino acids possessing a chiral center. This is a family of nineteen—all the amino acids except glycine.
Amino acids containing sulfur. This is a family of two—methionine and cysteine.
Amino acids containing a hydroxyl group. This is a family of two: serine and threonine.

Families are easy to come by and every amino acid is a member of more than one. In truth, we have been working with families all along, viewing each amino acid as a family of one: {A}, {V}, {L}, etc. The thing to note is that a family does not pop up from nowhere. Rather it derives, if tacitly, from our (1) specifying a condition, and (2) gathering parties from the eligible population that meet the condition. Let the terms set, family, group, and collection be interchangeable. In turn, a few principles from set theory provide the foundation for experiments. Recall from Chapter Two that probability and information are inextricably tied to set functions, as opposed to point functions.

Let us first count the available families. To explore how to do this, we imagine a small eligible population, say, A, V, and G. The possible subsets ↔ families would be:

{A, V, G}	{A, V}	{A, G}	{V, G}
{A}	{V}	{G}	

Yet families are predicated by conditions. If we specify that a member must contain bromine, the empty or vacuum set Ø would result—a family with *zero* members. Hence we count $7+1 = 8$ families given a population of three.

Then consider having four in the population: A, V, G, and M. But instead of forming the subsets by trial and error, we attempt a shortcut using binary labels. All subsets spring from a parent collection ↔ *the* eligible population. The components in a subset can subsequently be tagged: 1 indicates presence while 0 notes absence. For example, the parent set and corresponding labels present as:

Parent	A	V	G	M
Labels	1	1	1	1

The subset is then {A, V, G, M} which equates with the parent. Then consider:

Parent	A	V	G	M
Labels	1	1	1	0

The subset is {A, V, G} as indicated by the integer labels. Finally consider:

Parent	A	V	G	M
Labels	1	0	0	0

The subset must be {A}. It should be clear that the largest subset *is* the parent with labels 1, 1, 1, 1 while smallest is empty (vacuum set) with labels 0, 0, 0, 0. This means that there are $2^0+2^1+2^2+2^3 = 15$ subsets having one or more elements. By including the empty set Ø, we count $15 +1 = 16 = 2^4$ subsets total. Note that in assembling subsets, we pay no attention to order: {A, V, G} means the same as {G, V, A}.

We are directed to a counting formula. There are 20 amino acids from which to form families. In turn, there are

$$2^{20} = 1,048,576$$

families, including the empty one. There are

$$2^{20} - 1 = 1,048,575$$

families having one or more members. Families are subsidiary to the conditions *we* prescribe. Note that when presented with certain families sans clues, it is generally not easy to identify the prescribing conditions. This touches yet again on information asymmetry and its ramifications for proteins and biopolymers in general.

Then what families merit attention? Fortunately, these make for short lists, drawing from the million-plus. This is because a *partition* of the eligible population is required for base structure analysis. Partitions present *classes* of sets which are prescribed simultaneously *and* are mutually exclusive: a component in one set does not occupy another. This hearkens back to ideas in Chapter Two concerning disjoint sets and the law of total probability. But just as critical, the prescribing conditions must reflect the diversity of interactions expressed by the amino acids. From set theory, partitions accommodate an integer-labeling scheme. An [*i*, *j*, *k*] partition refers to a *class* of three mutually exclusive families: subsets with *i*, *j*, *k* denoting the number of elements in each. In base structure analysis, the integers always sum to 20. Note that prior to digression, we had been employing families *and* partitions all along. Exploring a protein in A, not-A ↔ A, A^c terms makes use of two families and a [1, 19] partition. Viewing the amino acids strictly by their identity applies a [1, 1, 1, 1, 1, 1, 1, 1, 1, 1, 1, 1, 1, 1, 1, 1, 1, 1, 1, 1] partition to a base and sequence. And the contact between partitions and amino acids does not begin with proteins: barring overlap, the coding, and non-coding regions of a genome offer a partition of two mutually exclusive sets.

In base structure analysis, the most insightful partitions concentrate on the α-carbon substituents of the amino acids, for these structures tune the folding and chemical functions in conjunction with the solvent environment (cf. Appendix One). One can then look broadly at the folding propensities of the amino acids. The result is a [7, 5, 8] partition with traditional family names:

External	{R, K, H, D, E, N, Q}
Internal	{L, V, I, F, M}
Ambivalent	{A, C, G, P, S, T, W, Y}

Another partition marks greater chemical distinctions leading to eight mutually exclusive subsets. This is a [2, 3, 5, 2, 3, 2, 1, 2] partition:

Acidic	{D, E}
Basic	{K, R, H}
Aliphatic	{A, G, I, L, V}
Amide	{N, Q}
Aromatic	{F, W, Y}
Hydroxyl	{S, T}
Imino	{P}
Sulfur-containing	{C, M}

Another partition marks fewer chemical distinctions, the result being an [8, 7, 2, 3] partition as detailed in reference [7]:

Polar	{G, S, T, C, Y, N, Q}
Non-Polar	{A, V, L, I, P, F, W, M}
Acidic	{D, E}
Basic	{K, R, H}

Still another strategy presents a [5, 2, 3, 5, 3, 2] partition:

Non-polar aliphatic	{G, A, V, L, I}
Non-polar "unusual"	{P, M}
Aromatic	{F, Y, W}
Polar	{S, C, T, N, Q}
Basic	{H, K, R}
Acidic	{D, E}

A partition strategy can be minimalist, for example [5, 15]:

Net Charged	{D, E, H, R, K}
Neutral	{A, V, L, I, P, F, W, M, G, S, T, C, Y, N, Q}

The α-carbon attachments are not everything. Thus a strategy can be grounded (arbitrarily) on molecular weight; the following presents a [2, 15, 3] partition:

Lightweights	{A, G}
Middleweights	{S, V, C, D, N, L, Q, K, I, T, P, E, M, F, H}
Heavyweights	{R, Y, W}

There are obviously many more partition arrangements to consider, but the lessons should be clear. For the families of a partition, the union must recover the 20 amino acids while intersections form an empty set ∅. More formally, the amino acids prescribed by the genetic code confer a sample space of 20 elements. If ∅ is added to the mix, then the possible conditions prescribe 2^{20} subsets total and a Boolean algebra of intersections, unions, complements, and differences. Note that a partition generally allows many possibilities, the exception being [1, 1, 1, 1, 1, 1, 1, 1, 1, 1, 1, 1, 1, 1, 1, 1, 1, 1, 1, 1]. For example, *the* [8, 7, 2, 3] ↔ (n, p, a, b) partition allows possibilities that number:

$$\frac{(8+7+2+3)!}{8!\,7!\,2!\,3!} = 997{,}682{,}400$$

This tells us that when we apply a *particular* [8, 7, 2, 3] partition to a structure analysis, we are leveraging information in amount:

$$I = \log_2(997{,}682{,}400) \approx 29.9 \text{ bits}$$

The information attests to the specialness of the short-list partitions and derives from our knowledge of the amino acids, their Angstrom-scale structures and chemistry. The shortfalls in applying the [1, 1, 1, 1, 1, 1, 1, 1, 1, 1, 1, 1, 1, 1, 1, 1, 1, 1, 1, 1] partition stemmed in part from the absence of leverage. Chemical intuition tells us that the prescribing conditions need to be substantive and not split hairs. At the same time, prescriptions are not necessarily hard and fast. For example, while the α-carbon substituents of H, K, R, D, and E may be readily categorized under suitable pH conditions, those of G and P are probably not (small, medium, etc.?).

Partitions ultimately draw coarse-grained pictures of protein sequences and base structures. If sequences are like biological sentences (cf. Preface), partitions illuminate the sentences in reduced-information terms. Metaphorically speaking, they emphasize the sense and grammar without getting bogged down in nuance. We can best see this through respellings. For lysozyme and its non-polar (n), polar (p), acidic (a), and basic (b) components, we have for the primary structure:

bnnabpannbpnbbnpnappbpnpnnpnnpnnbnapppppbnppppnpabppappnnpnppbpnppap
bpnpnnpnpbnpppnnnpapnnannnpnbbnnbanppnbnnnnnbpbpppbanbppnppppn

The corresponding base is follows:

$$n_{47}p_{52}a_{11}b_{20}$$

Application of the [7, 5, 8] $\leftrightarrow$ [external (e), internal (i), ambivalent (a)] to lysozyme gives us:

eiieeaeiaeaieeiaieaaeaiaiaaaiaiaeaeaaaaaeaaaaaaaeeaaeaaiieiaaeaaaaeaeaaaaiaaaeiaaaaii
eeaiaeaiaaaeeiieeaeaieaaiaaeaeaeaeeieeaieaaai

The base is follows:

$$a_{67}e_{37}i_{26}$$

Under suitable pH conditions, application of the minimalist [15, 5] $\leftrightarrow$ [uncharged (u), net-charged (c)] gives

cuuccucuucuuccuuucuucuuuuuuuuuuuuucucuuuuucuuuuuuuuccuucuuuuuuuuucuuuucucuuuu
uuuucuuuuuuuuucuuucuuuuuuccuuccuuuucuuuuuucucuuccucuuuuuuuu

The base is as follows:

$$u_{31}c_{99}$$

There is a loss of information in transitions from twenty- to fewer-letter spellings, and this may seem an imprudent thing. But there are two features to keep in mind. The first is that partitions are a mainstay in molecular structure analysis. Partitions are indeed among the most penetrating notions in chemistry. Consider the formula graph for sequence GA = GLY − ALA:

The *real* di-peptide hosts a variety of bond lengths, angles, and strengths. Even so, we instinctively coarse-grain the C–C, C–H, C–N, N–H, and C=O contacts and group them over a class of five mutually exclusive sets. In so many discussions of structure and reactivity, we do not split hairs, say, among the different C–H bonds. We only discriminate them from the other bond-types such as C=O and C–C. Thus when we direct partitions to protein sequences and bases, we are really following long-standing chemical suit: we view amino acids less by extended details, and more by fundamental type. There is a loss of information when applying the partitions. But multiple applications [7, 5, 8], [2, 3, 7, 8], etc., compensate by providing new perspectives. They let us view a structure from different vantage points.

The second thing to note is the simplification. By applying partitions to organic formula graphs, we represent molecules like acetone and alanine in less cumbersome and more transparent terms. The same is true for proteins. Several representations can be explored as projects while experiments occupy the next section.

C   PARTITION APPLICATIONS AND EXPERIMENTS

Partitions set the stage (finally!) for more counting and grouping experiments. We illustrate results of applying the third of the Section B strategies featuring a [8, 7, 2, 3] partition. This partition is discussed at length by Lehninger and other authors in classic biochemistry texts [7]. The partition is fundamental and serves as a gateway that will figure prominently down the line. Experiments applying additional partitions follow by similar procedure.

We imagine that base structures simply reflect the evolutionary "choice" of family sizes: the non-polar (n), polar (p), acidic (a), and basic (b) sets. The base for lysozyme contains more non-polar components (n: A, V, L, I, P, F, W, M) than acidic (a: D, E) because, well, the non-polar family has more members to contribute. This retains a probabilistic view of protein structure and invites layers of interpretation, for example, lysozyme hosts more n- than a-components because the non-polar interactions require more fine-tuning options. Availability and need go hand-in-hand.

Partitions reduce the information in sequences and base structures, yet the binomial and multinomial probabilities maintain as the analytical tools. Our warm-ups in Section A focused on alanine. For a new experiment, we consider not alanine per se, but rather the family to which it belongs, namely the non-polars {n: A, V, L, I, P, F, W, M}. We re-imagine the source pool just as before where every amino acid has equal representation. However, we regard drawing *any* member of the n-family a "success." The draw probability p becomes 8/20 given nature's stipulation of family size—the countable measure for the set of non-polars. The "failure" probability becomes 12/20 given the complement set: {(p, a, b): G, S, T, C, Y, N, Q, D, E, K, R, H}. There are 47 non-polar units in lysozyme; the probability of observing 47 successes in 130 random independent draws is as follows:

$$prob(k = 47, N = 130) = \frac{130!}{47!(130-47)!} \times \left(\frac{8}{20}\right)^{47} \cdot \left(\frac{20-8}{20}\right)^{130-47}$$

This works out to be:

$$prob(k = 47, N = 130) \approx 0.0485$$

The corresponding surprisal is as follows:

$$-\log_2(0.0485) \approx 4.37 \text{ bits}$$

The multinomial distribution applied to the total base in abnp-terms gives:

$$prob(k = 47, l = 52, m = 11, j = 20, N = 130) = \frac{130!}{47!52!11!20!} \times \left(\frac{8}{20}\right)^{47} \cdot \left(\frac{7}{20}\right)^{52} \cdot \left(\frac{2}{20}\right)^{11} \cdot \left(\frac{3}{20}\right)^{20}$$

The reader should verify that

$$prob(k = 47, l = 52, m = 11, j = 20, N = 130) \approx e^{-7.79} \approx 4.14 \times 10^{-4}$$

This is equivalent to one chance in 2,416 and surprisal 11.24 bits.

Now it may look as though another blind alley is being traveled that will end in disappointment—we find the probability low and the surprise high. But recall that all base structures are improbable and surprising and the reduced-information ones are no exception. Thus the sea of possibilities needs the customary attention. The results of surveys to approximate the distribution of structures appear in Figure Five. The point for lysozyme is noted by the large open circle. Note how the point *now* places in the infield instead of far right field. The $N = 130$ structures which conform to expectations *now* show surprisals near 10.0 bits. The base for lysozyme in abnp-terms is only a little more surprising: ca. half a standard deviation, at 11.2 bits. This signals that the lysozyme base, while not the most probable one, falls in the crowd of bases consonant with family-weighted selection.

Just as critical, lysozyme is less surprising than a typical variant. Apparently the sequence and base are idiosyncratic with respect to the components as individuals. However, it is unremarkable

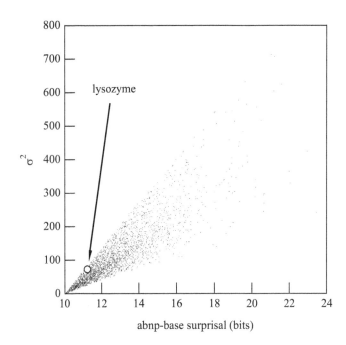

FIGURE FIVE Distribution of surprisals and deviations from expected for $N = 130$ family-weighted collections. The individual draw probabilities are proportional to the cardinality of the non-polar (n), polar (p), acidic (a), and basic (b) sets. The point for lysozyme is placed by the large open circle.

as an object with family representatives and room for family substitutions. This presents a vital clue about protein design via natural selection. If we look at the biopolymers a certain probabilistic way, aspects of their composition are not so surprising after all. This sets the stage for correspondences down the line.

Consider the significance further. Lysozyme's parts list is the way it is because it supports a biological catalyst with high specificity. Observe that evolution has made the families available more or less in proportion to biochemical needs. There are eight non-polar amino acids and two acidic ones available, not the other way around. Through the lens of [8, 7, 3, 2] partitions, the preparation solutions can be imagined as follows:

```
ppppppnbpppnnappnbbnbpnapanpbbnbnnaapapppaaannanpba
pnnbnbppppbnabpnpnnnabpbpnbnnabpnnnnnnpnpabnpnnbap
bpnnppnppananbnnnnnbppnppbbpbbnnppnpappnbnnppnppppb
nnnnnnpbappbbpbpbnnppppppnnnbnpbbpnnnbannapanabnnpb
nnnnbnnnnnapbappbnnnnnbbnbpppabnnpnnnnnpananppppp
nanbnnnanpnnnnbnppnnpppnpbpnnbpppbapnnpapnanbnnbnb
pnbnnnnnapnnppppannpnppbbapnpanbpppbnnbaapbpppbppp
nnnnannbnabppbpnpnanappbannnpbnppbnpnnpnpnppnnnppn
bpnbpbnnpbnpnnnnbbnabppnppappnnpnbnapppnapbpnnpnpa
pnpnpbbpbnannpnnanpnppnnbnppbnbnppnbannpapbnppbnpp
annanpnpppannnbnnpapnnbnnbappbpnppbnbnbpnnbnanpppn
npnnnannnpnnpnpnppppbbpappnpaanbbpnnnapnpbpnappbab
pnpnpnpnbnbnannpnnnnppnpnpppapnppnpnppbapnppnnnpnp
pppnnppppppnnapbanbnanappnppnbpnnbppanpppppnnnbpabp
pannnnnbpnpnpnnapppnnpnnpnbnpbaappnnppnpnpannnnbnnp
npppbpppbnnnnnabnpnpnnpnpnnnnnnnnnpnnnnpnapbnnapanp
```

ppbnnapnnabnnnnnabbppnpnppnppnpnpnpnnapppnbnnnbpp
bnpbabbbappbnnnpbbpbnppppnpbpnbnnnnnbpbnnpppbbapnp
bnnnnnbpnppppnbpppppnanannppabanpnppnpanpnnnnppppn
nnapnbppnpnnbppbpppanpbnbnpnpnbbnpnpbnppnnnpnppann

Then most any sector such as in bold approximates a working template, especially globular-type systems. Reservoirs *do not* look like:

apppbpbapaaabappappppapnpaaabpbppaaaaapapaaappappa
apppanaanapbapapapabpaaaapapanpaaapppaappaaanapana
panpaaaapbapapanppaanpbbnabaappapnaabbaapabanpppab
pbpaaapaapapnaaaaaappapaaaapapapanpapanabaanppnnaa
bappanaaaapnapapanpaapppnabpaaaappnabnpaaaaanpppp
ppbppaabaaappaabbpaabbpppabbaaaaapaapabnnabapapaaba
baaappapnaanppbappabnnppnaaaabaanababpnapaabbppapa
anbappapppapnapaabnppaapbpaappanbapnnppppapppbnapa
abnnbpanapaappabannnpapabbpppappaaabpppbbabanapbpb
appnaaapaapbanbnapbabbbbppanapabaapbaapanpnpbappppn
bpapbpppnbpaapapppaaapappappappaapppanappapnbappbb
abppaaaapabaannaabbabpapappbbapabbpppaabannpanapab
npnaaabaababpnpbpbbpaapaabppnaaaapaababbpnnppbbppa
abanabpbpbaabnpbanpanppaaaabpnnnnbpaaaaannabpnanpa
bpaaappaapaaapappbbpppappapabbaanbapappnpnapaanpaa
ppnpppnappaabpapaapppppababnpapappaappaabapapppbbb
pbappppnpapbpaapbpaapppapanabaanappaabaaapppapppppp
apaanpapnaappnaabpanpaaapabnapbaabpapppappaaapabap
pnapaabbaaabapabpapabaapaapapbappaabpbpannapppbppp
apbbapbabbpappbppapbaanpppaababbpbpbpbabbbpapaapap

If this were the case, too much of the building stock would be superfluous and get in the way. Further, the ion concentrations would exceed normal biological conditions, and protein manufacture would require too much trouble to marshal the right components. It can be shown that the base structures of diverse systems; for example, prions, heat shock proteins, and cytochromes, conform to the abnp-selection weights. Other globular systems better reflect three-family weights, for example, internal, external, and ambivalent. Partitions do not address matters of fine-tuning: lysozyme contains 9V, why not 8V or 10V? But they go a long way toward illuminating the coarse-tuning protocols.

We can flip things around to appreciate partitions from another perspective. When we assembled $N = 130$ bases via random draws, the structures themselves were variable while the solution-draw probabilities were held constant. Yet we can just as easily hold a base constant, and examine the effects of changing the draw probabilities.

We illustrate results of another experiment with lysozyme. In assembling the base in family terms by random draws, what if three of twenty amino acids were acidic while only two were basic? Then the base probability in family terms would be:

$$prob(k = 47, l = 52, m = 11, j = 20, N = 130) = \frac{130!}{47!52!11!20!}$$

$$\times \left(\frac{8}{20}\right)^{47} \cdot \left(\frac{7}{20}\right)^{52} \cdot \left(\frac{2+1}{20}\right)^{11} \cdot \left(\frac{3-1}{20}\right)^{20}$$

But what if four of the available amino acids were basic and six were polar? Then the probability would be

$$prob(k = 47, l = 52, m = 11, j = 20, N = 130) = \frac{130!}{47!52!11!20!}$$

$$\times \left(\frac{8}{20}\right)^{47} \cdot \left(\frac{7-1}{20}\right)^{52} \cdot \left(\frac{2}{20}\right)^{11} \cdot \left(\frac{3+1}{20}\right)^{20}$$

The notion is that by holding a base fixed and varying the family-selection weights, the weights themselves present as random variables. This process can be applied to multiple probability sets, for example,

$$\left\{ prob_{acid} = 4/20, prob_{base} = 2/20, prob_{polar} = 7/20, prob_{non\text{-}polar} = 7/20 \right\}$$

$$\left\{ prob_{acid} = 1/20, prob_{base} = 5/20, prob_{polar} = 8/20, prob_{non\text{-}polar} = 6/20 \right\}$$

$$\left\{ prob_{acid} = 3/20, prob_{base} = 3/20, prob_{polar} = 4/20, prob_{non\text{-}polar} = 10/20 \right\}$$

The restrictions (for simplicity) are that each entry is an integer multiple of 1/20. The results of examining several thousand randomly-assembled sets are shown in Figure Six. The horizontal axis marks the base information in bits. The vertical axis marks the sum of deviation from expectation (as in previous figures), viz.

$$\sigma^2 = \sum_{j=1}^{20} \left(k_j - \langle k \rangle \right)^2$$

The large filled circle marks the natural (canonical) partition weights: 2/20 (acid), 3/20 (base), 7/20 (polar), and 8/20 (non-polar) applied to lysozyme. Note how it falls near home plate on the

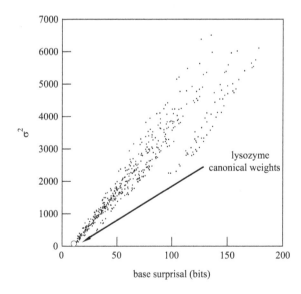

FIGURE SIX Distribution of surprisals and deviations from expected for lysozyme with family base $a_{11}b_{10}n_{47}p_{52}$. The draw probability sets have been taken as random variables. The large open circle identifies the result of applying the canonical set: $\{ prob_{acid} = 2/20, prob_{base} = 3/20, prob_{polar} = 7/20, prob_{non\text{-}polar} = 8/20 \}$.

far lower left. The lysozyme base in abnp-terms supports an active protein. And we learn from the experiments that base $a_{11}b_{20}n_{47}p_{52}$ is almost the least surprising according to family-selection weights. Similar results are obtained for multiple globular systems.

D   BASE STRUCTURES AND VARIATIONS

The experiments with lysozyme point to a general framework for discerning base structures. It is that while structures reflect the randomness of sequences, their abnp-units manifest roughly in proportion to the family sizes established by evolution. The framework is attractive on logistical grounds. It implies the design and manufacture of proteins to be efficient enterprises: organisms make the most of the amino acids underpinned by the genetic code; nature makes the component availability commensurate with biochemical needs. The framework accommodates chemical fungibility over the near and long term. With long sequences the rule, there is ample room for tweaking by substitutions.

But (again!) there is need for caution. Archetypes pose a miniscule fraction of the sequences on record and yet to be discovered. Hence their illumination of a framework can only be anecdotal. Fortunately, we need not look far for solid footing. Complete sets of sequences (aka proteomes) encoded by the genomes of organisms are available as NCBI and SwissProt downloads. Results which are emblematic appear in Figure Seven. The vertical axis marks the abnp-fractions encoded by the human genome while the horizontal axis marks the family fractions: 2/20 {a: D, E}, 3/20 {b: K, R, H}, 7/20 {p: G, S, T, C, Y, N, Q}, and 8/20 {n: A, V, L, I, P, F, W, M}. The scaling is linear with a best-fit-line slope of 0.966 and correlation coefficient R^2 of 0.989. The *measured* abnp-fractions place at 0.104, 0.153, 0.396, and 0.347 along the vertical axis. These values round nicely to 0.100, 0.150, 0.400, and 0.350 marked along the horizontal axis. This is an important picture, illustrating that multiple proteins—not just a few archetypes—would have pointed to the same general framework. We did not have to be especially lucky choosing lysozyme as a gateway. Pick an amino acid at random encoded by the human or other genome (cf. Chapter Five). On average, and to good approximation, 0.100, 0.150, 0.400, 0.350 offer the probabilities for correctly guessing

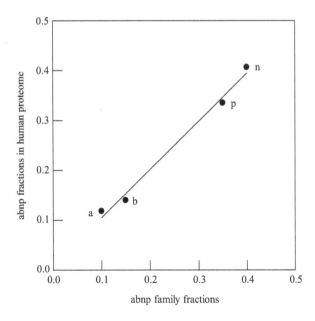

FIGURE SEVEN   Results of human proteome analysis. The measured abnp-fractions are plotted against the fractional abnp-family sizes: 2/20, 3/20, 8/20, and 7/20. The trend line follows from least squares analysis.

the abnp-family type in advance. Just as important, we have the average information I gained upon learning a family identity. The Shannon formula presents as (cf. Chapter Two):

$$I = -\left(\frac{2}{20}\right)\log_2\left(\frac{2}{20}\right) - \left(\frac{3}{20}\right)\log_2\left(\frac{3}{20}\right) - \left(\frac{8}{20}\right)\log_2\left(\frac{8}{20}\right) - \left(\frac{7}{20}\right)\log_2\left(\frac{7}{20}\right)$$

$$\approx 1.80 \text{ bits}$$

It can be shown that approximately the same number of bits attaches to each symbol in four-letter spellings of lysozyme, ribonuclease A, and myoglobin. Chapter Five expands upon this theme with discussion of multiple proteomes.

Then in the immediate term, we have guidance for interpreting base structures in fundamental family terms, for example,

$\mathbf{a_{10}b_{15}n_{40}p_{35}}$	$\mathbf{a_{20}b_{30}n_{80}p_{70}}$	$\mathbf{a_{15}b_{24}n_{62}p_{53}}$	$\mathbf{a_{25}b_{38}n_{105}p_{82}}$
$a_{35}b_{40}n_{15}p_{10}$	$a_{70}b_{80}n_{20}p_{30}$	$a_{53}b_{62}n_{24}p_{15}$	$a_{82}b_{105}n_{138}p_{25}$

All of the above represent templates for stable proteins. Each underpins a four-dimensional information vector (cf. Chapter Two), for example,

$$\mathbf{a_{10}b_{15}n_{40}p_{35}}: \vec{V}^{(4D)} = \sqrt{\frac{10}{100}}\hat{x}_a + \sqrt{\frac{15}{100}}\hat{x}_b + \sqrt{\frac{40}{100}}\hat{x}_n + \sqrt{\frac{35}{100}}\hat{x}_p$$

$$a_{35}b_{40}n_{15}p_{10}: \vec{V}^{(4D)} = \sqrt{\frac{35}{100}}\hat{x}_a + \sqrt{\frac{40}{100}}\hat{x}_b + \sqrt{\frac{15}{100}}\hat{x}_n + \sqrt{\frac{10}{100}}\hat{x}_p$$

Every template can be expanded as a sequence in four- or twenty-letter terms as discussed in Chapter Eight. But note how the vectors for the templates in bold font are nearly collinear. The structures they represent conform to the framework of natural proteins, viz.

The near-uniformity of directions brings to mind the composition vectors of ideal solution phases in equilibrium (cf. Chapter Two). In contrast, the templates in regular typeface present as unnatural and maybe xeno-biotic. Their abnp-weights are out of kilter and their information vectors point in wrong directions, viz.

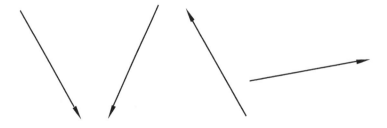

But what else does the framework teach? To dig deeper, we need to consider bases in terms of sets and complements. For example, if ~40% of a canonical base should consist of non-polar units, then ~60% (the complement) should be otherwise. If acidic units should be ~10% of a base, then ~90% should be non-acidic. If ~25% of a base represents units with net charge, then ~75% of the base should be electrically neutral. And so forth. These are extensions of the framework in conjunction with binary partitions [8, 12], [2, 18], [5, 15], etc.

Then revisit the probability of observing k successes in N independent random draws. If we imagine base construction where the draw probability p is itself variable, we can inquire how conditions can *maximize* the *total* probability. The total probability is represented by:

$$prob(k, N - k, p) = \frac{N!}{k!(N-k)!} \times p^k \cdot (1-p)^{N-k}$$

We explore the conditions of maximum $prob(k, N - k, p)$ by forming the partial derivative, taken with respect to p, and equating with zero, that is,

$$\left(\frac{\partial prob(k, N-k, p)}{\partial p} \right)_{N,k} = 0$$

The differentiation leads to

$$\left(\frac{\partial prob(k, N-k, p)}{\partial p} \right)_{N,k} = 0 = \frac{N!}{k!(N-k)!} \times kp^{k-1} \cdot (1-p)^{N-k} - \frac{N!}{k!(N-k)!} \times p^k \cdot (N-k)(1-p)^{N-k-1}$$

whereupon

$$\frac{N!}{k!(N-k)!} \times kp^{k-1} \cdot (1-p)^{N-k} = \frac{N!}{k!(N-k)!} \times p^k \cdot (N-k)(1-p)^{N-k-1}$$

The factorial terms cancel while algebra and further cancellations lead to the takeaway:

$$p = \frac{k}{N}$$

In words, the mathematics tells us the following. Given *our* take on the randomness of sequences, and the side-chain functions of the amino acids, we *expect* the abnp-fractions of proteins and their sources to *match* one another, for we have no reason to do otherwise. The bases we anticipate are subsequently the *most likely* ones from Brownian processes. Collectively, the processes *maximize* the probability of a base structure such as for lysozyme, and *minimize* our surprise upon finding it in nature.

The conditions of maximum probability and minimum surprise do not illuminate the biochemical functions—why is lysozyme KVFER… a hydrolase and not, say, a transferase? Nor do they bring physiological insights—why is lysozyme part of the immune system machinery and not, say, metabolic? What the conditions *do* carry is some hard thermodynamic currency. If the family mole fractions for a protein stay close to the preparatory solution, then the chemical potential gradients will tend toward minimal, just as the temperature and pressure gradients (cf. Appendix Two). We imagine the site of protein manufacture (highlighted in bold) surrounded by the source pool:

nnnnnnpbappbbpbpbnnpppppnnnbnpbbpnnnbannapanabnnpb
nnnnnbnnnnnapbappbnnnnnbbnbpppabnnpnnnnnpanannpppp
nanbnnnanpnnnnbnppnnpppnpbpnnbpppbapnnpapnanbnnbnb

```
pnbnnnnnapnnpppppannpnppbbapnpanbpppbnnbaapbpppbppp
nnnnannbnabppbpnpnanappbannnpbnppbnpnnpnpnpppnnnppn
bpnbpbnnpbnpnnnnbbnabppnppappnnpnbnapppnapbpnnpnpa
pnpnpbbpbnannpnnanpnppnnbnppbnbnppnbannpapbnppbnpp
annanpnpppannnbnnpapnnbnnbappbpnppbnbnbpnnbnanpppn
npnnnannnpnnpnpnpppppbbpappnpaanbbpnnnapnpbpnappbab
pnpnpnpnbnbnannpnnnnppnpnppppapnppnpnppbapnppnnnpnp
pppnnppppppnnapbanbnanappnpppnbpnnbppanpppppnnnbpabp
pannnnnbpnpnpnnapppnnpnnpnbnpbaappnpppnpnpannnnbnnp
npppbpppbnnnnnabnpnpnnpnpnnnnnnnnpnnnnpnapbnnapanp
ppbnnapnnabnnnnnabbppnpnppnpppnpnpnpnnnapppnbnnnbpp
bnpbabbbappbnnnpbbpbnppppnpbpnbnnnnnbpbnnpppbbapnp
bnnnnnbpnppppnbpppppnanannppabanpnppnpanpnnnnppppn
nnapnbppnpnnbppbpppanpbnbnpnpnbbnpnpbnppnnnpnppann
```

With thorough mixing, entropy production is limited to that incurred in peptide bond formation, water exclusion, and protein folding. Entropy is generated during synthesis, but in amounts restrained to some floor value. The system and surroundings as a whole approach—but (fortunately!) never quite reach—maximum probability states [17]. In turn, the rate of entropy production is minimized, as is the workings of steady state systems: they maintain the critical gradients and iron out the superfluous ones (cf. Appendix Two). The system is further resistant to thermodynamic fluctuations, in line with le Chatelier's principle. Point being made: protein sequences and bases appear cryptic with casual inspection. But with exploration, probability and information show them to reflect thermodynamic traits of their operating environments. These are the organisms whose existence depends on proteins.

Yet frameworks need to be probed for variations. There is a 50/50 chance of a coin landing heads. Yet we are unsurprised to observe four heads and six tails in ten tosses. Then consider the templates of several archetypes presented in integer and fractional terms. First, we have the respective structures for bovine ribonuclease A, human myoglobin, and a keratin (cf. sequences at beginning of Section A):

$$a_{10}b_{18}n_{37}p_{59} \leftrightarrow a_{0.0806}b_{0.145}n_{0.298}p_{0.475}$$
$$a_{22}b_{31}n_{62}p_{39} \leftrightarrow a_{0.142}b_{0.201}n_{0.403}p_{0.253}$$
$$a_0 b_0 n_{102} p_{67} \leftrightarrow a_{0.000}b_{0.000}n_{0.604}p_{0.396}$$

The sequence and template for a prion are represented as follows:

MRKHLSWWWLATVCMLLFSHLSAVQTRGIKHRIKWNRKALPSTAQITEAQVA
ENRPGAFIKQGRKLDIDFGAEGNRYYEANYWQFPDGIHYNGCSEANVT
KEAFVTGCINATQAANQGEFQKPDNKLHQQVLWRLVQELCSLKHCEFWLERG
AGLRVTMHQPVLLCLLALIWLTVK

$$a_{14}b_{28}n_{77}p_{57} \leftrightarrow a_{0.0795}b_{0.159}n_{0.4375}p_{0.324}$$

For human alpha synuclein, we have:

MDVFMKGLSKAKEGVVAAAEKTKQGVAEAAGKTKEGVLYVGSKTKEGVVHG
VATVAEKTKEQVTNVGGAVVTGVTAVAQKTVEGAGSIAAATGFVKKDQL
GKNEEGAPQEGILEDMPVDPDNEAYEMPSEEGYQDYEPEA

$$a_{24}b_{16}n_{55}p_{45} \leftrightarrow a_{0.171}b_{0.114}n_{0.393}p_{0.321}$$

A phospholipase is represented as follows:

NLYQFKNMIQCTVPSRSWWDFADYGCYCGRGGSGTPVDDLDRCCQVHDNC
YNEAEKISGCWPYFKTYSYECSQGTLTCKGGNNACAAAVCDCDR
LAAICFAGAPYNDNDYNINLKARC

$$a_{13}b_{11}n_{36}p_{58} \leftrightarrow a_{0.110}b_{0.0932}n_{0.305}p_{0.492}$$

A viral polymerase appears as follows:

MERIKELRNLMSQSRTREILTKTTVDHMAIIKKYTSGRQEKNPALRMKWMMA
MKYPITADKRITEMIPERNEQGQTLWSKMNDAGSDRVMVSPLAV TWWN
RNGPMTNTVHYPKIYKTYFERVERLKHGTFGPVHFRNQVKIRRRVDINPGHA
DLSAKEAQDVIMEVVFPNEVGARILTSESQLTITKEKKEELQDCKISPLM
VAYMLERELVRKTRFLPVAGGTSSVYIEVLHLTQGTCWEQMYTPGGEVKNDDVD
QSLIIAARNIVRRAAVSADPLASLLEMCHSTQIGGIRMVDILKQNPTEE
QAVGICKAAMGLRISSSFSFGGFTFKRTSGSSVKREEEVLTGNLQTLKIRVHEGYE
EFTMVGRRATAILRKATRRLIQLIVSGRDEQSIAEAIIVAMVFSQEDCMIK
AVRGDLNFVNRANQRLNPMHQLLRHFQKDAKVLFQNWGVEPIDNVMGMIG
ILPDMTPSIEMSMRGVRISKMGVDEYSSTERVVVSIDRFLRVRDQRGNVLL
SPEEVSETQGTEKLTITYSSSMMWEINGPESVLVNTYQWIIRNWETVKI
QWSQNPTMLYNKMEFEPFQSLVPKAIRGQYSGFVRTLFQQMRDVLGTFDTAQ
IIKLLPFAAAPPKQSRMQFSSFTVNVRGSGMRILVRGNSPVFNYNKATKRL
TVLGKDAGTLTEDPDEGTAGVESAVLRGFLILGKEDRRYGPALSINELSNLAK
GEKANVLIGQGDVVLVMKRKRDSSILTDSQTATKRIRMAIN

$$a_{83}b_{116}n_{317}p_{243} \leftrightarrow a_{0.109}b_{0.153}n_{0.417}p_{0.320}$$

A second keratin—the first appeared at the beginning of the chapter—is represented as follows:

MGCCGCSGGCGSSCGGCDSSCGSCGSGCRGCGPSCCAPVYCCKPVCCCVPA
CSCSSCGKRGCGSCGGSKGGCGSCGCSQCSCCKPCCCSSGCGSSCCQCSCCKP
YCSQCSCCKPCCSSSGRGSSCCQSSCCKPCCSSSGCGSSCCQSSCCKPCC
SQSRCCVPVCYQCKI

$$a_{1}b_{13}n_{19}p_{136} \leftrightarrow a_{0.00592}b_{0.0769}n_{0.112}p_{0.804}$$

A phosphofructokinase is represented as follows:

MAAVDLEKLRASGAGKAIGVLTSGGDAQGMNAAVRAVTRMGIYVGAKVF
LIYEGYEGLVEGGENIKQANWLSVSNIIQLGGTIIGSARCKAFTTREGRRAA
AYNLVQHGITNLCVIGGDGSLTGANIFRSEWGSLLEELVAEGKISETTARTYSH
LNIAGLVGSIDNDFCGTDMTIGTDSALHRIMEVIDAITTTAQSHQRTFVLEVM
GRHCGYLALVSALASGADWLFIPEAPPEDGWENFMCERLGETRSRGSRLNIII
IAEGAIDRNGKPISSSYVKDLVVQRLGFDTRVTVLGHVQRGGTPSAFDRILSS
KMGMEAVMALLEATPDTPACVVTLSGNQSVRLPLMECVQMTKEVQKAMDDK
RFDEATQLRGGSFENNWNIYKLLAHQKPPKEKSNFSLAILNVGAPAAGMN
AAVRSAVRTGISHGHTVYVVHDGFEGLAKGQVQEVGWHDVAGWLGRGGS
MLGTKRTLPKGQLESIVENIRIYGIHALLVVGGFEAYEGVLQLVEARGRYEEL
CIVMCVIPATISNNVPGTDFSLGSDTAVNAAMESCDRIKQSASGTKRRVFIV
ETMGGYCGYLATVTGIAVGADAAYVFEDPFNIHDLKVNVEHMTEKMKTDIQ

RGLVLRNEKCHDYYTTEFLYNLYSSEGKGVFDCRTNVLGHLQQGGAPTPFD
RNYGTKLGVKAMLWLSEKLREVYRKGRVFANAPDSACVIGLKKKAVAFSP
VTELKKDTDFEHRMPREQWWLSLRLMLKMLAQYRISMAAYVSGELEHVT
RRTLSMDKGF

$$\mathbf{a}_{84}\mathbf{b}_{102}\mathbf{n}_{327}\mathbf{p}_{267} \leftrightarrow \mathbf{a}_{0.108}\mathbf{b}_{0.131}\mathbf{n}_{0.419}\mathbf{p}_{0.342}$$

Lastly, we have the sequence and template for a fibroin:

MKPIFLVLLVATSAYAAPSVTINQYSDNEIPRDIDDGKASSVISRAWDYVDDTDK
SIAILNVQEILKDMASQGDYASQASAVAQTAGIIAHLSAGIPGDACAAANVI
NSYTDGVRSGNFAGFRQSLGPFFGHVGQNLNLINQLVINPGQLRYSVGPALGC
AGGGRIYDFEAAWDAILASSDSSFLNEEYCIVKRLYNSRNSQSNNIAAYITAH
LLPPVAQVFHQSAGSITDLLRGVGNGNDATGLVANAQRYIAQAASQVHV

$$\mathbf{a}_{22}\mathbf{b}_{20}\mathbf{n}_{118}\mathbf{p}_{102} \leftrightarrow \mathbf{a}_{0.0840}\mathbf{b}_{0.0763}\mathbf{n}_{0.450}\mathbf{p}_{0.389}$$

We observe framework conformity (more or less) across diverse systems, so indicated in bold font. Yet it is easy to find deviations, some glaring as represented normal typeface. This should not come as a shock—why should a "10, 15, 35, 40% rule" apply across the boards, even approximately? Broadly speaking, abnp-selection weights are approached by enzymes, hormones, and transport proteins—although again, we observe deviations via the phospholipase, a hydrolytic enzyme. But fibrous proteins and connective tissue favor polar components and skimp on acidic and basic units. Clearly there is more to the story as can be explored by experiments.

To prepare the stage, observe that every base structure, viewed through canonical (Figure Seven) selection weights, expresses probability:

$$prob\left(N_a, N_b, N_n, N_p\right) = \frac{\left(N_a + N_b + N_n + N_p\right)!}{N_a!N_b!N_n!N_p!} \times \left(\frac{2}{20}\right)^{N_a} \cdot \left(\frac{3}{20}\right)^{N_b} \cdot \left(\frac{8}{20}\right)^{N_n} \cdot \left(\frac{7}{20}\right)^{N_p}$$

This underpins surprisal:

$$S\left(N_a, N_b, N_n, N_p\right) = -\log_2 prob\left(N_a, N_b, N_n, N_p\right)$$

N_a, N_b, N_n, N_p count the abnp-representatives in a base. But note there to be *two* information terms in play. The factorial expression:

$$\Omega = \frac{\left(N_a + N_b + N_n + N_p\right)!}{N_a!N_b!N_n!N_p!}$$

counts the distinguishable isomers (permutations) of the primary structure which are consonant with the base (cf. Chapter One). Then *given* a base structure in family terms, *the* sequence in family terms conferred by evolution, gene transcription, and synthesis poses surprisal:

$$S(\Omega) = \log_2\left[\frac{\left(N_a + N_b + N_n + N_p\right)!}{N_a!N_b!N_n!N_p!}\right]$$

The surprisals of the base template and sequence are then related as follows:

$$S(\Omega) = -S\left(N_a, N_b, N_n, N_p\right) - \log_2\left\{\left(\frac{2}{20}\right)^{N_a} \cdot \left(\frac{3}{20}\right)^{N_b} \cdot \left(\frac{8}{20}\right)^{N_n} \cdot \left(\frac{7}{20}\right)^{N_p}\right\}$$

The above equation captures the interdependent, reciprocal relation between protein bases and sequences, given canonical family-selection weights. It describes how the information in one structure level is connected to the other.

The multinomial and reciprocal equations then motivate further experiments. The multinomial equation models not just selection probabilities, but also estimates of variations. For a given component, say, acidic, the *expected* variation in relation to the average is given by λ_a:

$$\lambda_a^{(ex)} = \frac{\sigma_a}{\langle N_a \rangle} = \frac{\sqrt{Np_a \cdot (1 - p_a)}}{Np_a}$$

$$= \frac{1}{\sqrt{N}} \cdot \frac{\sqrt{p_a \cdot (1 - p_a)}}{p_a} = \frac{1}{\sqrt{N}} \cdot \frac{\sqrt{\left(\frac{2}{20}\right) \cdot \left(\frac{18}{20}\right)}}{\left(\frac{2}{20}\right)}$$

$$= \frac{3}{\sqrt{N}}$$

The expressions for λ_b, λ_n, λ_p arrive just as easily. An experiment then queries whether the expressions hold for natural proteins: the binomial and multinomial equations submit that base structure λ should scale as $N^{-1/2}$. Just as critical, the reciprocal relation establishes a locus of points: $S(\Omega)$ as a function of $S(N_a, N_b, N_n, N_p)$. An experiment subsequently tests whether proteins composed of N units place points inside the locus or outside.

Going forward, we recall that lysozyme presents $N = 130$ units. Then lacking other information, we *expect* λ_a for the same size of base structure to be:

$$\lambda_a^{(ex)} = \frac{\sqrt{130 \cdot \left(\frac{2}{20}\right)\left(\frac{18}{20}\right)}}{130 \cdot \left(\frac{2}{20}\right)} \approx \frac{3.42}{13} \approx 0.263$$

But we subsequently look to other $N = 130$ sequences of, say, the human proteome. There are quite a few, for example, the *uncharacterized* proteins:

MFPGSLSRGRRAAVEMAWLPGSCARVAFAAGAAARYWTAWQGSAGPNPAAVA
EAHGSLFCGRATSARAWSLRRPGPGSPAHSGGVQTRENWIAYPLQSAEDGVA
TRLQIREESASCLAAEYWSQEPAMRF

MARRHCFSYWLLVCWLVVTVAEGQEEVFTPPGDSQNNADATDCQIFTLTP
PPAPRSPVTRAQPITKTPRCPFHFFPRRPRIHFRFPNRPFVPSRCNHRFPFQPFYW
PHRYLTYRYFPRRRLQRGSSSEES

MLAPLFLCCLRNLFRKLISFQPPQLGRTNMHYSKLPRTAIETEFKQNVGPPPKDL
TAEVYFPSIKSRSHLPAVFYNQYFKHPKCVGEYGPKNGAERQIEERKVLPTT
MMFSMLADCVLKSTPIPILGVAM

.

.

.

By the usual symbol counting and grouping for *all* $N = 130$ sequences of the proteome, we *observe*:

$$\lambda_a^{(obs)} = \frac{6.47}{14.1} \approx 0.458$$

The same efforts yield for basic components (b: K, R, H):

$$\lambda_b^{(ex)} = \frac{\sqrt{130 \cdot \left(\frac{3}{20}\right) \cdot \left(\frac{17}{20}\right)}}{130 \cdot \left(\frac{3}{20}\right)} \approx \frac{4.07}{19.5} \approx 0.209$$

$$\lambda_b^{(obs)} = \frac{6.33}{20.5} \approx 0.309$$

For non-polar components (n: A, V, L, I, P, F, W, M), we find:

$$\lambda_n^{(ex)} = \frac{\sqrt{130 \cdot \left(\frac{8}{20}\right) \cdot \left(\frac{12}{20}\right)}}{130 \cdot \left(\frac{8}{20}\right)} \approx \frac{5.59}{52.0} \approx 0.107$$

$$\lambda_n^{(obs)} = \frac{6.41}{54.4} \approx 0.118$$

Lastly, for polar components (p: G, S, T, C, Y, N, Q), we find:

$$\lambda_p^{(ex)} = \frac{\sqrt{130 \cdot \left(\frac{7}{20}\right) \cdot \left(\frac{13}{20}\right)}}{130 \cdot \left(\frac{7}{20}\right)} \approx \frac{5.44}{45.5} \approx 0.120$$

$$\lambda_p^{(obs)} = \frac{6.21}{41.0} \approx 0.152$$

We observe that the averages (denominators) conform closely to expectations of the multinomial probability and canonical selection weights. However, the variations (numerators) present as non-conformist. The observations for $N = 130$ are neither flukes nor isolated. Figure Eight presents data for the human proteome for $N \leq 1000$. The points for $\lambda_a^{(obs)}$ consistently fall above ones for $\lambda_a^{(ex)}$. Points for the latter are placed by assembling random bases using canonical abnp-weights. The curve traces $\lambda_a^{(ex)}$ for the multinomial equation in the limit of infinite-size populations.

Similar λ-illustrations are the case for basic, non-polar, and polar components. The lessons are striking: the *observed* variations considerably *exceed* those expected of the 10, 15, 35, 40% rule taken at face value. More to the point, the variations for large N-proteins speak to (1) canonically-weighted selection at *small N*, and (2) *combinations of small N-samples of like-deviation, that is, "like" is merged with "like."* In other words, pick one of the measured points in Figure Eight toward the far right. Then draw an arrow (horizontal dotted) from the point to the curve. The drop-down arrow points to N consonant with observed λ_a *and* random assembly.

We arrive at a vital frame of the story. Multiple bases *do conform* to canonical weights on average and their abnp-information vectors approach collinear. However, the variations are at odds with base units conferred independently of one another. This is unapparent from sequences and bases viewed casually or in isolation: the components always look unselected and unrelated to each other. Of course, we know otherwise from folding characteristics and biochemical functions; these attest to enormous selection of materials and interactions via evolution. However, selected combinations are manifest even at the composition levels of proteins. The interdependence of bases and sequences further emerges from viewing extended sets.

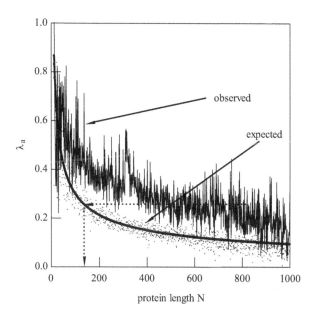

FIGURE EIGHT   Results of human proteome analysis. Observed λ_a consistently falls above values expected from canonical selection weights. The black curve traces expected λ_a for infinitely large populations as prescribed by the binomial distribution. The horizontal and vertical dotted arrows are discussed in the text.

We can look at things another way. A sequence confers a base structure and a subsequence does so for a partial base. Then consider lysozyme in two equal-size pieces distinguished by bold and normal typeface.

KVFERCELARTLKRLGMDGYRGISLANWMCLAKWESGYNTRATNYNAGDR
STDYGIFQINSRYWCNDGKTPGAVNACHLSCSALLQDNIADAVACAKRVVRD
PQGIRAWVAWRNRCQNRDVRQYVQGCGV

Question: what is the Kullback–Leibler penalty K_l for assuming the abnp-base for the boldface half to be echoed by the second half? It can be shown that the penalty is slight at ca. 0.02 bits. Apparently the family composition for one section of the protein carries substantial information about another—we incur only small penalties for assuming as much. To put things another way, lysozyme's base carries information that is internally mutual and reflexive. Compare the situation to lysozyme rendered *false*:

KVFERCELARTLKRLGMDGYRGISLANWMCLAKWESGYNTRATNYNAGDRS
TDYGIFQINSRYWCDAHQIYVHLRYDSRVDEHHRHKWLKWDPITPCEV
SLIANTATMAPLLYPHMDPYTVLKWFPRFG

Here the penalty can be shown to be about seven times greater. The two halves are ill-fitted beginning with their base information in family terms. Note that the reflexive properties of sequences and bases are not overly strong: the penalties for assumptions about 50–50 true–true systems are about a standard deviation less than for 50–50 true–false. But the mutual information is robust and holds even for proteins that deviate from canonical selection weights. Thus if a sequence and base are skewed in one part of a protein, other parts tend to follow suit. These properties assist in the analysis of protein sets discussed in Chapter Five. They further guide the practice of writing sequences in Chapter Eight.

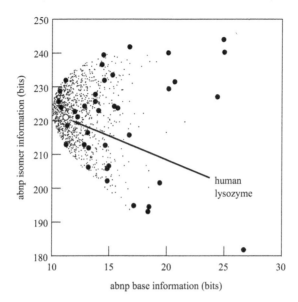

FIGURE NINE   Results of human proteome analysis. The vertical and horizontal axes mark sequence and base surprisals $S(\Omega)$ and $S(N_a, N_b, N_n, N_p)$. The small circles are placed by random bases constructed according to abnp canonical weights. The filled circles are placed by $N = 130$ proteins of the human proteome. The large open circle is placed by human-encoded lysozyme C.

There is more to take home from the surprisal terms: sequence $S(\Omega)$ in relation to base $S(N_a, N_b, N_n, N_p)$. The results in Figure Nine are typical, pertaining to $N = 130$ proteins encoded by the human genome (cf. above for three examples). The background of small circles mark the point locus established by the interdependent, reciprocal relation. The large open circle point corresponds to lysozyme while the filled circles are placed by other $N = 130$ proteins. We observe that the majority of points land in the most probable $\leftrightarrow$ background zone. However, a sizable minority places outside the zone and the results are much the same for other choices of N and organism proteomes. The minority proteins amplify the lesson of Figure Eight: while canonical abnp-weights apply to diverse proteins, the deviations exceed ones consistent with random, independent draws. This again bears witness that sequences and bases are highly selected, notwithstanding their appearances on paper and computer screens. This would be a trite lesson were keratins (having periodic sequences) the majority party in the protein universe. It is of major significance given the signature aperiodicity of sequences and the diversity of proteomes.

Another facet is illuminated by conditional probability (cf. Chapter Two). Pick a base component, say, acidic. For *given* $N = 130$, the probability of a particular N_a will depend on the sequence encoded by an organism: $0 \leq N_a \leq N$. For comparison, the binomial equation offers the probability for independent, random events and canonical draw weights, viz.

$$prob(N_a \mid N) = \frac{N!}{N_a!(N - N_a)!} \times \left(\frac{2}{20}\right)^{N_a} \left(\frac{20 - 2}{20}\right)^{N - N_a}$$

These values can be compared with measures observed through experiments with natural systems. For example, if a proteome presents, say, fifty $N = 130$ sequences, and we find three to demonstrate $N_a = 10$, then we have measure:

$$prob(N_a = 10 \mid N = 130) = \frac{3}{50} = 0.060$$

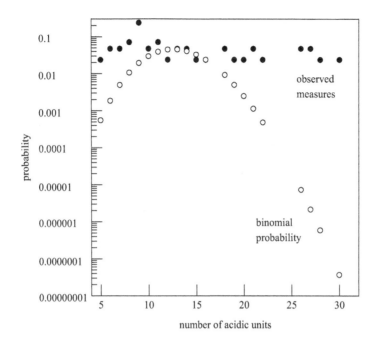

FIGURE TEN Results of human proteome analysis. The filled circles are placed by observed fractional weights for $N = 130$ proteins, whose acidic components are numbered on the horizontal axis. The open circles are placed by the binomial distribution.

The measure compares with the binomial value:

$$prob\left(N_a = 10 \mid N = 130\right) = \frac{130!}{10!(130-10)!} \times \left(\frac{2}{20}\right)^{10} \left(\frac{20-2}{20}\right)^{130-10} \approx 0.086$$

Figure Ten shows the contrast between the binomial values and experiment: open and filled circles, respectively. The mismatches should not surprise the viewer given data of the previous two figures. But note how the observed fractions are nearly uniform across the N_a range. Similar results are observed for other base components and choice of N. The takeaway is as follows. The general framework for base structures is reflective of thermodynamic systems under conditions of near-maximum probability and minimum entropy production. Figure Ten shows one example of how such reflection extends to sub-populations. Base structures are geared to express canonical weights *on average*. But there is more going on: bases are *grouped* in systems so as to push their diversity to the maximum.

E   SEQUENCES, VARIATIONS, AND STRUCTURE RECIPROCITY

The various classes of organic compounds present structural themes and variations, for example, aromatics:

The graphs and components (C_7H_8, C_8H_{10}, …) express reciprocal relations in their information. The graphs establish the components, but the components restrict what graphs or higher-level structures (bond angles, lengths, etc.) are allowed. This reprises notions in Chapter One and Section A of the present chapter regarding molecular components and structural isomers.

Proteins present structural themes just as much with reciprocal relations. In the previous section, we examined the base variations for given N. How do the variations look for a given biochemical function? We consider archetypes that have served as gateways.

For example, the following represent sequences and bases for lysozyme encoded by soft-shell turtles, sheep, canines, heliobacter pylori, and E. coli, respectively:

KIYEQCEAAREMKRLGLDGYDGYSLGDWVCTAKHESNFNTGATNYNRGDQS
TDYGIFQINSRWWCNDGKTPNAKNACGIECSELLKADITAAVICAKRVVRD
PNGMGAWVAWTKYCKGKDVSQWIKGCKL

$$A_{12}V_6L_6I_7P_2F_2W_6M_2G_{13}S_6T_7C_8Y_6N_9Q_4D_9E_6K_{12}R_6H_1 \leftrightarrow a_{15}b_{19}n_{43}p_{53}$$

KVFERCELARTLKRFGMDGFRGISLANWMCLARWESSYNTQATNYNSGDRSTD
YGIFQINSHWWWCNDGKTPGAVNACHIPCSALLQDDITQAVACAKR
VVSDPQGIRAWVAWRSHCQNQDLTSYIQGCGV

$$A_{12}V_7L_7I_7P_3F_4W_6M_2G_{10}S_{10}T_7C_8Y_4N_8Q_8D_8E_3K_4R_9H_3 \leftrightarrow a_{11}b_{16}n_{48}p_{55}$$

KIFERCELARTLKNLGLAGYKGVSLANWVCLAKWESNYNTRATNYNPGSKSTD
YGIFQINSRYWCNDGKTPRAVNACHISCSALLQDDITQAVACAKRVVSDPN
GIRAWVAWRAHCENRDVSQYVRNCGV

$$A_{14}V_{10}L_8I_6P_3F_2W_5M_0G_8S_9T_6C_8Y_6N_{12}Q_4D_6E_4K_7R_{10}H_2 \leftrightarrow a_{10}b_{19}n_{48}p_{53}$$

MILVASFLIVDSEGFSLSVYTDKTGHPTIGYGYNLSVYSYESRRITKAYGLLTDILK
KNHRALLSYEWYKNLDAMRRMVILDLSYNLGLNGLLKFKQFIKAIEDKNY
ALAVERLQKSPYFNQVKKERQGI

$$A_7V_7L_{18}I_9P_2F_5W_1M_3G_8S_{10}T_5C_0Y_{11}N_7Q_4D_6E_6K_{12}R_7H_2 \leftrightarrow a_{12}b_{21}n_{52}p_{45}$$

MKMSYWALILTFIACVAGGLVWSANHYHGKFLEEQKRADAAEQRADSTEAITA
NVLRTMAITNIIQEANQHAKQQIALESQRTQEDIKVAVADDDCASRPV
PAAAADRLRKYANSLRPGSGSSVTSQPDG

$$A_{22}V_7L_8I_8P_4F_2W_2M_3G_6S_{10}T_7C_2Y_3N_5Q_9D_8E_7K_6R_8H_3 \leftrightarrow a_{15}b_{17}n_{56}p_{42}$$

Discerning structural themes in proteins is more challenging than for small organics: random strings baffle more than icons and pictures! Such challenges are ordinarily met via bioinformatics and modeling software: BLAST, MSA, molecular dynamics, and more. Among other things, alignments, gaps, and homologies are established while folded configurations are explored via potential functions and simulations.

Base-sequence relations shine complementary light. They point to general trends of what evolution is "trying to do" and "trying to avoid" in conferring structures and functions. As usual, the quotation marks acknowledge that evolution does not try to do anything. Sequences and bases just happen, and some happen to be more robust or at least tolerable across species than others.

The sequence confers the base while the base limits the sequence. For example, the lysozyme encoded by soft-shell turtles underpins sequence surprisal:

$$S(\Omega) = \log_2\left[\frac{(N_a + N_b + N_n + N_p)!}{N_a!N_b!N_n!N_p!}\right]$$

$$= \log_2\left[\frac{(15+19+43+53)!}{15!19!43!53!}\right] \approx 227 \text{ bits}$$

This applies to the protein viewed in abnp-partition terms. Then the base-sequence relation asserts:

$$S(N_a, N_b, N_n, N_p) = -S(\Omega) - \log_2\left\{\left(\frac{2}{20}\right)^{N_a} \cdot \left(\frac{3}{20}\right)^{N_b} \cdot \left(\frac{8}{20}\right)^{N_n} \cdot \left(\frac{7}{20}\right)^{N_p}\right\}$$

$$= -S(\Omega) - \log_2\left\{\left(\frac{2}{20}\right)^{15} \cdot \left(\frac{3}{20}\right)^{19} \cdot \left(\frac{8}{20}\right)^{43} \cdot \left(\frac{7}{20}\right)^{53}\right\}$$

$$\approx -227 + 239 = 12 \text{ bits}$$

The surprisals place a *state point* for the turtle lysozyme with x,y-coordinate (12 bits, 227 bits). The same approach can be directed to a collection of lysozymes and indeed proteins in general.

How do we interpret state point placements? It is that bases and sequences present three paradigms at the extremes. Think of these as different phase types in thermodynamic systems with everyday analogs: materials as a rule offer three distinct phases, solid, liquid, and gas, while solutions present three modes of behavior—adherence to, positive, and negative deviations from ideality. Figure Eleven illustrates the paradigms by revisiting the point locus of Figure Nine. The first paradigm—our choice of labels is arbitrary—corresponds to proteins which place $S(N_a, N_b, N_n, N_p)$,

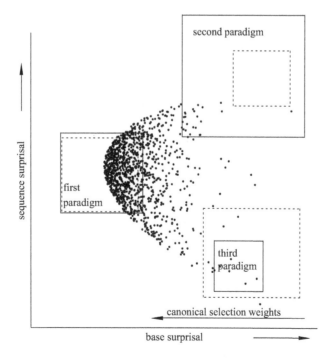

FIGURE ELEVEN Paradigms of base-sequence relationships. The vertical and horizontal axes mark surprisals $S(\Omega)$ and $S(N_a, N_b, N_n, N_p)$ as in Figure Nine. The room for sequence possibilities is depicted by the solid-line boxes. The room allotted for conservative substitutions is depicted schematically by the dotted-line boxes.

$S(\Omega)$ coordinates toward the far left. States in this region apply to proteins which closely reflect canonical selection weights. A base structure controls the room for the sequence possibilities. The allotted room is indicated schematically by the solid-line box so labeled. Volume-wise, the room falls in-between those for the other two paradigms. Upon encountering first-paradigm proteins, our surprise is minimal (or should be) given their (comparatively) high base probability.

The second and third paradigms apply to proteins that express non-canonical selection weights. There are two ways to deviate just as in liquid solutions. In the second paradigm, the base probability trends lower and the surprisal higher. The deviations carry critical effect by *increasing* the sequence possibilities—the number of ways to order the amino-acid units. In steering molecules along the second paradigm, evolution not only confers the biochemical functions, but also signals an extra degree of rarefaction. It is as if the proteins were *ultra*-custom-fabricated for the task at hand. When we encounter second-paradigm proteins, our surprise is high on *both* base and sequence-order accounts. The second paradigm corresponds to the upper regions of Figure Eleven, the solid-line box indicating the size (schematically speaking) of the sequence space.

The third paradigm also deviates from canonical abnp-weights and is no less critical. Here evolution establishes the biochemical functions, but restricts somewhat the space for amino-acid arrangements, again as depicted by the solid-line box. Our surprise is relatively high for the base structure, although less so for the sequence-order compared with the first and second paradigms. It is as if limits have been imposed on the custom tailoring; perhaps the task at hand does not require as much as the other paradigms. It goes without saying that all base and sequence details start with the genes and environments underpinning the proteins.

The paradigms mark the extremes of base-sequence interdependence. And they carry vital information about more than one type of space. We refer to the room allotted for abnp-conservative substitutions. Importantly, this room grows or shrinks in a direction *opposite* the sequence-order possibilities. This is indicated schematically by the dotted-line boxes in Figure Eleven. For consider an $N = 100$ base with canonical weights: $a_{10}b_{15}n_{40}p_{35} \leftrightarrow a_{10\%}b_{15\%}n_{40\%}p_{35\%}$. The base is first-paradigm with respective measures for the sequence and substitution spaces:

$$\frac{N!}{N_a!N_b!N_n!N_p!} = \frac{100!}{10!15!40!35!} \approx 2.3 \times 10^{51}$$

$$2^{N_a} \cdot 3^{N_b} \cdot 8^{N_n} \cdot 7^{N_p} = 2^{10} \cdot 3^{15} \cdot 8^{40} \cdot 7^{35} \approx 7.4 \times 10^{75}$$

Compare these values to second-paradigm cases: these originate when one or both minority components are increased at the expense of the majority. For the example of $a_{10+5}b_{15+5}n_{40-5}p_{35-5} \leftrightarrow a_{15\%}b_{20\%}n_{35\%}p_{30\%}$ we have for measures:

$$\frac{N!}{N_a!N_b!N_n!N_p!} = \frac{100!}{(10+5)!(15+5)!(40-5)!(35-5)!} \approx 1.1 \times 10^{55}$$

$$2^{N_a} \cdot 3^{N_b} \cdot 8^{N_n} \cdot 7^{N_p} = 2^{10+5} \cdot 3^{15+5} \cdot 8^{40-5} \cdot 7^{35-5} \approx 1.0 \times 10^{71}$$

The third paradigm manifests when one or both minority components are diminished in favor of the majority, for example, $a_{10-5}b_{15-5}n_{40+5}p_{35+5}$. The measures become:

$$\frac{N!}{N_a!N_b!N_n!N_p!} = \frac{100!}{(10-5)!(15-5)!(40+5)!(35+5)!} \approx 2.2 \times 10^{45}$$

$$2^{N_a} \cdot 3^{N_b} \cdot 8^{N_n} \cdot 7^{N_p} = 2^{10-5} \cdot 3^{15-5} \cdot 8^{40+5} \cdot 7^{35+5} \approx 5.2 \times 10^{80}$$

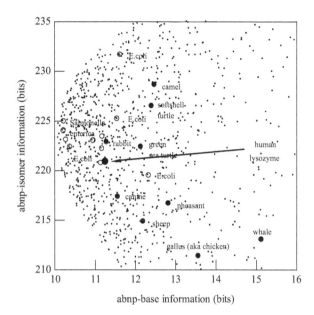

FIGURE TWELVE Sequence-versus-base information in abnp-terms for lysozymes. The state points for various species are labeled. The open circles are placed by bacteria. The filled circles are placed by the lysozymes of higher organisms. The backdrop of points represent $N = 130$ random structures according to canonical selection weights.

Where do archetypes place their state points? Figure Twelve addresses matters by showing the $S(N_a, N_b, N_n, N_p)$, $S(\Omega)$ coordinates for multiple lysozymes. The backdrop is the locus of points for $N = 130$ structures with canonical selection weights. The open circles refer to lysozymes encoded by bacteria while the filled circles are placed by the lysozymes of several higher organisms. The placement for human-encoded lysozyme KVFER…GCGV is marked indicated via the arrow. The horizontal axis refers to the base surprisal $S(N_a, N_b, N_n, N_p)$ in abnp-family terms. The vertical axis marks the sequence surprisal $S(\Omega)$.

We see that evolution is directing lysozymes primarily toward the first and third paradigms. In turn, the lysozymes for higher organisms express bases that are more abnp-rarefied, compared with bacteria. The base surprisal for the human-encoded version is the lowest among the higher organisms. Lysozymes encoded by chickens and whales pose sequences and bases that are most rarefied: their lysozymes manifest the lowest volumes of sequence space. By the same token, evolution is maximizing the room for conservative substitutions. It is as if the proteins are conferred with an eye toward future needs and updates. What is evolution seemingly avoiding? That would be forcing the enzymes toward the second paradigm which ramps up the custom tailoring of sequences but restricts (somewhat) the room for conservative substitutions.

Figure Thirteen presents $S(N_a, N_b, N_n, N_p)$, $S(\Omega)$ for multiple ribonucleases of the A-variety. As in the previous figure, the backdrop is provided by canonical selection weights. The placement for human ribonuclease A is noted by the large filled circle. We learn that just as with lysozymes, evolution is directing the ribonucleases across the first and third paradigms while staying clear of the second paradigm. Yet again, the room for conservative updates is maximized.

Figure Fourteen presents $S(N_a, N_b, N_n, N_p)$, $S(\Omega)$ for myoglobins, the backdrop provided by family-selection weights. The placement for human-encoded myoglobin (cf. beginning of chapter) is noted by the arrow. Here we observe markedly different behavior from the previous two figures. Evolution is directing the myoglobins for mammals (filled circles) almost exclusively along the second paradigm. They indeed mark and follow a corridor at the top of the state diagram. In contrast, sharks (open circles) having ancient myoglobins place state points in the first-paradigm region.

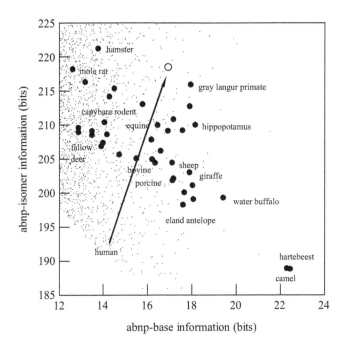

FIGURE THIRTEEN   Sequence-versus-base information in abnp-terms for ribonucleases. The state points for various species are labeled. The small filled circles are placed by $N = 124$ random structures according to canonical selection weights.

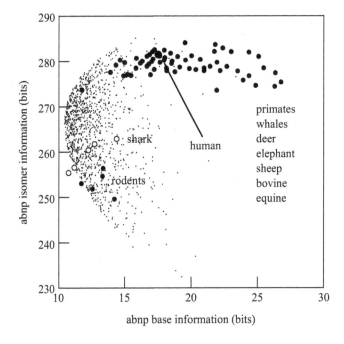

FIGURE FOURTEEN   Sequence-versus-base information in abnp-terms for myoglobins. The data points for various species are labeled: filled and open circles are placed by mammals and sharks, respectively. The placement of human myoglobin is noted by the arrow while the background reflects $N = 154$ random structures according to canonical selection weights.

What is evolution seemingly avoiding? The myoglobins are largely kept away from bases which restrict the sequence room but maximize the space for conservative substitutions. It is as if evolution were shying from rocking the boat via substitutions.

The major points of Chapter Three are as follows:

1. All proteins present base structures by their amino-acid compositions $\leftrightarrow$ parts lists. Bases are conferred by evolution and gene expression and intersect with all the levels of structure: sequences, α-helices, β-sheets, 3D folds, etc. Virtually all bases are thermodynamically stable, and 20 components allow for truly many possibilities. Probability and information offer lenses for exploring the design and selection traits.

2. The base structures for archetypes are ill-correlated with genetic-code biases toward the amino acids. Further, bases evince only slight preferences for lower mass components over the heavies. This signals material- and entropy-economy as a minor factor in base design and selection, as viewed in 20-component terms.

3. Base structures trend toward canonical selection weights with respect to non-polar, polar, acidic, and basic components. The weights are tied to the family sizes established by evolution. To be sure, there are deviations from canonical. These speak further to choices of base structures in composing natural sets. This is examined in greater detail in Chapter Five.

4. There is reciprocity between the base and sequence information: a sequence confers the base while the base restricts the sequences. The interdependence offers a means for discriminating evolutionary trends in protein families. The trends concern both the space available for amino-acid order and the room for updates by conservative substitutions.

EXERCISES

1. The following sequence represents a keratin molecule. Re-write the sequence in maximum-condensed terms. What is the (highly non-canonical) base structure?

   MGCCGCSEGCGSGCGGCGSGCGGCGSGCGGCGSSCCVPVCCCKPVCCCV
   PACSCSSCGSCGGSKGGCGSCGGSKGGCGSCGCSQCSCYKPCC
   CSSGCGSSCCQSSCCKPCCCQSSCCKPCCCSSGCGSSCCQSSCCNPCCSQ
   SSCCVPVCCQCKI

2. Groups of atoms form the bases of simple organics. Is $C_{11}H_{24}$ a valid base?

3. The following sequence describes a carbonic anhydrase:

   MRALVLLLSLFLLGGQAQHVSDWTYSEGALDEAHWPQHYPACGGQRQSP
   INLQRTKVRYNPSLKGLNMTGYETQAGEFPMVNNGHTVQISLPSTMRMT
   VADGTVYIAQQMHFHWGGASSEISGSEHTVDGIRHVIEIHIVHYNSKYKSYDIAQ
   DAPDGLAVLAAFVEVKNYPENTYYSNFISHLANIKYPGQRTTL
   TGLDVQDMLPRNLQHYYTYHGSLTTPPCTENVHWFVLADFVKLSRTQVW
   KLENSLLDHRNKTIHNDYRRTQPLNHRVVESNFPNQEYTLGSEFQFYLHKIE
   EILDYLRRALN

   a. What is the base structure?
   b. How selected is the base by the alanine content as measured by the surprisal? Take the amino acids to be equally probable.
   c. Which component appears most selected? Least?
   d. Use the multinomial distribution to compute the total surprisal.

4. Consider the base structure:

$$A_7V_2L_4I_{10}P_8F_{11}W_5M_7G_3S_5T_6C_{10}Y_8N_{12}Q_7D_6E_1K_6R_5H_7$$

Imagine the structure resulting from independent, random draws from an equal-representation pool of amino acids at 10^{-12}s per structure. Estimate the time to obtain the base.

5. Consider the probability and surprisal for the lysozyme base according to the multinomial distribution. To how many base structures do the probability and surprisal apply?

6. Consider the base of ribonuclease A obtained through independent random draws weighted by the genetic code. What is the surprisal in bits?

7. Compare the lysozyme base to the source-pools featured in the experiments.
   a. What is the angle in radians between 20D vectors allied with the two systems?
   b. Let the same question apply to 4D vectors determined by abnp-components.

8. Revisit the sequence for human-encoded myoglobin.

MGLSDGEWQLVLNVWGKVEADIPGHGQEVLIRLFKGHPETLEKFDKFKHL
KSEDEMKASEDLKKHGATVLTALGGILKKKGHHEAEIKPLAQSHATKHKIPVKYL
EFISECIIQVLQSKHPGDFGADAQGAMNKALELFRKDMASNYKELGFQG

   a. What is the base structure in abnp-partition terms?
   b. Use the multinomial distribution and canonical selection weights to compute the surprisal in bits.
   c. Compute the volume of space allowed for abnp-conservative substitutions.

9. An amino-acid site is selected at random in a protein of interest and found to be non-polar.
   a. What is the probability that the site belongs to the ambivalent family?
   b. What is the probability that the site belongs to the internal family?
   c. Do the answers change if we are given knowledge of the protein sequence?

10. An amino-acid site is selected at random and found to belong to the polar family.
   a. What is the probability that the site belongs to the ambivalent family?
   b. What is the probability that the site belongs to the external family?
   c. Do the answers change if the protein sequence is specified?

11. Tyrosine contains a phenol unit (cf. Chapter One). Phenols are characterized as weak Bronsted–Lowry acids. Should tyrosine be viewed as a member of the family of acidic amino acids?

12. Consult the literature regarding amino acids that manifest due to post-genomic processing. If these are added to the standard 20, how many amino-acid families are conferred by nature?

13. Calculate the bits per symbol for the standard and emoji-containing representations of lysozyme. Take the probability weight of each symbol type to be proportional to its occurrence in the sequence.

14. Estimate the probability of observing the myoglobin base through independent, random draws. Take the draw probability for each amino acid to be 1/20. What is the surprisal in bits?

15. The special point in far right field of Figure One applies not just to lysozyme, but rather to an extended class of bases. How many bases are members of this class?

16. Consider the following mutually exclusive families of amino acids:

GASPVHFRYW   ETCILNMDKQ

Discuss possible conditions that prescribe the families.

17. Prescribe two partitions of the amino acids and discuss their usefulness in protein structure analysis.

18. What is the surprisal for ribonuclease A in bits viewed in abnp-partition terms and canonical selection weights?

19. Consider the lysozyme base and sequence in abnp-terms. Suppose there were eight different acidic amino acids available and two non-polars. Calculate the number of distinguishable sequences. How does it compare the number allowed by eight non-polar amino acids and two acidic ones?

20. The following are respective sequences for prion, heat shock, and cytochrome proteins:

MRKHLSWWWWLATVCMLLFSHLSAVQTRGIKHRIKWNRKALPSTAQITEAQVA
ENRPGAFIKQGRKLDIDFGAEGNRYYEANYWQFPDGIHYNGCSEANVTKE
AFVTGCINATQAANQGEFQKPDNKLHQQVLWRLVQELCSLKHCEFWLER
GAGLRVTMHQPVLLCLLALIWLTVK

MTERRVPFSLLRGPSWDPFRDWYPHSRLFDQAFGLPRLPEEWSQWLGGSSWPG
YVRPLPPAAIESPAVAAPAYSRALSRQLSSGVSEIRHTADRWRVSLDVNHFAPD
ELTVKTKDGVVEITGKHEERQDEHGYISRCFTRKYTLPPGVDPTQVSSSLSP
EGTLTVEAPMPKLATQSNEITIPVTFESRAQLGGPEAAKSDETAAK

MAAAAATLRGAMVGPRGAGLPGARARGLLCGARPGQLPLRTPQAVSLSSK
SGLSRGRKVILSALGMLAAGGAGLAVALHSAVSASDLELHPPSYPWSHRG
LLSSLDHTSIRRGFQVYKQVCSSCHSMDYVAYRHLVGVCYTEDEAKALAEE
VEVQDGPNEDGEMFMRPGKLSDYFPKPYPNPEAARAANNGALPPDLSYIVR
ARHGGEDYVFSLLTGYCEPPTGVSLREGLYFNPYFPGQAIGMAPPIYNEVLEF
DDGTPATMSQVAKDVCTFLRWAAEPEHDHRKRMGLKMLLMMGLLLPLVYAM
KRHKWSVLKSRKLAYRPPK

Which steers closest to canonical abnp-selection weights? Which steers farthest?

21. Consider information in the abnp-respellings of lysozyme, ribonuclease A, and myoglobin, and that of the keratin in Exercise One. Take the probability weight of each symbol type to be proportional to its occurrence. Which structure expresses the fewest bits per symbol? Which structure expresses the most bits per symbols?

22. Which of the following represents true-blue myoglobin? Address the question by comparing the bold- and normal-font sections of the following sequences:

MGLSDGEWQLVLNVWGKVEADIPGHGQEVLIRLFKGHPETLEKFDKFK
HLKSEDEMKASEDLKKHGATVLTALGGILKKKGHHEAEIKPLAQSHATKHKI
PVKYLEFISECIIQVLQSKHPGDFGADAQGAMNKALELFRKDMASNYKELGFQG

MGLSDGEWQLVLNVWGKVEADIPGHGQEVLIRLFKGHPETLEKFDKFKH
LKSEDEMKASEDLKKHGATVLTALGGILRAECQIPCDVSRGIYCTAKTTYGFN
NALFGFPNWEGHPAGMKNCTTFGGQVELYHVVLAYSQQDIGGCGQCWIVEQ

23. Can we always assume that the abnp-bases of the first and second halves of a sequence are more similar than not? Consider the following $N = 130$ sequence of the human proteome to answer this question:

MTTPRNSVNGTFPAEPMKGPIAMQSGPKPLFRRMSSLELVIAGIVENEWK
RTCSRPKSNIVLLSAEEKKEQTIEIKEEVVGLTETSSQPKNEEDIEIIPIQ
EEEEEETETNFPEPPQDQESSPIENDSSP

24. The distance between sets A and B is defined as follows:

$$d(A,B) = m(A \cup B - A \cap B)$$

What kinds of partitions of the 20 amino acids maximize the average distance between sets? The term $m(A \cup B - A \cap B)$ refers to the number of elements in the set so formed: the *countable* measure.

NOTES, SOURCES, AND FURTHER READING

Protein structure analysis using probability and information proceeds along three general fronts. The first is grounded in statistical thermodynamics and attends to 3D configurations (helices, folds, β-sheets) and phase transitions (folding, unfolding) subject to model potential functions. The second encompasses bioinformatics and attends to sequence variation and phylogeny plus biomedical applications. The third front applies probability theory such as discussed in Chapter Two. For example, *given* a base structure B, what is the probability of observing, in a linear stochastic process, specific runs R, charge clusters C, mass distributions M, and residue spacings S:

$$prob(R \mid B) \quad prob(C \mid B) \quad prob(M \mid B) \quad prob(S \mid B)$$

The probabilities are conditioned whereby low values point to *special R, C, M, and S* in the sense of evolutionary and biochemical significance. The third front, in tandem with bioinformatics, has long applied partition strategies—ways of classifying the amino acids. Even so, partitions and the respellings that go hand-in-hand typically receive glancing attention in the biochemistry curriculum—there is too much other territory to cover! The present chapter concentrated on the *B*-parts of probability terms such as above. This motivated experiments guided by the fundamental families of amino acids. The reader is encouraged to study Karlin [19–22], Pande [23–26], and White [27–29] for partition and respelling insights and principles. It is also time well spent to review the work of Fisher regarding the composition and limiting law properties of globular proteins [30]. We highly recommend the review by Randić and co-workers on the graphical properties of biopolymer sequences, both DNA and protein [31]. The thermodynamic facets of protein sequences and functions have been central to works by Yockey, Shakhnovich, Dill, and their co-workers [32–35]. We direct the reader to Durbin regarding the application of probability models to sequence analysis [36]. Proteins are famously given to unusual compositions as discussed by Wootton [37]. Lastly all molecules present information by their compositions. The base information in small organic compounds has been researched by the author [38].

1. Muraki, M., Harata, K., Sugita, N., Sato, K. 1996. Origin of carbohydrate recognition specificity of human lysozyme revealed by affinity labeling, *Biochemistry* 35, 13562.
2. Raines, R. T. 1998. Ribonuclease A, *Chem. Rev.* 98(3), 1045–1066.
3. Savino, C., Miele, A. E., Draghi, F., Johnson, K. A., Sciara, G., Brunori, M., Vallone, B. 2009. Pattern of cavities in globins: The case of human hemoglobin, *Biopolymers* 91, 1097.
4. Sawyer, R. H. 1987. *The Molecular and Developmental Biology of Keratins*, Academic Press, Orlando, FL.
5. Squire, J., Vibert, P. J. 1987. *Fibrous Protein Structure: A Volume Dedicated to Dr. Arthur Elliott*, Academic Press, London.
6. Darby, N. J., Creighton, T. E. 1993. *Protein Structure*, IRL Press at Oxford University Press, Oxford.
7. Lehninger, A. L. 1970. *Biochemistry*, Worth Publishers, New York.
8. Prigogine, I. 1957. *The Molecular Theory of Solutions*, North Holland Publishing Company, Amsterdam.
9. le Noble, W. J. 1974. *Highlights of Organic Chemistry*, Dekker, New York.
10. Parzen, E. 1960. *Modern Probability Theory and Its Applications*, Wiley, New York.
11. Wilf, H. S. 1978. *Mathematics for the Physical Sciences*, Dover, New York.
12. Jaynes, E. T. 1979. Where Do We Stand on Maximum Entropy? in *The Maximum Entropy Formalism*, M. Tribus, R. D. Levine, eds., MIT Press, Cambridge, MA.
13. Balakrishna, N. 2003. *A Primer on Statistical Distributions*, Wiley, Hoboken, NJ.
14. Hamming, R. W. 1991. *The Art of Probability for Scientists and Engineers*, Addison-Wesley, Redwood City, CA.
15. Bennett, C. H. 1982. Thermodynamics of computation—A review, *Int. J. Theo. Phys.* 21, 905.
16. White, A., Handler, P., Smith, E. L. 1972. The proteins I, in *Principles of Biochemistry*, McGraw-Hill, New York.

17. Yourgrau, W., Van der Merwe, A., Raw, G. 1966. *Treatise on Irreversible and Statistical Thermophysics: An Introduction to Nonclassical Thermodynamics*, Macmillan, New York.
18. Hill, T. L. 1986. *An Introduction to Statistical Thermodynamics*, Dover, New York.
19. Karlin, S., Buche, P., Brendel, V., Altschul, S. F. 1991. Statistical methods and insights for protein and DNA sequences, *Annu. Rev. Biophys. Biophys. Chem.* 20, 175–203.
20. Karlin, S., Ghandour, G. 1985. Multiple-alphabet amino acid sequence comparisons of the immuno-globulin kappa-chain constant domain, *PNAS USA* 82(24), 8597–8601.
21. Karlin, S., Ghandour, G. 1985. The use of multiple alphabets in kappa-gene immunoglobulin DNA sequence comparisons, *EMBO J.* 4(5), 1217–1223.
22. Karlin, S., Ghandour, G., Foulser, D. D., Korn, L. J. 1984. Comparative analysis of human and bovine papillomaviruses, *Mol. Biol. Evol.* 1, 357.
23. Pande, V. S., Grosberg, A. Y., Tanaka, T. 1994. Nonrandomness in protein sequences: Evidence for a physically driven stage of evolution, *PNAS USA* 91, 12972–12975.
24. Pande, V. S., Grosberg, A. Y., Tanaka, T. 2000. Heteropolymer freezing and design: Toward physical models of protein folding, *Rev. Mod. Phys.* 72, 259–314.
25. Pande, V. S., Bowman, G. R. 2010. Protein folded states are kinetic hubs, *PNAS USA* 107(24), 10890–10895.
26. Pande, V. S., Grosberg, A. Y., Tanaka, T. 1995. Phase diagram of heteropolymers with an imprinted conformation, *Macromolecules* 28(7), 2218–2227.
27. White, S. H. 1994. Global statistics of protein sequences: Implications for the origin, evolution, and prediction of structure, *Annu. Rev. Biophys. Biomolec. Struct.* 23, 407–439.
28. White, S. H. 1992. The amino acid preferences of small protein: Implications for protein stability and evolution, *J. Mol. Biol.* 227, 991–995.
29. White, S. H., Jacobs, R. E. 1993. The evolution of protein from random amino acid sequences I. Evidence from the lengthwise distribution of amino acids in modern protein sequences, *J. Mol. Evol.* 36, 79–95.
30. Fisher, H. F. 1964. A limiting law relating the size and shape of protein molecules to their composition, *PNAS USA* 51, 1285–1291.
31. Randić, M., Zupan, J., Balaban, A. T., Dražen, V.-T., Plavšić, D. 2011. Graphical representation of proteins, *Chem. Rev.* 111, 790–862.
32. Yockey, H. P. 1992. *Information Theory and Molecular Biology*, Cambridge University Press, Cambridge, UK.
33. Broglia, R. A., Shakhnovich, E. I., eds. 2001. *Proceedings of the International School of Physics, Course CXLV, Protein Folding, Evolution, and Design*, IIOS Press, Amsterdam.
34. Shakhnovich, E. 2006. Protein folding thermodynamics and dynamics: Where physics, chemistry and biology meet, *Chem. Rev.* 106(5), 1559–1588.
35. Dill, K. A. 1985. Theory for the folding and stability of globular proteins, *Biochemistry* 24(6), 1501–1509.
36. Durbin, R., Eddy, S., Krogh, A., Mitchison, G. 1998. *Biological Sequence Analysis: Probabilistic Models of Proteins and Nucleic Acids*, Cambridge University Press, New York.
37. Wootton, J. C. 1994. Sequences with "Unusual" amino acid compositions, *Curr. Opin. Struct. Biol.* 4, 413–421.
38. Graham, D. J., Schacht, D. 2000. Base information content in organic formulas, *J. Chem. Info. Compt. Sci.* 40, 942.

Four Base Structure Analysis and Constituent Numbers

Proteins express information by their constituent numbers N: the number of amino acids apart from identity. The chapter concentrates on the length properties of proteins that typically receive glancing attention during sequence and folded structure inspection. The tools of Chapter Two are directed (again!) to archetypal molecules and sets. The compressibility of a sequence is explored through information-conserving transforms. The discussion is followed by the application of probability functions to the N-distributions expressed by proteomes. The chapter closes with attention to the information asymmetry of distributions.

A SEQUENCES, BASES, AND COMPRESSIBILITY

The preceding chapter focused on the information in protein compositions. However, it did not consider perhaps the most rudimentary feature: the number of component units N irrespective of identity. Revisit the sequence and bases for human-encoded lysozyme, bovine ribonuclease A, and human-encoded myoglobin [1–3]:

KVFERCELARTLKRLGMDGYRGISLANWMCLAKWESGYNTRATNYNAGDRST
DYGIFQINSRYWCNDGKTPGAVNACHLSCSA<u>LL</u>QDNIADAVACAKR<u>VV</u>RDP
QGIRAWVAWRNRCQNRDVRQYVQGCGV

$$A_{14}V_9L_8I_5P_2F_2W_5M_2G_{11}S_6T_5C_8Y_6N_{10}Q_6D_8E_3K_5R_{14}H_1$$

KETAAAKFERQHMDSSTSAASSSNYCNQMMKSRNLTKDRCKPVNTFVHESLAD
VQAVCSQKNVACKNGQTNCYQSYSTMSITDCRETGSSKYPNCAYKTTQANKHIIV
ACEGNPYVPVHFDASV

$$A_{12}V_9L_2I_3P_4F_3W_0M_4G_3S_{15}T_{10}C_8Y_6N_{10}Q_7D_5E_5K_{10}R_4H_4$$

MGLSDGEWQLVLNVWGKVEADIPGHGQEVLIRLFKGHPETLEKFDKFKHLKS
EDEMKASEDLKKHGATVLTALGGILKKKGHHEAEIKPLAQSHATKHKIPVKYLEFI
SECIIQVLQSKHPGDFGADAQGAMNKALELFRKDMASNYKELGFQG

$$A_{12}V_7L_{17}I_8P_5F_7W_2M_4G_{15}S_7T_4C_1Y_2N_3Q_7D_8E_{14}K_{20}R_2H_9$$

The above present $N = 130$, 124, and 154, respectively. A phospholipase with $N = 118$ (cf. Chapter One) is presented as follows [4]:

SLLEFGKMILEETGKLAIPSYSSYGCYCGWGGKGTPKDATDRCCFVHDCCYGN
LPDCNPKSDRYKYKRVNGAIVCEKGTSCENRICECDKAAAICFRQNLNTY
SKKYMLYPDFLCKGELKC

$$A_6V_3L_9I_5P_5F_4W_1M_2G_{11}S_7T_5C_{14}Y_9N_6Q_1D_7E_7K_{13}R_5H_1$$

An alpha synuclein with $N = 140$ looks like [5]:

MDVFMKGLSKAKEGVVAAAEKTKQGVAEAAGKTKEGVLYVGSKTKEGVVHGV
ATVAEKTKEQVTNVGGAVVTGVTAVAQKTVEGAGSIAAATGFVKKDQLGKNE
EGAPQEGILEDMPVDPDNEAYEMPSEEGYQDYEPEA

$$A_{19}V_{19}L_4I_2P_5F_2W_0M_4G_{18}S_4T_{10}C_0Y_4N_3Q_6D_6E_{18}K_{15}R_0H_1$$

The list can be endless and questions come easily. $N = 130$ for lysozyme. Why not 131 or 129? The examples feature $N = 130, 124, 154, 118$, and 140. Is there a bias in nature toward N-even over odd?

As with other facets of protein structure, making sense of N is an uphill climb, for the complexities take root *before* the molecules are synthesized N- to C-terminal. Their information originates in the genome of an organism or virus. Yet we cannot discern the length of a DNA or RNA source (the exception is mRNA) from a protein structure. The information is stored in a gene at (typically) low density and transferred asymmetrically (cf. Chapter One) [6,7]. As for particular N-values, the throwaway explanation is that $N = 130$ for lysozyme because, well, *that* is what *works* catalytically. $N = 154$ succeeds for certain myoglobins and oxygen transport functions. And so forth. Such explanations are 100% valid in that evolution and biochemistry underpin all the properties. Can probability and information provide extra light?

The answer is mixed. Probability and information cannot justify N for a specific protein, but can illuminate trends and complexities: counting and grouping experiments point us to models for the N-distributions in natural systems. These give us a feel for the kinds of properties that determine sequence lengths.

We begin by examining the compressibility of sequences. $N = 130$ for lysozyme. Do we really need 130 symbols to represent the primary structure? We take for granted that the representations for organic structures are compressible. For example, the following apply to lysine:

The restrictions imposed by atomic valence and electron sharing allow using fewer symbols in the right-side figure. We are able to expand it at will to obtain the left side, based on chemistry training. Keratins with periodic sequences can be written in condensed terms (cf. Chapter Three). With so many proteins, however, sequence compression seems less than straightforward. Granted, we can add or subtract units arbitrarily, for example, in lysozyme:

$$KVFER...CGVG \leftrightarrow A_{14}V_9L_8I_5P_2F_2W_5M_2G_{11+1}S_6T_5C_8Y_6N_{10}Q_6D_8E_3K_5R_{14}H_1$$
$$\_VFER...CGV \leftrightarrow A_{14}V_9L_8I_5P_2F_2W_5M_2G_{11}S_6T_5C_8Y_6N_{10}Q_6D_8E_3K_{5-1}R_{14}H_1$$

But doing so alters the information and does nothing to explore N criteria. We can also appeal to any number of text-compression routines such as Huffman coding. The routines, however, teach more about the intricacies of codes and little about molecular information. This calls for more experiments.

Our procedure directs the Burrows–Wheeler transform to archetypes such as lysozyme [8,9]. As with Huffman coding, this is an established (if somewhat obscure) algorithm for text compression, typically aimed at storage economy: why keep 1,000 symbols in computer memory if only 900 are needed? Where text is compressible, the transform yields strings containing runs of identical symbols. For example, a transform might result in:

$$....CTVVVVVVVRAMLDSSSSS....$$

This can be rewritten as

$$....CTV_6RAMLDS_5....$$

with no compromise of the information. The new string presents 11 symbols in place of the original 19. Recovering the original means re-tracing steps: expanding the string and applying the Burrows–Wheeler algorithm in reverse. Importantly, the method succeeds when correlations are harbored by the source, for these render parts of the message redundant. The transform paves the way for pruning redundancies without sacrificing the message integrity. The transform depends critically on the order, number, and identities of sequence units.

Can the sequence for lysozyme be compressed? The question is more than academic, and not because of data storage. If the answer is yes, then $N = 130$ is not so hard and fast. The representation KVFER...GCGV contains correlations and redundancies, however obliquely. In communication and computation, correlations and redundancies minimize errors in the transmission and reception of messages. These would seem desirable traits in proteins given their shape, volume, and energy fluctuations (cf. Chapter One). If the answer is no, then KVFER...GCGV represents a large molecule that not only succeeds biochemically, but also has been engineered according to minimum coding protocols. The primary structure is bare bones whereby every constituent is vital. The minimization and correction of errors transpire through processes beyond the sequence, for example, involving the folded structure, solvent, and electrolyte conditions.

What do we expect for the answer? A perusal of lysozymes across species finds sequences with $N < 130$. For example, the following with $N = 129$ is encoded by chickens:

KVFGRCEL**AAA**MKRHGLDNYRGYSLGNWVCAAKFESNFNTQATNRNTDGST
DYGILQINSRWWCNDGRTPGSRNLCNIPCSALLSSDITASVNCAKKIVSDGNGMNA
WVAWRNRCKGTDVQAWIRGCRL

Even without Burrows–Wheeler, the above is *slightly* compressible, viz.

KVFGRCEL**A**$_3$MKRHGLDNYRGYSLGNWVCAAKFESNFNTQATNRNTDGST
DYGILQINSRWWCNDGRTPGSRNLCNIPCSALLSSDITASVNCAKKIVSDGNG
MNAWVAWRNRCKGTDVQAWIRGCRL

We can save one symbol (in bold underline) in the re-writing and observe the same in other archetypes. Then in thermodynamic terms, sequences seem closer to liquids than gases: gases are highly compressible at ambient pressures whereas liquids are not.

The Burrows–Wheeler transformation involves three stages. First, the *rotations* $R^{(0)} - R^{(N-1)}$ of the sequence in question are constructed. Each obtains by placing the last symbol at the head of the line. For human-encoded lysozyme, we have:

$R^{(0)}$: KVFERCELARTLKRLGMD...QGCGV
$R^{(1)}$: **V**KVFERCELARTLKRLGMD...QGCG
$R^{(2)}$: **GV**KVFERCELARTLKRLGMD...QGC
$R^{(3)}$: **CGV**KVFERCELARTLKRLGMD...QG

.

.

.

$R^{(N-1)}$: VFERCELARTLKRLGMD...QGCGV**K**

Second, the rotations are placed in lexicographic order. It is here where an experiment takes shape. The single-letter abbreviations for the amino acids are mnemonic-friendly, however they present no hard and fast order. For example, we could arbitrarily take the order to follow the Roman alphabet:

A C D E F G H I K L M N P Q R S T V W Y

Or we could follow the order of partitions, for example, the [8, 7, 2, 3] of Chapter Three:

AVLIPFWM	GSTCYNQ	DE	KRH
Nonpolar	Polar	Acidic	Basic

The point is that order is a fluid notion for components by themselves. Given 20 amino acids, there are indeed $20! \approx 2.43 \times 10^{18}$ ways to place a set of sequence rotations in lexicographic order.

In experiments, we first form the rotations for a sequence. We subsequently examine the ordering effects of *trial* lexicographic keys. These are typically formed from random permutations of AVLIPFWMGSTCYNQDEKRH. Each key so produced gives an ordered set, for example,

{G, A, D, E, P, L, W, K, H, S, Y, Q, F, I, C, R, V, M, T, N}

The set offers a lens through which to view the sequence for correlations and redundancies. Upon ordering the rotations, the third stage is to assemble a sequence using the last symbol of each rotation. This obtains the Burrows–Wheeler image that may—or may not be text-compressible. As an example, let the ordering be governed by

{A, V, L, I, P, F, W, M, G, S, T, C, Y, N, Q, D, E, K, R, H}

Then the first several rotations of lysozyme in order prove to be:

AVACAKRVVRDPQGIRAWVAWRNRCQNRDVRQYVQGCGVKVFERCELART
LKRLGMDGYRGISLANWMCLAKWESGYNTRATNYNAGDRSTDYGIFQINSRY
WCNDGKTPGAVNACHLSCSALLQDNIA**D**

AVNACHLSCSALLQDNIADAVACAKRVVRDPQGIRAWVAWRNRCQNRDVR
QYVQGCGVKVFERCELARTLKRLGMDGYRGISLANWMCLAKWESGYNTRATN
YNAGDRSTDYGIFQINSRYWCNDGKTP**G**

ALLQDNIADAVACAKRVVRDPQGIRAWVAWRNRCQNRDVRQYVQGCGVKVF
ERCELARTLKRLGMDGYRGISLANWMCLAKWESGYNTRATNYNAG
DRSTDYGIFQINSRYWCNDGKTPGAVNACHLSC**S**

AWVAWRNRCQNRDVRQYVQGCGVKVFERCELARTLKRLGMDGYRGISLANWM
CLAKWESGYNTRATNYNAGDRSTDYGIFQINSRYWCNDGKTPGAVNA
CHLSCSALLQDNIADAVACAKRVVRDPQGI**R**

AWRNRCQNRDVRQYVQGCGVKVFERCELARTLKRLGMDGYRGISLANWMCL
AKWESGYNTRATNYNAGDRSTDYGIFQINSRYWCNDGKTPGAVNA
CHLSCSALLQDNIADAVACAKRVVRDPQGIRAW**V**

.
.
.

Note how the rotations follow the order prescribed by the key: AV..., AL..., AW.... The terminal symbols have been noted in bold. These are pieced together to render the final image:

DGSRVNRVNLILCLWARKAYGDVSCEARHLTN**GG**QGTDIVANYKAWGPCYRQ
LQSDADCIERLNRKASNAMGSW**RR**AQRDN**GG**YVDAIYTCRQFPVRCLA**RR**MN
TQGCWFVAGALIT**KK**YDANESWVNV

The end product is a sea change from the original KVFER... GCGV. Think of it as an encrypted version of lysozyme! Not to worry. The *true* message KVFER...GCGV can always be recovered by back-transforming.

What to make of image DGSRVN...WVNV? It is that in this particular experiment, our efforts to find correlations and trim redundancies proved to no avail. The runs are the five dimers (di-peptides) noted by the bold underlines; this compares with two in the original (also noted by underlines). But the runs do nothing for compression: **GG** can be rewritten as **G**$_2$ requiring two symbols either way. The meagerness of runs is due to deficiencies of {A, V, L, ..., K, R, H} as a lens. Further experiments are necessary using other lexicographic keys.

These experiments have been carried out and the results show the sequences for lysozyme and other globular proteins to be *slightly* compressible. The possible ordering keys (20!) cannot be examined in any reasonable time. However, we get a sense of the compressibility through experiments with ~10^5 keys. Some resulting transforms (encryptions!) for lysozyme are as follows. The compressible parts are marked in bold underline; the lexicographic keys appear as ordered sets in bold.

CLRLINVNSDGLRVGVL**AAA**KSRNYVYICTDQRACRELINCLRFPVWCFDTAQR
TNMRGNGDQGRNQ**GGGG**WIVPDASDRYQLQCAWSRRGMACSETHLRAAW
GAYKRDVTIAWDVVNSYNE**KKK**NYAAC

{A, K, T, N, S, Q, E, P, D, Y, I, M, F, G, C, L, V, R, W, H}

KVDYITSAKNENVWCDVRGKYW**AAA**LVA**GGGG**NQDTIVWGFPVRLCNE
CRILCDQYRLPDSQ**AAA**NYKLDGCLNVRRSNVILGQDRNGNKRSATLHRECSA
ARSGAMWRGRRNMATQRQDIAYVTYCFWC

{R, H, V, K, I, P, F, M, Q, S, G, W, A, Y, T, L, C, D, N, E}

ASKRNYVYIDCTARQWGWASRRGMARLVNLCINSGDRVLRLCEINCFLRVPQ
NG<u>GGG</u>VALAQATNMRRGGNDQRGWCFLQPRYQDASDCIVDTSCEHLTR<u>A</u>
<u>AA</u>WYGKRDVCNYK<u>AAA</u>WNETIDVVNSYKK

{T, N, M, C, A, S, Q, I, K, D, Y, E, G, F, P, L, V, H, W, R}

The above sequences can be rewritten using $N = 126$ symbols, down from 130. The compressed versions would do nothing for storage purposes, however. We would have to include the lexicographic keys—20-symbol ordered sets—to find our way back to KVFER...GCGV.

The lesson is this. Protein sequences present base information by their N-values. The values are not hard and fast as evidenced by the slight variations across species. Compression studies echo this by showing values to be *nearly* hard and fast. The most pronounced effects are distortions of a sequence. Name a property such as the repeat dimers in lysozyme. The property is significantly skewed in the Burrows–Wheeler image. This indicates sequences to be thermodynamically closer to solids than to liquids: work to compress them leads primarily to distortions and defects. More to the point, a sequence manifests a stringent code economy not apparent during casual inspection. The structure of the protein is simultaneously rich and lean in information.

B   NATURAL SETS AND N-DISTRIBUTIONS

We are back to the question: what determines N for a protein? The immediate biochemical needs form only part of the answer. Recall that the design of one sequence involves others. Ribonuclease A and ribonuclease inhibitor were cited in Chapter One along with trypsin and its inhibitor. It seems impossible to comprehend the structure and mechanisms of one member of an enzyme-inhibitor pair independently of the other.

But there is much more going on. Inhibitors do not disrupt the workings of the organism as a robust system. They have affinity for their "assigned" catalysts and leave others alone in biochemical operations, Brownian motion, and so forth. This implies that protein structures at all levels somehow take into account *all* the chemistry of an organism. What determines N for lysozyme? The best, if fuzzy, answer would be factors that concern far-from-equilibrium systems as a whole. Then to make headway with N-information, we need to look not just at individual proteins but rather the distributions manifest in natural sets. This hearkens back to Chapter Three where we had to look beyond archetypes in isolation. N-properties received first attention by their abnp-weights and dispersion in sets. Here we explore properties further via N-distributions.

Databases provide thousands of sets of proteins a few clicks away at the NCBI and Swiss-Prot websites [10,11]. The N-information follows from counting the units in each sequence. The proteomes for viruses pose a few to several hundred N-values while bacteria sets offer several thousand. Higher organisms (e.g., mammals) feature tens of thousands of sequences and corresponding N. Unsurprisingly, the elements of N-sets are highly dispersed. Our experiments should commence by examining and comparing sets and looking for threads.

Let's start with small sets to make two points. We will then try to anticipate threads and patterns that apply to large sets. The genome of the human immunodeficiency virus (HIV-1) encodes ten proteins [12]. Each has a much researched structure and function: antisense protein (Asp), virion infectivity factor (Vif). And so forth. The corresponding N-set is presented as follows:

$${189, 1435, 206, 96, 82, 856, 192, 500, 116, 86}$$

This is base information. What does it teach?

Our first point is that N-sets, as with folded protein structures, are more easily appreciated in visual terms. Figure One presents a distribution: the number of encoded sequences on the vertical

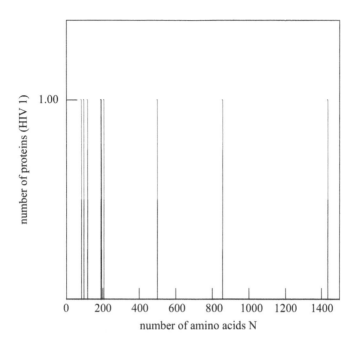

number of proteins (HIV 1)

1.00

0   200   400   600   800   1000   1200   1400

number of amino acids N

FIGURE ONE   The N-distribution for proteins encoded by the HIV-1 virus.

axis versus N on the horizontal. For the HIV set, each N proves unique. In turn, the distribution appears as ten vertical needles, each of height 1.

Observations about the set then come easily, for example,

There are more needles (set elements) at $N < 250$ compared with $N > 250$.
The needles are further apart at the higher N.
There is a bias toward N-even by way of eight of the ten elements.

Little more can be said, however, because with so few elements, there are scant patterns on which to latch. More importantly, viruses require more than one set of proteins given their dependence on host organisms. Thus the Figure One distribution offers only fragments of what has to be a long and complicated story. While it seems counterintuitive, we have to look to large sets to obtain lessons. The simplicity presented by small genomes and protein sets is deceptive. This is our second point about sequence and base N.

While exploring sets for N-information, the biochemical details of proteins are irrelevant. A set reads tediously as {164, 325, 120, ..., 220, 451}, typically with thousands of elements. Can we anticipate what a distribution plot might look like? Several possibilities should come to mind.

In one scenario, we reason there to be some minimum and maximum N for an organism. A molecule too small, say, an octa-peptide, is unable to fold and function like an enzyme. A molecule too large, say, $N = 100,000$, is too unwieldy and material-expensive to make. Besides, making large-N proteins generates more entropy than small-N, and an organism strives for states of minimum generation. One can expect assorted N between minimum and maximum, yet there seems no reason (for the moment) to favor one value over another. If these ideas are translated visually, we obtain Figure Two. The plot presents a window bounded by N_{min} and N_{max} and manifests N in-between with roughly equal weight. The plot is suggestive of a fine-tooth comb.

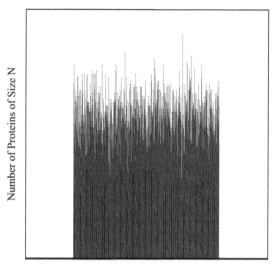

FIGURE TWO   One scenario for an *N*-distribution featuring a window bounded by minimum and maximum values. Disparate values manifest inside the window with roughly equal weight.

A second scenario builds on the first. We imagine there to be a window defined by N_{min} and N_{max} However, the organism exercises material, energy, and entropy economy by favoring low-*N* molecules. These properties are reflected in Figure Three. If the organism can succeed using small proteins, why should it expend for large ones?

The scenario can be taken further. Figure Four shows a window plus a more pronounced material, energy, and entropy economy.

As a fourth scenario, we picture the window differently. We imagine there to be some *average N* governed by myriad factors—organism type, environment, solubility, and more. We anticipate

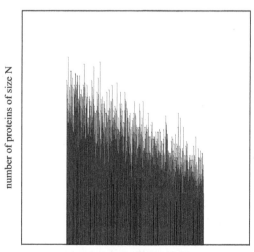

FIGURE THREE   A second plausible *N*-distribution. The window appears as in Figure Two. The distribution is biased, however, toward the lighter-weight proteins.

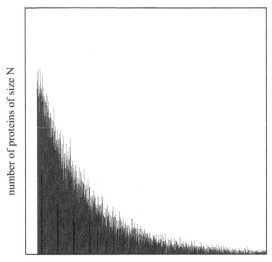

FIGURE FOUR   A third plausible *N*-distribution showing a window and pronounced material, energy, and entropy economy.

multiple factors to determine *N* for a task: catalysis, transport, signaling, etc. Combinations of factors predicate a normal distribution about the average as illustrated in Figure Five.

But then maybe an organism's proteins are needed in sizes small, medium, and large. A particular size is dictated by the biochemical function, for example, catalytic versus signaling versus connective tissue. Then the *N*-distributions might appear as in Figure Six.

It is easy to think of more scenarios. It is not so easy to anticipate if different scenarios apply to different organisms. Perhaps bacteria subscribe to Scenario Two while mammals embrace Five. What to do?

Here is where experiments shine light. We simply count all the *N*-values of a proteome and plot their distribution: the number of times a given *N* is observed. There is no end to experiments of

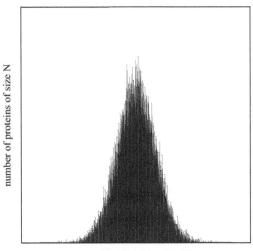

FIGURE FIVE   A fourth plausible *N*-distribution that follows normal-law statistics.

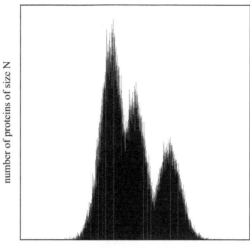

number of amino acids N

FIGURE SIX A fifth plausible N-distribution showing different-size groups dependent on the biochemical functions.

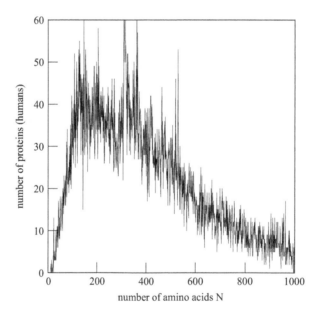

number of amino acids N

FIGURE SEVEN N-distribution observed for proteins encoded by the human genome.

this sort, thus we illustrate the results for two systems that are emblematic. The N-distribution for human-encoded proteins appears in Figure Seven while that for a strain of Yersinia Pestis bacteria is illustrated in Figure Eight.

The lesson is immediate and striking: radically different systems—in this case, one highly toxic to the other—manifest highly similar N-distributions! To be sure, the number of elements (set cardinality) is system-specific, hence the differences in distribution heights. But the plots for these sets of proteins and others show skewed, single-mode distributions peaked at nearly the same place with nearly the same width. None of our scenarios hit the nail on the head!

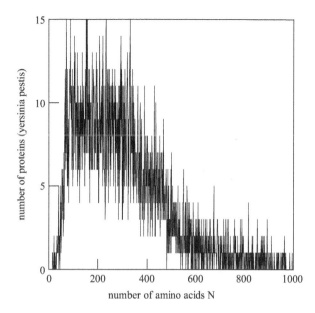

FIGURE EIGHT *N*-distribution observed for proteins encoded by Yersinia Pestis bacteria.

But partial credit is deserved on four accounts, and this is part of the lesson. First, the scenarios anticipated windows of viable *N* and this is the case for natural sets. Second, natural distributions are (largely) single-mode as imagined in four of the five scenarios. Third, natural sets are tilted toward lower *N*, not the other way. Thus, material, energy, and entropy economy is practiced by organisms. Lastly, there are *N*-clusters in several places as posited in Scenario Five. In short, while our imagination fell short, our rationales were not totally off the mark.

Most important is the correspondence: one manner of *N*-distribution applies to radically different systems. The distribution for human-encoded proteins expresses a somewhat greater width and longer tail compared with bacteria-encoded, and the apex of each distribution occurs at slightly different *N*. However, the principal features are the shapes-in-common and near-alignment. This is not a quirk of humans and bacteria, for the same shapes and alignments are observed when comparing the proteomes of other organisms, low and high.

The *N*-correspondences are surprising perhaps, but only if we look at systems myopically: humans look and act differently from bacteria, right? Yet data such as in Figures Seven and Eight are not at all startling when we consider the depth of shared properties: genetic codes, biochemical functions (e.g., metabolic and transcription), and more. From a thermodynamic perspective, we further note the role of mixing: systems subject to mixing evolve similar statistical distributions over time, transcending disparities of size, structure, and chemistry. The mixing of protein sources occurs multiple ways. Organisms are host to biomes, for example, and share evolutionary branches. This does not force the convergence of distributions. Rather it means that one system offers significant intensive-state information about another over the long term. This point is firmed via partitions and probability structures. If we sort and integrate the fractional populations in Figures Seven and Eight in mutually exclusive sets of uniform width ΔN, we obtain Figure Nine. Pick a protein at random encoded by the human genome. Figure Nine communicates that the probability of the sequence length N_r falling somewhere in the window $N \leq N_r \leq N + \Delta N$ is highly correlated with, and indeed almost matched, by a bacterial proteome. In effect, the probability structures of protein lengths are closely aligned. The exceptions to this statement lie at the extremes of low probability—the tail regions of Figures Seven and Eight. The significance of this property cannot be overemphasized. The genomes that underpin protein systems are markedly different size-wise, viz.

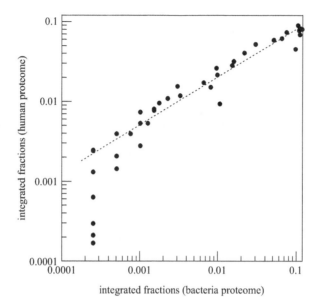

integrated fractions (bacteria proteome)

FIGURE NINE   Integrated fractional populations from Figures Seven and Eight. The fractions are assigned to mutually exclusive sets of width $\Delta N = 50$. The correspondence maintains at other choices of ΔN. The dotted line marks the correspondence region.

That their N-structures present highly mutual information reflects common biochemistry, mixing, and gradient control over millennia. Recall that mixing and gradient control figured prominently in Chapter Three concerning base structures and protein synthesis. The manifestations are encountered again in the most rudimentary of base properties, the protein sizes apart from composition.

Is there a bias toward N-even? This question was raised at the beginning of the chapter. It can be addressed by computing *N modulo 2* for each integer of a set, for example,

$$130 \bmod 2 = 0 \quad\quad 265 \bmod 2 = 1 \quad\quad 418 \bmod 2 = 0$$

We subsequently figure the percentage of zeroes versus ones. What we find is that biases are evident in some viral proteomes. With organisms, however, biases appear only slightly, for example,

	Percent even N	Percent odd-N
Humans	49.8	50.2
Yersinia bacteria	50.7	49.3

The even–odd character of N does not appear as a significant selection trait.

More importantly, we recall what initiated this chapter: lysozyme demonstrates $N = 130$; why not 129 or 131? The answer is that we cannot pin down the reasons for any particular sequence. However, we learn from proteomes that archetypal molecules have plenty of company: $N = 130$ lands near the distribution peak for organism-encoded sets. We further learn that $N > 1000$ is not favored—evolution confers comparatively few molecules this size and above.

We can take our observations further with probability and information. The N-distributions have distinctive shapes that can be approximated using real, continuous-variable functions. We illustrate two ways of thinking about them. Let us refer to a distribution as $f(N)$, the number of proteins of unit number N in a set.

Let us imagine having to design a protein from scratch for an organism. We do not want to rock the boat, much less sink the ship. Rather we want the protein to contribute a beneficial function, say, as a potent antibody to some invasive virus. How should we start? Well, we could take the easy road by tweaking a sequence already in place, say, by substituting a few valines for alanines. But that is neither fun nor in the spirit of the challenge.

Hence we ponder the beginnings of a few rough drafts:

MGVARHLW

KMGCYHLR

MTYWHIGG

There are eight units in each and we have a long way to go. We then contemplate what unit to add in position nine on the right $\leftrightarrow$ C-terminal end. Should it be F, L, or W? G perhaps? Only then do we realize the depths of the challenge, for we are way over our head. With each unit added, there is some enhancement or re-direction of the biochemical function. But adding a unit introduces new mechanisms for incompatibility. Consider the first of the drafts:

MGVARHLW**X**

The chemistry expressed by the units in place can be altered by attaching, say, aspartic acid (D) at site **X**. The catch is that D could be a deal-breaker now or somewhere down the line. It could compromise the folding, functions, and cross-link configurations. Worse, D at site **X** could introduce chemistry harmful to the organism. Adding a unit furthers the design, but increases the risk of side effects!

These ideas can be phrased in probability terms. Let the functional capacity C of a protein *we* design scale linearly with N: the larger the N, the more operations that the protein can handle. But let there be a small probability p that, with each unit added, the risk for conflicts with the organism chemistry increases. Then $1 - p$ is the probability that a unit furthers the design *and* maintains the compatibility. The probability should be close to 1, given extensive deliberations on our part. But it falls short due to our imperfect grasp, not just of the protein chemistry, but also the workings of the organism.

The drafts all feature $N = 8$. C for each draft then looks like:

$$C \propto N = aN = a \cdot 8$$

The coefficient a is specific to the protein under design and the environment in which it is made. The polypeptide is of benefit so long as the compatibility is not compromised. The probability of meeting these conditions with $N = 8$ is as follows:

$$(1-p) \cdot (1-p) \cdot (1-p) \cdot (1-p) \cdot (1-p) \cdot (1-p) \cdot (1-p) \cdot (1-p)$$

$$= (1-p)^8$$

The effective or practical capacity C' is then a product of terms:

$$C' = a8 \times (1-p)^8$$

In the general case, we have

$$C' = aN \times (1 - p)^N$$

It should be clear what is going on. In designing a protein from scratch, we enhance the biochemical capabilities by increasing N. But we boost the chances for failure: C' falls via the factor $(1 - p)^N$.

A sidebar prepares for the landing. Consider the Taylor expansion for e^x:

$$e^x = 1 + \frac{x}{1!} + \frac{x^2}{2!} + \frac{x^3}{3!} + \frac{x^4}{4!} + \ldots.$$

If x is positive and close to zero, only the first two terms are significant. Then

$$e^x \approx 1 + x, \quad 0 < x \ll 1$$

If we insert the approximation into the probability function, for $0 < p \ll 1$, we have:

$$C' = aN \times (1 - p)^N \approx aN \times (e^{-p})^N = aN \cdot e^{-Np}$$

Therein lies the lesson. It is *not* how to design proteins from scratch—we discuss this topic in Chapter Eight. Rather it points to the mountains that evolution—or present day peptide chemists—must scale to do so. Given a distribution of mountains, $f(N)$ for an organism set should track with C':

$$f(N) \approx A \cdot N e^{-pN}$$

The function $f(N)$ features prominently in the family of gamma distributions [13]. Connecting it to proteomes obtains by determining A and p from the distribution maxima. At maximum height, the first derivative of $f(N)$ is zero:

$$\left[\frac{df(N)}{dN} \right]_{N = N_{max}} = 0 = A \times e^{-Np} - A \cdot N \cdot p \times e^{-Np}$$

After cancellations and solving for p, we arrive at:

$$p = \frac{1}{N_{max}}$$

We obtain A by looking again at the maximum:

$$f(N_{max}) \approx A \cdot N_{max} \times e^{-N_{max}p} = A \cdot N_{max} \cdot e^{-1}$$

A is calculated from rearranging the above and multiplying through by e^1:

$$A = \frac{f(N_{max}) \cdot e^1}{N_{max}}$$

The distributions for human and bacteria proteins were showcase examples. By re-visiting Figures Seven and Eight, we estimate:

$$p_{\text{human}} = \frac{1}{N_{\text{max}}} \approx \frac{1}{220} \qquad\qquad p_{\text{bacteria}} = \frac{1}{N_{\text{max}}} \approx \frac{1}{190}$$

$$A_{\text{human}} = \frac{1}{N_{\text{max}}} \cdot f_{\text{max}} \cdot e^1 \approx \frac{1}{220} \cdot 40 \cdot 2.71 \qquad A_{\text{bacteria}} = \frac{1}{N_{\text{max}}} \cdot f_{\text{max}} \cdot e^1 \approx \frac{1}{190} \cdot 10 \cdot 2.71$$

Figure Ten then compares the probability function for human-encoded proteins. The reader should construct the analogous plot for bacteria proteins using the data of Figure Eight. Suffice to say that the gamma distribution captures the salient features of $f(N)$ in these and multiple other cases. It cannot do justice to the N-clusters in places, but the foundation is simple and furthers our thinking about sequences and base N. Most importantly, the probability model emphasizes that a protein is *not* of size N merely because, well, *that* is what works for folding and functions—end of story. Rather it reflects that N depends on properties concerning the organism as a whole, somehow balancing biochemical needs and mitigating conflicts. Note also that the model offers a basis for the distribution asymmetry. The favored lengths are in the low N-range: $50 < N < 700$, not the high-N sectors. This means that if we really were to design a protein from scratch, we should not aim for, say, $N = 5,000$ and bet on success. Molecules with N on the high side are encoded by systems but are sparse due to low solubility for openers, plus energy, entropy and material considerations. For example, the following describes a base structure with $N = 1487$:

$$A_{136}V_{33}L_{59}I_{35}P_{257}F_{24}W_7M_{20}G_{404}S_{56}T_{44}C_{19}Y_{11}N_{33}Q_{61}D_{67}E_{74}K_{65}R_{74}H_8$$

But the information applies to a collagen having no particular chemical functions. Broadly speaking, the chemically sophisticated proteins express N in the "Goldilocks" zone of distributions $f(N)$.

In the preceding chapter, probability and information gave us more than one way to view composition properties. The same is true for base structure N. Our first view contemplated proteins designed from

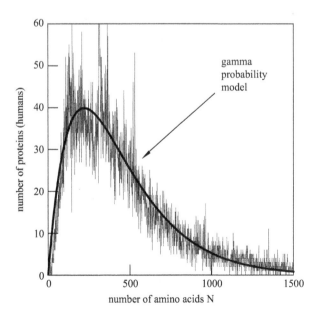

FIGURE TEN N-Distribution of proteins encoded by humans and that prescribed by the gamma probability function (solid curve).

scratch. A second approach imagines fashioning proteins from ones already found in nature. No elaborate engineering is called for. We rely on evolution and transcription chemistry for the heavy lifting.

We consider taking a large protein α which hosts, say, a few thousand units. We then imagine removing a small piece by cleaving at the site indicated by the arrow:

This yields a much smaller protein, which can be tested for folding, biochemistry, and possible benefits. This is not the only candidate, for we can cleave again to obtain a second molecule for testing, viz.

Note that we do not have to cleave near the C-terminal side or even be choosy about where to cleave. The object is to obtain lots of candidates that *might* carry useful functions. This is not unlike the search for pharmaceutical "hits" using combinatorial and other high throughput syntheses. Ultimately, many molecules are surveyed for biological activity and therapeutic benefit. In our experiment, we cannot cleave indefinitely, of course. Sooner or later, we arrive at a terminal protein (or peptide) having an extensive family tree. But why stop there? We can take other starters β, γ, δ, etc., and subject them to sequences of cuts. Or we can take another α and apply a different sequence of cuts.

Our search for candidates connects with probability. Let N_0 represent the number of units (length) in a starter protein; the subscript indicates the number of cuts having been applied. Following a cut, we obtain a second protein with $N_1 < N_0$. N for the candidate is then some fraction λ_0 of the parent. A second cut leads to $N_2 < N_1$. The candidate so obtained expresses N which is a fraction λ_1 of its parent. We keep track of things by writing:

N_0 = length of starter protein
N_1 = length after a single cut
$N_0 - N_1$ = length of first fragment = $\lambda_0 N_0$
N_2 = length after two cuts
$N_1 - N_2$ = length of second fragment = $\lambda_1 N_1$

In the general case, we have:

N_i = length after i-number of cuts
$N_{i-1} - N_i$ = length $i\underline{th}$ fragment = $\lambda_{i-1} N_{i-1}$

Then we note the fractions $\lambda_0, \lambda_1, \lambda_2, \ldots$ to present a distribution:

$$\frac{N_0 - N_1}{N_0} = \frac{\Delta N_{0,1}}{N_0} = \lambda_0$$

$$\frac{N_1 - N_2}{N_1} = \frac{\Delta N_{1,2}}{N_1} = \lambda_1$$

$$\vdots$$

$$\frac{N_{i-1} - N_i}{N_{i-1}} = \frac{\Delta N_{i-1,i}}{N_{i-1}} = \lambda_{i-1}$$

If we were to sum the fractions, the operation would look like

$$\sum_{i=1} \frac{N_{i-1} - N_i}{N_{i-1}} = \sum_{i=1} \frac{\Delta N_{i-1,i}}{N_{i-1}} = \lambda_0 + \lambda_1 + \lambda_2 + \dots$$

There are many λ_i terms given all the starter proteins, cutting scenarios, and candidates. In turn, the sums of fractions should follow normal-law probability. But now observe that if we were to approximate the summation via an integral, things would look like:

$$\sum_{i=1} \frac{N_{i-1} - N_i}{N_{i-1}} \approx \sum_i \frac{\Delta N_i}{N_i} \approx \int \frac{dN'}{N'} \approx \ln(N)$$

The fractions should be normal-law distributed—we have no reason to expect otherwise. But this implies that the *logarithms* of the protein sizes N should be distributed normally! In other words, the N-distributions for proteins obtained by our second design method should be log-normal.

We revisit the N-distributions for human-encoded proteins. Only this time we use a logarithm scale for the horizontal axis. Note the distinctive bell-shape of the distribution as illustrated in Figure Eleven. The plot for a Gaussian probability model has been included that matches the raw data quite well. The match is just as good for the bacteria N-distributions viewed in logarithm terms. The alignment is as striking as observed for the gamma probability model.

As with our first probability model, we do not learn how to design proteins, much less group compatible ones in sets. But we gain perspective of what evolution is doing and has more yet to do. What determines N for a protein? Our second perspective offers that N is governed by quasi-random factors in *stepwise* processes. It is *not* a sum of multiple factors which determines the information, but rather a product. If sums were the case, then the N-distributions we find in nature would be normal as considered in Scenario Four (cf. Figure Five).

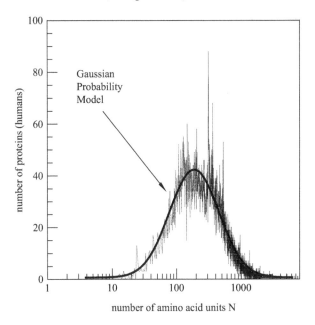

FIGURE ELEVEN *N*-Distributions of proteins encoded by humans plus normal probability plot. The horizontal axis is presented as a logarithmic scale.

C SEQUENCE *N* AND INFORMATION ASYMMETRY

Probability and information encourage (at least!) two ways of thinking about sequence and base *N*. Both spring from the correspondences of *N*-sets for different sources of proteins and make use of real-, continuous-variable models. Data wise, the gamma and log-normal functions seem equally adept at approximating our observations. They have company, too. Other mathematical approaches detailed in the literature apply as well to the *N*-distributions of proteomes [14–16].

The gamma and log-normal functions are a dead heat providing *N*-distribution fits. However, they carry subtly different takes on information properties. The gamma function attached to proteins imagined from the ground up. The suitability of a sequence for folding and biochemical functions was dependent on information in *both* the design and organism at every step. Schematically the scenario can be pictured as:

The system ↔ organism restricts the sequences allowed to proceed from the left to right side of the arrow. Where a design proves beneficial, or at least compatible, it is incorporated by the system to contribute new functions. In turn, the *new* system subsequently impacts designs down the road. Critically, information is conserved throughout the process. It is possible for an observer to work backward from the product state (right side of arrow) to reconstruct the initial state (left side). There is no asymmetry of information going from the design to incorporation stage.

The log-normal distribution aligns with a different story. Here a starter protein is sliced and diced to render fragments. These are candidates to be combined with the system where suitable. Information is *not* conserved, even if all the candidates pose compatible folds and biochemical operations. Schematically, we can think of the process as:

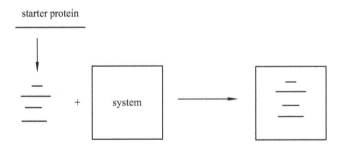

It is *not* possible for an observer to work backward from the product state to reconstruct the initial. Each fragment ↔ candidate constitutes a bona fide sequence having N- and C-terminal groups and sites in-between, for example,

GPLGSHMSTESND
GVSETLLAWRHIDFWTSE
HNPDLNATLSDPCTQNDITHAEEDLEV
SFPNPVKAS
FKIHDGQEDLESMTGTS

But information in the fragments does not include their places in the starter! The transfer of information is *asymmetric* going from an initial to final state.

Both probability functions of Section B model the N-distributions of natural systems. However, only the log-normal function captures at least some of the flavor of asymmetric information transfer. Things work as follows.

Probability functions hold information that includes moments at all orders [17]. If $f(x)$ is a normalized probability density, the nth-order moment arrives via the integral:

$$< x^n > = \int_{-\infty}^{\infty} x^n \cdot f(x)\, dx$$

But this is not the only route, for the moments can obtain from computing a *single integral* followed by nth-order derivatives. We refer to the moment generating function $m(t)$, viz.

$$m(t) = \int_{-\infty}^{\infty} e^{tx} \cdot f(x)\, dx$$

The alternative function $m(t)$ is the Laplace transform of $f(x)$ and is a wellspring for the moments, viz. [18]:

$$\left\langle x^n \right\rangle = \left[\frac{d^n m(t)}{dt^n} \right]_{t=0}$$

We then visit the family of gamma distributions in normalized terms:

$$f(x) = \frac{1}{\Gamma(\alpha+1)\beta^{\alpha+1}} \cdot x^{\alpha} \cdot e^{-x/\beta}$$

In Section B, the designed-from-scratch proteins made appeal to family members with $\alpha = 1$. The denominator term in the above then refers to the integral:

$$\Gamma(\alpha+1=1+1=n=2) = \int_{0}^{\infty} e^{-x} x^{n-1}\, dx$$

For $n = 2$, we have:

$$\Gamma(2) = \int_{0}^{\infty} e^{-x} x^{1}\, dx$$

Integrating by parts shows $\Gamma(2) = 1$. The normalized probability function becomes:

$$f(x) = \frac{1}{\beta^2} \cdot x \cdot e^{-x/\beta}$$

Then the moment generating function becomes:

$$m(t) = \frac{1}{\beta^2} \int_{0}^{\infty} e^{tx} x \cdot e^{-x/\beta}\, dx = \frac{1}{\beta^2} \int_{0}^{\infty} x \cdot e^{x\left(t-\frac{1}{\beta}\right)}\, dx$$

The above is also amenable to integration by parts whereby we obtain:

$$m(t) = \frac{1}{\left(1 - \beta t\right)^2}$$

The moments follow straightaway via derivatives, for example,

$$\left\langle x^1 \right\rangle = \left[\frac{dm(t)}{dt} \right]_{t=0} = \left[\frac{dm\left(1 - \beta t\right)^{-2}}{dt} \right]_{t=0}$$

$$= \left[\frac{\beta}{(1 - \beta t)^3} \right]_{t=0} = \beta$$

The property to note is that for the gamma distribution, information shuttles equivalently in two directions. The distribution holds information for the moments *and* the moment generator $m(t)$. In turn, the *existence* of $m(t)$ means that the moments at *all* orders, viz.

$$\left\langle x^1 \right\rangle, \left\langle x^2 \right\rangle, \left\langle x^3 \right\rangle, \ldots$$

constitute information equivalent to the gamma distribution. In effect, the gamma function underpins information symmetrically regarding the moments—and vice-versa.

The log-normal function presents a curious case of asymmetry. For illustrative purposes, let random variable x be normally distributed with mean zero and variance one. The log-normal distribution applies to variable $y = e^x$ or $\ln(y) = x$. All the moments spring from integrals in the usual way:

$$\left\langle y^n \right\rangle = \left\langle e^{nx} \right\rangle = \frac{1}{\sqrt{2\pi}} \int_{-\infty}^{+\infty} e^{nx} e^{-x^2/2} \, dx$$

Rearranging the integrand leads to:

$$\left\langle e^{nx} \right\rangle = \frac{1}{\sqrt{2\pi}} \int_{-\infty}^{+\infty} \exp\left[nx - \frac{x^2}{2} \right] dx$$

$$= \frac{1}{\sqrt{2\pi}} \int_{-\infty}^{+\infty} \exp\left[\frac{-x^2 + 2nx}{2} \right] dx$$

$$= \frac{1}{\sqrt{2\pi}} \int_{-\infty}^{+\infty} \exp\left[\frac{-x^2 + 2nx - n^2}{2} + \frac{n^2}{2} \right] dx$$

$$= \frac{1}{\sqrt{2\pi}} \int_{-\infty}^{+\infty} \exp\left[\frac{-(x - n)^2}{2} + \frac{n^2}{2} \right] dx$$

$$= e^{n^2/2}$$

The log-normal probability function is thereby a source for moment information. But the moments do not reciprocate because the log-normal distribution lacks a finite generating function. If we try computing $m(t)$, we confront:

$$m(t) = \left\langle e^{ty} \right\rangle = \left\langle e^{te^x} \right\rangle = \frac{1}{\sqrt{2\pi}} \int\limits_{-\infty}^{+\infty} e^{te^x - x^2/2} \, dx$$

The exponent in the integrand can be expanded as a series (cf. Taylor expansion in Section B). We then have the inequality:

$$\left\langle e^{te^x} \right\rangle \geq \frac{1}{\sqrt{2\pi}} \int\limits_{0}^{+\infty} \exp\left[t \cdot \left(1 + x + \frac{x^2}{2!} + \frac{x^3}{3!} \right) - \frac{x^2}{2} \right] dx$$

The exponential tends to infinity as x goes to infinity dragging $m(t)$ with it [19]. Lacking a finite moment generating function, the moments *do not* offer information commensurate with the probability function. In effect, the log-normal function expresses information for the moment representation, albeit asymmetrically. It is not possible to work backward from moments to reconstruct the function.

We encounter issues of information symmetry in several places of protein structure and chemistry. With only modest adjustments, the design experiments of Section B can be initiated via polynucleotides DNA and RNA and taken to completion using the gamma and log-normal functions. In the gamma approach, instead of contemplating amino-acid strings as rough drafts, we ponder codon sequences that underpin the drafts, for example, AUG GGU GUU GCU AGA CAU UUA UGG → MGVARHLW. Adding a codon enhances or re-directs the biochemical functions in MGVARHLW, but doing so increases the risk of side effects in the organism. This is the same scenario as our first approach to protein design, only initiated farther upstream. In the log-normal approach, instead of fragmenting starter proteins, we could slice and dice starter DNA and RNA, the fragments of which could be transcribed to make candidate proteins.

Information asymmetry is part-and-parcel to the codons and transcription leading to proteins. The gamma and log-normal distributions approximate the N-distributions we observe in nature. However, only the log-normal retains some of the flavor of the asymmetric information transfer. This is not cause for dismissing the gamma description. Indeed the probability function emphasizes the wealth of factors that must be taken into account in protein design. But so does the log-normal!

The major points of Chapter Four are:

1. Proteins present information at the base level via their constituent numbers. As with compositions, this information is conferred by gene expression and connects with all manner of biological structure and function. All possibilities for N offer stable proteins, although some would seem more capable and robust than others due to folding, solubility, and other requirements. Proteins having N-too-small pose sparse folding and functional capabilities. N-too-large pushes against the grain of material, energy, and entropy economy.

2. Protein sequences exhibit sparse compressibility by information-conserving transforms such as Burrows–Wheeler. The transforms point to a stringent code economy; there appear few superfluous units in the representations of natural sequences.

3. As with amino-acid composition, N is case-specific for a host, biochemical function, and operating environment. Even so, the N-distributions across proteomes express striking correspondences. These reflect biochemical and genomic properties in common, and the

effects of careful gradient control. For systems to manifest otherwise would be at odds with mixing via evolution and the need to limit entropy production.

4. We are directed to probability functions by contemplating proteins designed from the ground up, or as the fragments of large ones. Both approaches offer more or less equivalent experiment-fitting models via the gamma and log-normal functions. The functions diverge, however, on matters of information symmetry having to do with moments at low and high orders.

EXERCISES

1. The first ten units of lysozyme appear as KVFERCELAR. Construct the Burrows–Wheeler transform for each using AVLIPFWMGSTCYNQDEKRH as the lexicographic key.
2. The maxima for the N-distributions for human and bacteria proteins were observed at different N. Discuss reasons for the difference.
3. A study was made of the even/odd properties several viral protein sets. The number of even-N and odd-N proteins is noted. Comment on the significance.

Protein set	Number of even	Number of odd
Simian immunodeficiency	3	5
Simian T lymphocyte	3	3
Strep Phage	38	30
SARS	9	5
Influenza A H1N1 (5b)	6	6
Salmonella Phage	30	41
Rubella	1	1
Rabies	4	1
Fowl adenovirus	20	19
Measles	2	6
Sudan Ebola	6	2
Variola	97	100
Ebola	6	3
Vesicular stomata virus	1	5
Human parvo virus	3	3
Human papilloma	2	4
Torque teno virus	2	4
Hepatitis C	0	2

4. The N-distribution for proteins encoded by honeybees is illustrated in Figure Twelve.
   a. What are suitable values of A and p from the gamma distribution? Construct a plot of the distribution using these values. How well does the plot match the data?
   b. Construct a plot of $f(N)$ versus natural log(N). Does the plot resemble a bell-shape curve?
5. The N-distribution for proteins encoded by Bos taurus (aka cows) appears as in Figure Thirteen.
   a. How different are the A, p-values for the gamma distribution from Exercise Four?
   b. Construct the log-normal representation of $f(N)$.
6. The N-distribution for the variola virus is plotted in Figure Fourteen. Compare the distribution to the ones for organisms. Discuss similarities and differences.

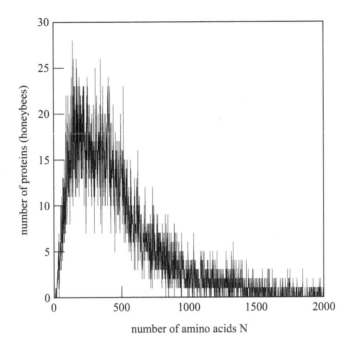

FIGURE TWELVE *N*-distribution observed for proteins encoded by Apis mellifera (aka honeybees).

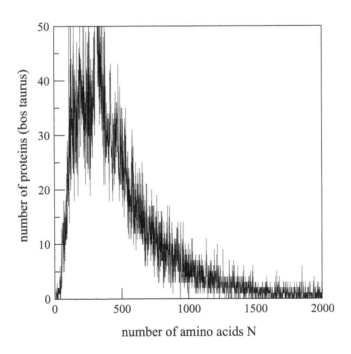

FIGURE THIRTEEN *N*-distribution observed for proteins encoded by Bos taurus (aka cows).

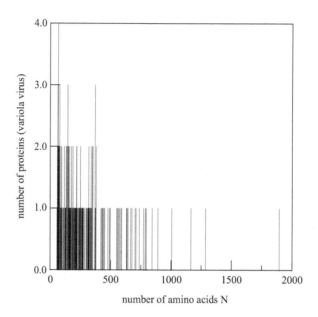

FIGURE FOURTEEN N-distribution for proteins encoded by variola virus.

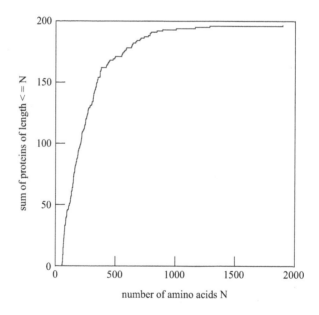

FIGURE FIFTEEN Integrated N-distribution for proteins encoded by variola virus.

7. The length distribution data were illustrated as $f(N)$ versus N. Construct sketches for the integrals of $f(N)$. Figure Fifteen illustrates the integrated distribution for variola proteins of the previous exercise. Does the plot resemble your sketches?

8. Consider a sequence investigated for its compressibility via the Burrows–Wheeler transform. Let the transform algorithm for a given lexicographic key require one second of computer time. How long does it take to examine the transforms for all possible keys?

9. Revisit the respective sequences for lysozyme and three Burrows–Wheeler transforms:

KVFERCELARTLKRLGMDGYRGISLANWMCLAKWESGYNTRATNYNAGDRS
TDYGIFQINSRYWCNDGKTPGAVNACHLSCSALLQDNIADAVACAKRVVRDPQ
GIRAWVAWRNRCQNRDVRQYVQGCGV

CLRLINVNSDGLRVGVLAAAKSRNYVYICTDQRACRELINCLRFPVWCFDTAQR
TNMRGNGDQGRNQGGGGWIVPDASDRYQLQCAWSRRGMACSETHLR
AAWGAYKRDVTIAWDVVNSYNEKKKNYAAC

KVDYITSAKNENVWCDVRGKYWAAALVAGGGGNQDTIVWGFPVRLCNE
CRILCDQYRLPDSQAAANYKLDGCLNVRRSNVILGQDRNGNKRSATLHRECSA
ARSGAMWRGRRNMATQRQDIAYVTYCFWC

ASKRNYVYIDCTARQWGWASRRGMARLVNLCINSGDRVLRLCEINCFLRVPQN
GGGGVALAQATNMRRGGNDQRGWCFLQPRYQDASDCIVDTSCE
HLTRAAAWYGKRDVCNYKAAAWNETIDVVNSYKK

Discuss enzyme characteristics that are highly distorted via the transforms.

NOTES, SOURCES, AND FURTHER READING

Size distributions have been central to molecular structure analysis for over a century. The properties of small organics are discussed at length by Morawitz and Kauffman [20,21]. Regarding even, odd characteristics, the carbon contents of fatty acids hold significant information about sources and synthetic pathways [22]. Proteins motivate the same intense interest regarding N-properties, especially in relation to biochemical functions, evolution of genomes and operating environments [23]. Probability models for N-distributions have included the gamma and log-normal distributions discussed in this chapter. Further contact has been made by taking into account the frequency of start- and stop-codons in DNA. The reader is directed to the writings of White for in-depth perspectives [23–26]. Chothia further presents length fundamentals that determine protein structure and function [27]. Shakhnovich and Gutin bring together properties of lengths, thermodynamics, and evolution [28]. Interestingly, the log-normal distribution featured prominently in a model for crushed rocks originated by Kolmogorov [29]. The model has contributed to the understanding of noise and growth phenomena in diverse systems. To the author's knowledge, Fisher was the first researcher to address quantitatively the scaling and length properties of proteins [30].

1. Muraki, M., Harata, K., Sugita, N., Sato, K. 1996. Origin of carbohydrate recognition specificity of human lysozyme revealed by affinity labeling, *Biochemistry* 35, 13562.
2. Chatani, E., Hayashi, R., Moriyama, H., Ueki, T. 2002. Conformational strictness required for maximum activity and stability of bovine pancreatic ribonuclease A as revealed by crystallographic study of three Phe120 mutants at 1.4 A resolution, *Protein Sci.* 11, 72–81.
3. Makino, M., Sugimoto, H., Sawai, H., Kawada, N., Yoshizato, K., Shiro, Y. 2006. Crystal structure of human cytoglobin at 1.68 Angstroms resolution, *Acta Crystallogr., Sect. D: Biol. Crystallogr.* 62, 671–677.
4. White, S. P., Scott, D. L., Otwinowski, Z., Gelb, M. H., Sigler, P. B. 1990. Crystal structure of cobra-venom phospholipase A2 in a complex with a transition-state analogue, *Science* 250(4987), 1560–1563.
5. Mirecka, E. A., Shaykhalishahi, H., Gauhar, A., Akgul, S., Lecher, J., Willbold, D., Stoldt, M., Hoyer, W. 2014. Sequestration of a [beta]-hairpin for control of [alpha]-synuclein aggregation, *Angew. Chem. Int. Ed. Engl.* 53(16), 4227–4230.
6. Gilbert, W. 1978. Why genes in pieces? *Nature* 271, 501. doi:10.1038/271501a0.
7. Peters, C. W. B., Kruse, U., Pollwein, R., Grzeschick, K.-H., Sippel, A. E. 1989. The human lysozyme gene, *Eur. J. Biochem.* 182, 507–516.
8. Berstel, J., Lauve, A., Reutenauer, C., Saliola, F. c2009. Part I: Christoffel words, in Combinatorics on Words: Christoffel Words and Repetitions in Words, American Mathematical Society, Providence, RI.

9. Burrows, M., Wheeler, D. J. 1994. A block-sorting lossless data compression algorithm, *Technical Report* 124, 48.
10. www.ncbi.nlm.nih.gov/
11. www.uniprot.org/
12. Voyles, B. A. 1993. *The Biology of Viruses*, Mosby-Year Book, St. Louis, MO.
13. Balakrishnan, N. 2003. *A Primer on Statistical Distributions*, Wiley, Hoboken, NJ.
14. Tiessen, A., Pérez-Rodriguez, P., Delaye-Arredondo, L. J. 2012. Mathematical modeling and comparison of protein size distribution in different plant, animal, fungal and microbial species reveals a negative correlation between protein size and protein number, thus providing insight into the evolution of proteomes, *BMC Res. Notes* 5, 85. doi:20.1186/1756-0500-5-85.
15. Brocchieri, L., Karlin, S. 2005. Protein length in eukaryotic and prokaryotic proteomes, *Nucleic Acids Res.* 33(10), 3390–3400.
16. Ramirez-Sanchez, O., Pérez-Rodriguez, P., Delaye, L., Tiessen, A. 2016. Plant proteins are smaller because they are encoded by fewer exons than animal proteins, *Genomics Proteomics Bioinformatics* 14(6), 357–370.
17. Dudley, R. M. 1989. *Real Analysis and Probability*, Wadsworth & Brooks/Cole, Pacific Grove, CA.
18. Itô, K. 1984. *Introduction to Probability Theory*, Cambridge University Press, New York.
19. Romano, J. P., Siegel, A. F. 1986. *Counterexamples in Probability and Statistics*, Wadsworth & Brooks/Cole, Monterey, CA.
20. Kauffman, S. A. 1993. *The Origins of Order: Self-Organization and Selection in Evolution*, Oxford University Press, New York.
21. Morowitz, H. J. 1979. *Energy Flow in Biology: Biological Organization as a Problem in Thermal Physics*, Ox Bow Press, Woodbridge, CT.
22. White, A., Handler, P., Smith, E. L. 1972. *Principles of Biochemistry*, chap. 5, McGraw-Hill, New York.
23. White, S. H. 1992. The amino acid preferences of small proteins: Implications for protein stability and evolution, *J. Mol. Biol.* 227, 991–995.
24. White, S. H., Jacobs, R. E. 1993. The evolution of proteins from random amino acid sequences I. Evidence from the lengthwise distribution of amino acids in modern protein sequences, *J. Mol. Evol.* 36, 79–95.
25. White, S. H. 1994. The evolution of proteins from random amino acid sequences II. Evidence from the statistical distributions of the lengths of modern protein sequences, *J. Mol. Evol.* 38(4), 383–394.
26. White, S. H. 1994. Global statistics of protein sequences: Implications for the origin, evolution, and prediction of structure, *Annu. Rev. Biophys. Biomolec. Struct.* 23, 407–439.
27. Chothia, C. 1984. Principles that determine the structure of proteins, *Ann. Rev. Biochem.* 53, 537–572.
28. Shakhnovich, E. I., Gutin, A. M. 1990. Implications of thermodynamics of protein folding for evolution of primary sequences, *Nature* 346, 773–775.
29. Shlesinger, M. F., West, B. J. 1988. 1/f versus $1/f^\alpha$ Noise, in *Random Fluctuations and Pattern Growth. Experiments and Models*, H. E. Stanley, N. Ostrowsky, eds., Kluwer Academic Publishers, London.
30. Fisher, H. F. 1964. A limiting law relating the size and shape of protein molecules to their composition, *PNAS USA* 51, 1285–1291.

Five   Base Structures and Proteomic Sets

Proteins express information not just individually, but also collectively in sets. Accordingly, set principles and applications have entered several discussions up to now. The present chapter delves more deeply by way of proteomes: complete sets of sequences encoded by the genomes of organisms and viruses. The examples add to the probability structures and correspondences illustrated in preceding chapters. Probability and information further draw out protein set topologies, both external and internal. These are abstractions which bring to light shape and growth correspondences of sets.

A   SEQUENCES, BASE STRUCTURES, AND SETS

We have examined the sequences and bases of archetypal proteins, focusing on the compositions and unit numbers N. Yet we never traveled far before set properties rose to the surface. It was too restrictive to view amino acids solely by their identity A, V, L, …, K, R, H. We had to consider families and the information leveraged by partitions and respellings. Likewise, archetypes like lysozyme could not be analyzed as individuals in a vacuum; we had to include families of the archetypes to appreciate variations. We were further called to inspect sequences having N in common, but different biochemical functions. Like all natural products, proteins cannot be viewed independently of one another or their sources and environments. Moreover, the notions of probability and information go hand in hand with set functions. This has been a critical theme in places.

The first lessons from sets concerned partitions and selection weights: multiple proteins, especially globular ones, express acidic (a), basic (b), nonpolar (n), and polar (p) components in proportions that match (more or less) those of the 20 amino acids: ~10%, ~15%, ~40, and ~35%. The proportions are not confined to individual sequences but rather extend to the sets conferred by evolution (cf. Figure Seven of Chapter Three). These are the proteomes which themselves express information at the base level. Just as for individual sequences, such information is in place prior to gene transcription and protein synthesis. It is the foundation for proteins and is maintained by an organism throughout its life. Proteomes express base information which is transferred through generations. What can we learn from it?

To begin, the base for *any* set of proteins is straightforward to assemble and explore. For example, three sequences—epidymis secretory ($N=246$), 14-3-3 theta ($N=245$), and HLA Class I antigen ($N=365$)—encoded by the human genome list as follows:

MTMDKSELVQKAKLAEQAERYDDMAAAMKAVTEQGHELSNEERNLLSVAYKNV
VGARRSSWRVISSIEQKTERNEKKQQMGKEYREKIEAELQDICNDVLE
LLDKYLIPNATQPESKVFYLKMKGDYFRYLSEVASGDNKQTTVSNSQQAY
QEAFEISKKEMQPTHPIRLGLALNFSVFYYEILNSPEKACSLAKTAFDEAIAEL
DTLNEESYKDSTLIMQLLRDNLTLWTSENQGDEGDAGEGEN

MEKTELIQKAKLAEQAERYDDMATCMKAVTEQGAELSNEERNLLSVAYKN
VVGGRRSAWRVISSIEQKTDTSDKKLQLIKDYREKVESELRSICTTVLELLDKY
LIANATNPESKVFYLKMKGDYFRYLAEVACGDDRKQTIDNSQGAYQEAFDI
SKKEMQPTHPIRLGLALNFSVFYYEILNNPELACTLAKTAFDEAIA
ELDTLNEDSYKDSTLIMQLLRDNLTLWTSDSAGEECDAAEGAEN

MAVMAPRTLVLLLSGALALTQTWAGSHSMRYFFTSVSRPGRGEPRFIAVGYVD
DTQFVRFDSDAASQRMEPRAPWIEQEGPEYWDGETRKVKAHSQTHRVDLGT
LRGYYNQSEAGSHTVQRMYGCDVGSDWRFLRGYHQYAYDGKDYIALKED
LRSWTAADMAAQTTKHKWEAAHVAEQLRAYLEGTCVEWLRRYLENGKE
TLQRTDAPKTHMTHHAVSDHEATLRCWALSFYPAEITLTWQRDGEDQTQDT
ELVETRPAGDGTFQKWAAVVVPSGQEQRYTCHVQHEGLPKPLTLRWEPSS
QPTIPIVGIIAGLVLFGAVITGAVVAAVMWRRKSSDRKGGSYSQAASSDSAQGSD
VS**LTACKV**

By symbol counting and grouping, the respective bases work out to be:

$$A_{21}V_{11}L_{24}I_{10}P_5F_6W_2M_8G_{10}S_{19}T_{12}C_2Y_{11}N_{14}Q_{15}D_{14}E_{30}K_{20}R_{10}H_2$$
$$A_{24}V_{10}L_{27}I_{12}P_4F_6W_2M_6G_9S_{16}T_{16}C_5Y_{11}N_{12}Q_{10}D_{18}E_{26}K_{19}R_{11}H_1$$
$$A_{38}V_{26}L_{26}I_9P_{15}F_9W_{12}M_8G_{28}S_{26}T_{29}C_5Y_{15}N_2Q_{21}D_{21}E_{22}K_{13}R_{27}H_{13}$$

If all three bases are grouped together to make a set Π, a *collective* base obtains, namely:

$$A_{21+24+38}V_{11+10+26}L_{24+27+26}I_{10+12+9}P_{5+4+15}F_{6+6+9}W_{2+2+12}$$
$$M_{8+6+8}G_{10+9+28}S_{19+16+26}T_{12+16+29}C_{2+5+5}Y_{11+11+15}N_{14+12+2}$$
$$Q_{15+10+21}D_{14+18+21}E_{30+26+22}K_{20+19+13}R_{10+11+27}H_{2+1+13}$$

It is as if we had combined the three sequences upfront to render a single with $N = 246 + 245 + 365 = 856$, viz.

$$\textbf{MTMDKS.....MEKTEL......LTACKV}$$
$$A_{83}V_{47}L_{77}I_{31}P_{24}F_{21}W_{16}M_{22}G_{47}S_{61}T_{57}C_{12}Y_{37}N_{28}Q_{46}D_{53}E_{78}R_{48}H_{16}$$

The collective follows from counting and grouping symbols, just as we would for one sequence by itself.

Combining sequences grows sets, the most illustrious being proteomes, henceforth denoted by $\mathbf{\Pi}$ in boldface. The N-distribution for human-encoded $\mathbf{\Pi}$ was featured in Chapter Four and the abnp-fractions appeared in Figure Seven of Chapter Three. We explore new territory by joining the sequences and focusing on the total composition. After counting, and grouping symbols, we obtain:

$$A_{1,025,558} \; V_{890,696} \; L_{1,412,028}I_{608,324} \; P_{921,885} \; F_{515,649} \; W_{172,541} \; M_{339,405}G_{944,215} \; S_{1,210,578} \; T_{750,752} \; C_{338,036}Y_{372,202}$$
$$N_{565,120} \; Q_{676,096} \; D_{668,928}E_{1,011,509} \; K_{791,932} \; R_{796,528} \; H_{426,748}$$

The base for human-encoded $\mathbf{\Pi}$ is rapidly assembled by computer and most tediously otherwise [1]. Every primary structure contributes to the base irrespective of its biological function or population in the organism. The weights are instead governed by the genome whereby a large transcription product such as a collagen contributes more than a small one, for example, a ubiquitin. There are tradeoffs: a large-N sequence adds more mass to a base compared with small-N, but is encoded sparsely by the genome. This is one of our lessons from N-distributions.

Every proteome $\mathbf{\Pi}$ presents a base so conferred by evolution. The N-distributions for Yersinia Pestis bacteria were included in the previous chapter. The base for one strain assembles as follows:

$$A_{118,\,976} \; V_{93,\,335} \; L_{140,\,569} \; I_{78,963}P_{65,\,761} \; F_{48,276} \; W_{18,668} \; M_{33,574}G_{95,508} \; S_{86,889} \; T_{68,243} \; C_{17,704}Y_{45,965} \; N_{55,515} \; Q_{68,187}$$
$$D_{65,371} \; E_{80,856} \; K_{55,024} \; R_{71,536} \; H_{29,400}$$

The bases for viral proteomes are just as readily constructed. The respective bases for strains of H1N1 influenza, listeria phage, and HIV 1 are as follows:

$$A_{274}V_{264}L_{390}I_{311}P_{187}F_{186}W_{79}M_{189}G_{313}S_{363}T_{300}C_{87}Y_{133}N_{255}Q_{186}D_{210}E_{359}K_{296}R_{322}H_{85}$$
$$A_{437}V_{569}L_{651}I_{595}P_{270}F_{391}W_{107}M_{240}G_{496}S_{451}T_{548}C_{71}Y_{346}N_{471}Q_{318}D_{570}E_{555}K_{676}R_{365}H_{160}$$
$$A_{238}V_{235}L_{324}I_{254}P_{209}F_{96}W_{99}M_{80}G_{264}S_{211}T_{227}C_{82}Y_{94}N_{172}Q_{214}D_{140}E_{257}K_{245}R_{227}H_{90}$$

At this writing, over 1,000 organism and 7,000 viral proteomes are available from NCBI, each presenting a unique set of sequences and collective base. By sampling proteomes across the spectrum, the immediate lessons reprise ones for individual proteins: (1) sets of sequences bewilder with casual inspection, and (2) the bases for sets are highly irregular and idiosyncratic. Compare leucine (L) and tryptophan (W) in the above examples whence the former substantially exceeds the latter. These and other disparities point to selection mechanisms at work. It would be most improbable to observe the component disparities from independent random draws.

The base structures of Π-sets prompt the same inquiries as individual proteins. For example, we questioned in Chapter Three whether a component amount reflected weights in the genetic code: the code favors some amino acids by apportioning more codons. The lysozyme base was weakly supportive of codon-amino-acid correlations and multiple other proteins follow suit. Matters take a turn, however, when it comes to proteomes, or at least the sizable ones. Definitive correlations emerge and strengthen; evolution *does* apportion codons roughly in line with biochemical needs [2]. Figure One shows the fraction of each amino-acid type encoded by humans versus the corresponding fractions for codons. The linear correlation coefficient R^2 is respectable at 0.628—roughly triple that for lysozyme and other archetypes in Chapter Three. The base of the human proteome expresses about six times more leucine than tryptophan. The number of codons specific to leucine and tryptophan is six and one, respectively (cf. Appendix V).

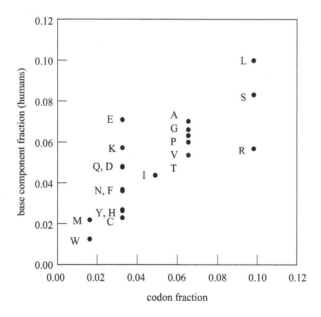

FIGURE ONE Fraction of each amino acid in human proteins versus the fraction of codons specific to the amino acid.

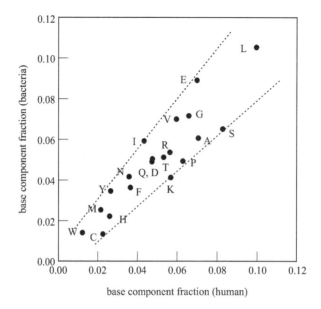

FIGURE TWO   Base component fractions for bacteria proteomes versus human. The dotted lines follow the dispersion of fractions left-to-right ascendant.

Correlations are affirmed across the spectrum. This is especially apparent when the base information of different sets is combined on graphs. For example, Figure Two plots the component fractions for Yersinia Pestis bacteria versus the fractions for human-encoded proteins.

The plot shows that disparate sources of proteins follow similar protocols at the base: R^2 for the data is just less than 0.80. More pointedly, we observe a critical correspondence for collections of proteins. Recall from Chapter Two that for molecular solutions, the component fractions do double duty: they reflect the intensive thermodynamic states *and* random draw probabilities. For proteomes, there is more to appreciate, for the component fractions ↔ probabilities apply to far-from-equilibrium systems *and* are near-universal. When exploring proteome bases, it is as if we were tracking systems that share the same intensive-state recipe with modest variations. Pick a protein constituent at random encoded by the human genome. Figure Two teaches that the probability of the constituent being, say, leucine, is almost matched by a bacterial genome. Consider the probability structures in reverse: pick a window, say, 0.080–0.120. The figure tells us that much the same components are allied with the window, irrespective of the proteome. This is remarkable and echoes themes of the previous chapter. We observed how the *N*-distributions of radically disparate sources are closely correspondent.

Figure Two contains another lesson indicated by the dotted lines. These track the dispersion of the component fractions. We observe how the fractions deviate for different sources: the deviations are roughly proportional to the fractions themselves. This is a subtle point which acquires further currency at the higher information levels introduced in Chapter Six. For now, we can appreciate that the fraction correspondences are remarkable across species, and yet unsurprising from a wide-lens perspective. The genomes underpinning proteomes evince marked size disparities, viz.

That the component fractions correspond reflects the significant sharing of codes, biochemical functions, and ancestral trees. In addition, there are the effects of mixing and gradient-control over time—processes critical to steady-state (or nearly so) systems. A genome does not evolve in a vacuum and its proteome reflects as much.

The correspondences extend to the genomes themselves. Below is represented part of the genetic material for an organism via the horizontal line; the blocks mark protein coding regions. If we confine attention to regions in, say, the dotted circle, only poor correlations between base components and codons can be realized (cf. Figure Three of Chapter Three). In contrast, the base for a proteome is governed by the coding regions in total. These number thousands of codons (typically) and enable correlations to find their form.

In Chapter Three we inquired about material, energy, and entropy economy whereby the correlations for individual sequences proved slight. The correlations strengthen somewhat when we turn to Π-sets. Figure Three shows the fraction of amino acids encoded by the human genome versus molecular weight.

The trend is nominally linear with negative slope: organisms *do* favor the lighter, fewer-atom amino acids over the heavies, but only by a little, that is, $R^2 \approx 0.326$. This is consistent with the N-distributions skewed toward lower weight sequences. By the same token, Figure Three tells us that that neither component fractions nor weights offer information that is substantially mutual. Pick a protein unit at random encoded by a genome and measure the molecular weight with a resolution of, say, 15%. Figure Three tells us that the measurement says little about the component identity. If the weight were, say, 120 ± 18 g/mol, does the component correspond to P, V, D, T, I, N, or C? The chances of a correct guess are about one in seven.

We are letting components and figures do the talking. To be sure, the base sizes of proteomes vary wildly across nature. But the structures prove memorably correspondent intensive-state-wise.

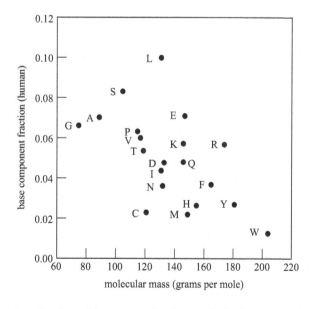

FIGURE THREE   Fraction of amino acids versus molecular weight for human proteins.

As another case in point, Figure Four shows the fractions for human-encoded proteins versus drosophila (aka fruit-flies). The linear scaling is striking with $R^2 \approx 0.961$ and the slope of the regression line is 1.01 ± 0.049. The fractions constitute probability sets whereby the set for the higher mammal closely tracks the insect, and vice-versa. In effect, the sets express information at the base level that is highly mutual. Who would have suspected this based on the appearances of humans and fruit-flies?

But what if the source is inanimate? Probability correspondences extend further when we compare organism and agricultural sets. For example, Figure Five compares the component fractions of human- versus rice-encoded proteins. Humans and rice grains are radically different systems. Even so, their proteomes present base structures that are highly correspondent. Pick a protein unit at random encoded by the human genome. The probability that the unit is, say, proline, is tracked closely by the rice genome. Alternatively, pick a probability window, say, 0.050–0.070, and gather components whose fractions fall somewhere in the window. One component set offers significant information about the other.

The takeaway is that to learn about the base probabilities for proteome B, we need only look at A—most *any* A— to get a good idea. In effect, the sets conferred in nature express composition information considerably beyond their borders. We revisit the Shannon formula of Chapter Two:

$$I = -\sum_i prob_i \log_2 prob_i$$

Let a component identity constitute a state such that the summation involves 20 terms. Then to compute I for a single unit drawn at random from a proteome, we consider:

$$I = -prob_A \log_2 prob_A - prob_V \log_2 prob_V - prob_L \log_2 prob_L - \cdots$$

For the human proteome, I works out to be ~4.18 bits. This is the information allied with a *single* amino-acid unit, given our knowledge of the base structure. Importantly, the bits are very close for other

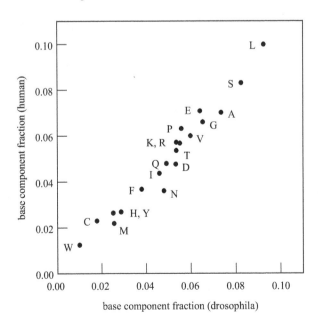

FIGURE FOUR   Amino-acid fractions for humans versus drosophila proteins. Note the minimal dispersion compared with Figure Two.

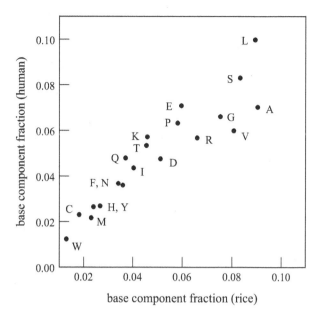

FIGURE FIVE Amino-acid fractions for humans versus rice proteins.

organisms. What do **Π**-sets share in common? We witnessed in Chapter Four how the *N*-distributions are closely correspondent. The correspondences extend to the bits per component unit.

The Kullback–Leibler formula provides complementary light. Suppose we assumed that the $prob_i$ for human-encoded sets were tracked by, say, drosophila. How grievous is the error? We look to:

$$K_I = + \sum_i prob_i^{(\text{human})} \log_2 \left(\frac{prob_i^{(\text{human})}}{prob_i^{(\text{fruitflies})}} \right)$$

The summation works out to be less than 0.010 bits—a miniscule penalty for a seemingly radical assumption. In visual terms, we have that the information vectors for the two systems at base level are closely aligned, viz.

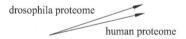

drosophila proteome

human proteome

The pointing details for one arrow carry strong indicators for the other. Proteomes are the subject at hand, but it is as if we were comparing the equilibrium phases of ideal solutions (cf. Chapter Two). The information vectors for the phases are closely aligned, irrespective of details such as volume, temperature, and pressure.

Clearly base information provides a framework for viewing proteomes. In place of integer subscripts, we employ the component probabilities such as for human proteomes so as to represent a template:

$A_{0.0710} V_{0.0617} L_{0.0978} I_{0.0421} P_{0.0638} F_{0.0357} W_{0.0119} M_{0.0235} G_{0.0654} S_{0.0838} T_{0.0520} C_{0.0234} Y_{0.0258} N_{0.0391}$

$Q_{0.0468} D_{0.0463} E_{0.0701} K_{0.0548} R_{0.0552} H_{0.0296}$

Then what do the base structures look like for, say, canine, and equine sets? Grapes? To good approximation, they are simple multiples of the template, followed by rounding. There seems no best template to use, as the base vectors align closely for proteomes far and wide. In turn, the penalties weighed by K_1 are nearly trivial for assuming one base composition to be matched by another.

But we need to dial back and qualify statements. Evolution is not restricted to a one playbook for individual sequences, for example, enzymes versus keratins, and the same is true for collections. We illustrated Π-measures for several organisms. What about encoders at the threshold of life?

The component probabilities expressed by viral proteomes present a mixed bag. For example, Figure Six shows the measures for human-encoded proteins versus HIV 1. The correlation coefficient is significant at $R^2 \approx 0.703$.

The data are followed by Figure Seven comparing the measures for human- and variola-encoded proteins: the correlations are diminished considerably with $R^2 \approx 0.292$.

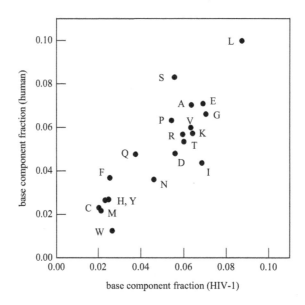

FIGURE SIX   Amino-acid fractions for humans versus HIV 1 proteins.

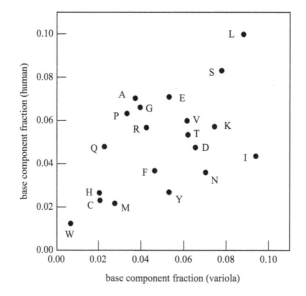

FIGURE SEVEN   Amino-acid fractions for humans versus variola proteins.

The lesson is that viral and organism component probabilities correspond significantly in some cases, but diverge in others. This is in spite of the near-uniformity of the bits per component. We list several examples:

Humans	4.184 bits
Honeybees	4.194 bits
Drosophila	4.187 bits
Grape	4.177 bits
Yersinia Pestis	4.165 bits

Variola	4.147 bits
Herpes	4.030 bits
Rabies	4.173 bits
H1N1 Influenza	4.200 bits
H5N1 Influenza	4.204 bits

Of significance are the higher values for viruses such as influenza. They suggest less selection bias toward the amino acids and greater accommodation of mutations.

B   SET STRUCTURES, PARTITIONS, AND INFORMATION ECONOMY

Partitions and families bring additional correspondences to light. Recall from Chapter Three that the bases for individual sequences can be expressed in reduced-information terms. This leverages information by the choice of partition of the 20 amino acids. The same partition and respelling strategies apply to proteomes. For example, four of the bases from Section A present as follows:

Humans	$b_{2,050,965}a_{1,687,145}p_{4,893,549}n_{5,842,828}$
Yersinia Pestis	$b_{156,572}a_{148,031}p_{440,329}n_{593,000}$
Variola	$b_{7495}a_{6461}p_{18,865}n_{21,513}$
HIV 1	$b_{562}a_{392}p_{1264}n_{1535}$

The bases are expressed at lower resolution and the diversity of subscripts makes things look complicated. However, the representations take us to plots as in Figure Eight. The vertical axis marks the natural abnp-family-fractions for the 20 amino acids while the horizontal axis marks the fractions observed for the four proteomes. Again, the linear correlations speak to the correspondences of otherwise disparate sets. Thus to learn about the abnp-measures for system B, we need only look at A for cues. But more than that: we can bypass A and view the family measures conferred by evolution: 2/20 (a), 3/20 (b), 7/20 (p), 8/20 (n). We encounter again themes of Chapter Three. Why do 8 out of 20 amino acids encoded by the genetic code carry nonpolar side chains? Answer: a Π-set requires about 40% of the amino-acid units to be nonpolar. Π-sets in nature do not present 60% (or 20%, 32%, etc.) components as nonpolar. If this were the case, there would be a gross mismatch between system requirements and component availability. Supply would not be commensurate with demand.

We can take matters farther by averaging the family-fractions for multiple Π, organism and viral. Results are illustrated in Figure Nine based on 60 systems (cf. Appendix Six). The error bars do not devalue the correspondences among base structures in family terms. Pick a protein unit at random encoded by a genome selected at random. Figure Nine communicates that the probability of the unit family being, say, acidic is closely matched by the acidic-set measure quite apart from genome identity. This is significant and far-reaching. Realizing the base structures for proteomes entails

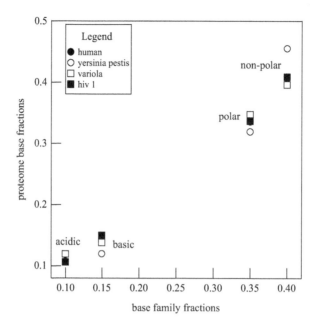

FIGURE EIGHT Comparison of base structure abnp-family-fractions for diverse proteomes: human, bacteria, and viral. The horizontal axis marks the fractional sizes of the family sets. In rendering the figure, the filled circle for human proteome data is obscured by the symbols for HIV 1 and variola.

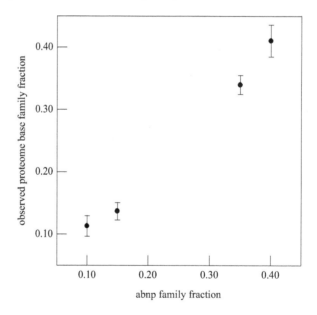

FIGURE NINE Family component fractions averaged over sixty proteomes: 23 organisms and 37 viruses listed in Appendix Six. The error bars mark one standard deviation about the average. The horizontal axis marks the abnp-fractions for the 20 amino acids.

the discrimination and counting of thousands of symbols. Yet we can gauge the family-fractions ↔ probabilities well in advance simply by inspecting the library of 20 amino acids.

The correspondences touch upon other issues. We saw in Figure One how amino-acid-codon correlations strengthen when it comes to Π-sets. Stronger still are the correlations leveraged by partitions. If we plot the family fractional measures for the Figure Eight proteomes versus the codon

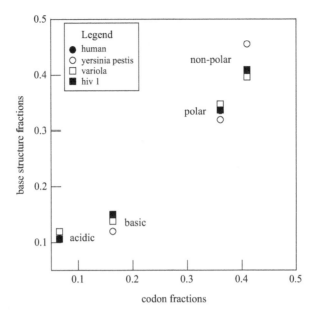

FIGURE TEN Comparison of base structure family-fractions for diverse proteomes: human, bacteria, and viral. The horizontal axis marks the fraction of codons allied with each family. In rendering the figure, the filled circle for human proteome data is obscured by the symbols for HIV 1 and variola.

fractions for each family type, we obtain Figure Ten. The linear correlation coefficients R^2 exceed 0.950. The measures affirm that proteomes require that ca. 40% of the amino acids carry nonpolar side chains. The percentage of codons so dedicated approximately approaches this at $25/61 \approx 0.389$.

The correspondences acquire still firmer footing through thermodynamic analysis. Suppose that we were to label the acidic, basic, polar, and nonpolar families, respectively, as states "1", "2", "3", and "4". Then imagine the experiment whereby we choose an amino-acid unit at random encoded by a genome. We subsequently view the unit to note the family type and write down integer 1, 2, 3, or 4 in response. The record of our experiment would look (numbingly) like:

$$2, 4, 2, 1, 3, 4, 4, 1, 2, 1, 1, 4, 3, 2, 1, 2, 4, 2....$$

Question: what is the average of all the numbers? This is trivial, right? For, say, human-encoded $\mathbf{\Pi}$, the measures from Figure Eight tell us:

$$\text{Average} = 0.1146(1) + 0.1417(2) + 0.3381(3) + 0.4037(4) = 3.0037$$

But now let us suppose that a colleague performs the experiment and communicates *only* the average 3.0037. Question: could we re-construct the fractions or at least supply reasonable estimates? This is an unusual query because it looks to connecting the family base structure—four pieces of information—with a single entity.

The answer is mostly affirmative and shines light on the thermodynamics of proteins and their operating environments. The Gibbs formula establishes that for a system under conditions of maximum entropy, the following holds for expectation values [3,4]:

$$\langle i \rangle = \frac{\sum\limits_{i=1}^{4} i \cdot e^{-\lambda i}}{\sum\limits_{i=1}^{4} e^{-\lambda i}}$$

The denominator is the partition function of the system, a sum of Boltzmann factors:

$$\sum_{i=1}^{4} e^{-\lambda i} = e^{-\lambda \cdot 1} + e^{-\lambda \cdot 2} + e^{-\lambda \cdot 3} + e^{-\lambda \cdot 4}$$

The argument in each exponential term contains λ, a best-fit parameter. The numerator features a sum of i-terms weighted by Boltzmann factors. To be sure, there are all kinds of combinations of 1, 2, 3, 4 which can express an average of 3.0037. But if we determine λ on the basis of maximum entropy conditions, we arrive at $\lambda \approx -0.4230$. It follows that

$$\sum_{i=1}^{4} e^{-\lambda i} = Z \approx 12.845$$

The fractional measures for each state arrive as follows:

$$prob_1 = \frac{e^{-\lambda \cdot 1}}{Z} \approx 0.1188 \quad prob_2 = \frac{e^{-\lambda \cdot 2}}{Z} \approx 0.1814$$

$$prob_3 = \frac{e^{-\lambda \cdot 3}}{Z} \approx 0.2769 \quad prob_4 = \frac{e^{-\lambda \cdot 4}}{Z} \approx 0.4227$$

Note how the measures do not stray too far from the family measures conferred by evolution. Figure Eleven shows the above measures along with observed for human-encoded proteins. The correspondence is modest for two of the measures, close for one, and spot on for another. The takeaway is that while Π-sets are governed by all manner of evolutionary complexity, they express characteristics

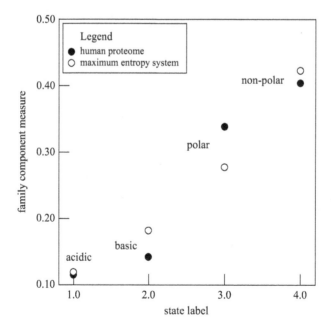

FIGURE ELEVEN   Comparison of base structure family-fractions for the human proteome (filled circles) and that computed by maximum entropy methods (open circles).

of *near*-maximum entropy systems. The selection and maintenance of components scale with the availability of fundamental families, their types and sizes.

The lesson travels farther still by revisiting the Angstrom-scale structures of the amino acids (cf. Chapter One). It is clear that the building blocks for proteins present different amounts of complexity as individuals, for example,

Cysteine   Proline   Tyrosine

Greater complexity means greater information in the transmission and receipt of messages. For organic compounds, the information hinges on the number and diversity of electronic sites. And processing-wise, information is registered at the Angstrom level via thermal collisions of molecules in solution. Every event entails a local trapping of electronic structure details—charge density and spatial distribution—of the colliding parties. Brownian computation models have been developed for such trapping events [5–10]. The models look quantitatively to the sets of collisions allowed by the covalent bond networks expressed in molecules. For example, it should be apparent that tyrosine (Y) above offers a more extended and diverse network than proline (P). It follows that tyrosine offers greater diversity in thermal collision events and correspondingly more information.

Figure Twelve shows the component fractions for the human-encoded proteome versus high-order electronic information of the amino acids [5]. The linear correlation coefficient R^2 is fair to modest at 0.545. We are nonetheless provided further insights on why some amino acids, for example, tryptophan (W), present sparingly in Π-sets, compared with, say, leucine (L) and serine (S). These reflect back to the component-codon relationships of Figure One and the N-distributions in Chapter Four. Regarding codons, why are six specific to leucine and only one to tryptophan? Regarding N-properties, why are smaller proteins encoded more favorably than larger ones? Figure Twelve shows how information economy enters into the answers to these questions. The principles are thermodynamic and computational: high information messages require more work to generate, transmit, and process than low-information ones [11]. We can even think in binary terms, for example, 100010100100101111 ↔ 18 bits is a more work-demanding message than 101 ↔ 3 bits. Whether digital or Brownian, efficient computation makes the most of low-information words and operations. Just as critical, high information words and operations are employed sparingly as they carry higher price tags. Proteins are Brownian computers directing systems far from equilibrium; their Π–sets attest to programming economy required for the long haul.

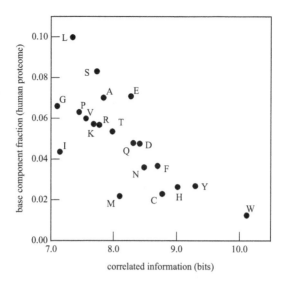

FIGURE TWELVE   Amino-acid fractions of the human-encoded proteome versus high-order correlated information of the amino acids. The data are taken from reference [5]. The linear correlation coefficient R^2 is 0.545.

C   SET STRUCTURES AND SHAPE PROPERTIES

Protein biochemistry centers around 3D folded configurations in solution—sizes, shapes, and surfaces. Importantly, the base structures for proteomes also carry topological features, only here things are more abstract. Recall that the base structures for individual proteins can be articulated by information vectors (cf. Chapter Two, Section D). When the vectors for individuals are grouped in sets, we acquire the luxuries of high-dimension pictures. As with the folded structures for individual proteins, natural sets favor some representations over others. These add to our list of correspondences.

To see how things work, revisit the three human-encoded sequences (MTMDKS…, MEKTEL…, MAVMA…) presented at the beginning of the chapter. The base structures for each underpin 20D information vectors as follows:

$$\vec{V}_{\text{epidymis}} = \sqrt{\frac{21}{246}} \cdot \hat{x}_{\text{A}} + \sqrt{\frac{11}{246}} \cdot \hat{x}_{\text{V}} + \sqrt{\frac{24}{246}} \cdot \hat{x}_{\text{L}} + \cdots + \sqrt{\frac{2}{246}} \cdot \hat{x}_{\text{H}}$$

$$\vec{V}_{\text{theta}} = \sqrt{\frac{24}{245}} \cdot \hat{x}_{\text{A}} + \sqrt{\frac{10}{245}} \cdot \hat{x}_{\text{V}} + \sqrt{\frac{27}{245}} \cdot \hat{x}_{\text{L}} + \cdots + \sqrt{\frac{1}{245}} \cdot \hat{x}_{\text{H}}$$

$$\vec{V}_{\text{antigen}} = \sqrt{\frac{38}{365}} \cdot \hat{x}_{\text{A}} + \sqrt{\frac{26}{365}} \cdot \hat{x}_{\text{V}} + \sqrt{\frac{26}{365}} \cdot \hat{x}_{\text{L}} + \cdots + \sqrt{\frac{13}{365}} \cdot \hat{x}_{\text{H}}$$

There is a distance between each vector measured by an angle, for example,

$$\theta_{\text{epidymis, theta}} = \cos^{-1}\left\{ \sum_{i=1}^{i=20} \left(prob_i^{(\text{epidymis})} \right)^{1/2} \cdot \left(prob_i^{(\text{theta})} \right)^{1/2} \right\}$$

$$= \cos^{-1}\left\{ \sqrt{\frac{21}{246}} \cdot \sqrt{\frac{24}{245}} \cdot \hat{x}_{\text{A}} \cdot \hat{x}_{\text{A}} + \sqrt{\frac{11}{246}} \cdot \sqrt{\frac{10}{245}} \cdot \hat{x}_{\text{V}} \cdot \hat{x}_{\text{V}} + \cdots + \sqrt{\frac{2}{246}} \cdot \sqrt{\frac{1}{245}} \cdot \hat{x}_{\text{H}} \cdot \hat{x}_{\text{H}} \right\}$$

$$\approx 0.107 \text{ radians} \approx 6.13°$$

The three sequences together make a set and likewise, their vectors and angles. We will refer to angle sets using Greek letter Ξ. Then the Ξ elements for the three bases can be shown to be:

$$\Xi : \left\{ \theta_{\text{epidymis, theta}}, \theta_{\text{epidymis, antigen}}, \theta_{\text{theta, antigen}} \right\} \leftrightarrow \left\{ 6.13°, 17.6° \; 17.6° \right\}$$

If a Π-set contains Y number of elements, then affiliated Ξ hosts $D = \dfrac{Y \times (Y-1)}{2}$ number of angles. Each angle θ falls *somewhere* in the window:

$$0 \leq \theta \leq 90° \leftrightarrow \frac{\pi}{2} \text{ radians}$$

The placement depends on the information in one base structure and how close it is to another.

Here is where the topology acquires form, for each Ξ presents a characteristic volume and diameter. The former is measured by the product of angles and applies to a "shape" of dimension D, viz.

$$\text{Vol}(\Xi) = \prod_{k=1}^{D} \theta_k$$

The volume will ring bells with some readers because it is closely related to the geometric mean of a numerical set [12]. The volume can be as small as zero, as would be the case if the vectors of two or more bases were collinear. Yet the volume of a set of base structures can never exceed $\left(\dfrac{\pi}{2} \text{ radians} \right)^D$: the topology space can only be so large. The upper limit applies to sets where all the vectors point at right angles. As for the diameter, this equates with the square root of the sum of squares of the angles [13]:

$$\text{Diam}(\Xi) = \left(\sum_{k=1}^{D} \theta_k^2 \right)^{1/2}$$

Our spatial imaginations are limited. Even so, things are not challenging for the set under discussion given that $Y=3$, $D=3$. The shape can be imagined as a box with unequal sides and six faces:

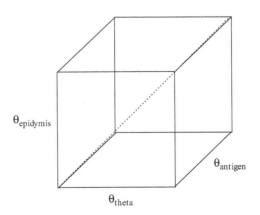

The volume follows from the product of the side lengths:

$$\text{Vol}(\Xi) = \theta_{\text{epidymis}} \cdot \theta_{\text{theta}} \cdot \theta_{\text{antigen}}$$

In turn, diagonal along the interior traced by the dotted line establishes the diameter:

$$\text{Diam}(\Xi) = \left[\theta_{\text{epidymis}}^2 + \theta_{\text{theta}}^2 + \theta_{\text{antigen}}^2 \right]^{1/2}$$

For Ξ in question, it can be shown that the volume and diameter equal 0.0101 radians³ and 0.447 radians, respectively.

Proteomes are rich with sequences and base structures, hence their shapes and sizes are quite beyond visualization, for example, if $Y \approx 20K$ for a Π–set, then $D \approx 2 \times 10^8$. Viruses are a different story, however, as their Π–sets contain as few as one element, for example, avian carcinoma virus. If $Y = 1$ then $D = 0$ whence the volume and diameter are both zero. Multiple cases, however, present $Y = 3$, $D = 3$ whereby boxes are the imagination-accessible shape.

What do we learn from volumes and diameters? Well, if all they did was remind us that Π–sets host different numbers of elements, this would be no revelation. We already knew this! Rather set topologies bring to light additional, deep-rooted correspondences and a subtle feature along the way. First the feature: it is that Π–sets generally do not express duplicate base structures. If they did, the information vectors for two or more structures would be collinear with zero radians separating them. The volume of a set would collapse to zero as it follows from a product of angles. While perhaps taken for granted, this characteristic places a finer point on the information economy in protein systems. Why should, say, a bacterium encode a particular enzyme more than once if life can succeed with a single encoding? Code economy was shown in Chapter Four to be a salient trait of individual sequences. The economy extends to the sets hosting individuals.

The topological correspondences deal with size and scaling, for the shapes of otherwise diverse Π–sets grow at near-universal rates. For example, Figure Thirteen shows the logarithm of set

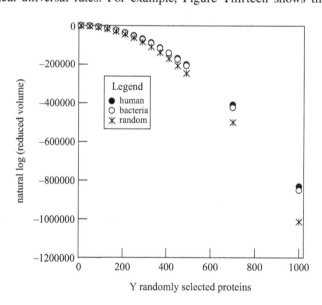

FIGURE THIRTEEN Volume scaling of natural Π–sets. The graph was constructed by averaging over randomly selection populations of bacterial (Yersinia Pestis) and human protein sequences. The 20D vectors for each base were assembled and the angles were computed using the inverse cosine formula. Each angle θ was then rescaled as $\theta' = \theta/(\pi/2)$. The rescaling rendered the set volume dimensionless, hence the term *reduced* volume. The error bars are less than the symbol widths. The asterisk symbols are placed by randomly generated strings having the same N-distributions as natural Π–sets.

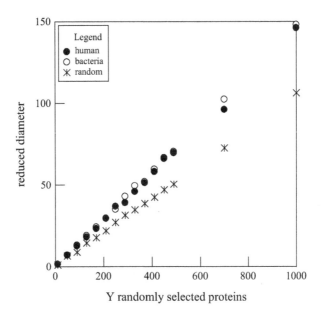

FIGURE FOURTEEN   Diameter scaling of natural **Π**–sets. The diameters were computed in reduced units as with the volumes of the previous figure. The error bars are less than the symbol widths. The asterisk symbols correspond to randomly generated strings having the same N-distributions as natural **Π**–sets.

volumes as a function of Y number of randomly selected proteins from human and bacterial **Π**–sets (filled and open circles, respectively). The symbols are closely aligned from low to higher Y. The alignment is all the more impressive given that different populations sampled lead to error bars less than the symbol widths. Figure Fourteen affirms matters by showing the diameters as a function of Y. The topological correspondences are as significant as the ones for component fractions. They further testify to the sharing of biochemical functions and evolutionary tracks; ditto for the mixing of systems and gradient controls over time.

There is another lesson for the taking, for the asterisks in the figures are placed by random strings of amino acids having the same N-distributions as **Π**-sets. These emphasize how natural **Π**-sets express topologies significantly expanded compared with abiotic sets of random independent sequences. We can think of the contrasts as follows:

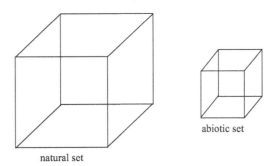

Importantly, the ratio of true/false volumes grows exponentially with Y *and* at near-universal rates for proteomes across the spectrum. How does a Y=20K **Π**-set differ from a Y=4,000 set? The reasons include that the expansion factor over random sets is orders of magnitude larger. Note that the Y- and D-measures are modest for viral proteomes, compared with organisms. For example, the base structure of a strain of H1N1 influenza was listed in Section A. It obtains from a $Y = 12$ set with corresponding Ξ having $D = 66$. The volume and diameter can be shown to be:

$$\text{Vol}_{\text{H1N1}} \approx 8.67 \times 10^{-9} \text{ radians}^{66} \quad \text{Diam}_{\text{H1N1}} \approx 0.695 \text{ radians}$$

The exponent allied with the radian units drives home the high dimension of the topological structure. Information vectors, volumes, and diameters draw out the topology of a $\mathbf{\Pi}$-set. In addition, the vectors provide a means for measuring the distances between sets. Revisit the $Y = 3$, $D = 3$ set (MTMDKS..., MEKTEL..., MAVMA...) $\leftrightarrow N = 246 + 245 + 365 = 856$. The base structure presented as follows:

$$A_{83}V_{47}L_{77}I_{31}P_{24}F_{21}W_{16}M_{22}G_{47}S_{61}T_{57}C_{12}Y_{37}N_{28}Q_{46}D_{53}E_{78}R_{48}H_{16}$$

A 20D information vector then looks like:

$$\vec{V}_{\text{whole set}} = \sqrt{\frac{83}{856}} \cdot \hat{x}_A + \sqrt{\frac{47}{856}} \cdot \hat{x}_V + \sqrt{\frac{77}{856}} \cdot \hat{x}_L + \cdots + \sqrt{\frac{16}{856}} \cdot \hat{x}_H$$

The vector points in a direction guided by three proteins collectively. It projects at an angle with respect to the vector prescribed by another set. The angle is a distance measure satisfying the vector space criteria illustrated in Chapter Two.

As an example, consider the proteomes for killer whales (*Orcinus orca*) and ghost sharks (*Callorhinchus milii*). The respective base structures assemble as follows:

$$A_{990,791} \; V_{842,744} \; L_{1,377,912} \; I_{601,814}P_{892,113} \; F_{505,612} \; W_{165,913} \; M_{300,097}G_{922,676} \; S_{1,168,945}T_{734,250} \; C_{301,147}Y_{370,660}$$
$$N_{505,926} \; Q_{663,650} \; D_{684,265} \; E_{994,418} \; K_{800,064} \; R_{804,712} \; H_{357,997}$$

$$A_{882,682} \; V_{855,106} \; L_{1,310,535} \; I_{701,115} \; P_{749,306} \; F_{497,179} \; W_{153,853} \; M_{330,982} \; G_{820,944} \; S_{1,210,064}T_{792,474} \; C_{294,714}Y_{395,910}$$
$$N_{618,730} \; Q_{697,863} \; D_{732,444} \; E_{1,023,197} \; K_{907,325} \; R_{748,851} \; H_{363,120}$$

$N = 13,985,706$ in the first structure while $N = 14,086,394$ in the second. The information vectors express in the usual way:

$$\vec{V}_{\text{orca}} = \sqrt{\frac{990,791}{13,985,706}} \cdot \hat{x}_A + \sqrt{\frac{842,744}{13,985,706}} \cdot \hat{x}_V + \sqrt{\frac{1,377,912}{13,985,706}} \cdot \hat{x}_L + \cdots + \sqrt{\frac{357,997}{13,985,706}} \cdot \hat{x}_H$$

$$\vec{V}_{\text{shark}} = \sqrt{\frac{882,682}{14,086,394}} \cdot \hat{x}_A + \sqrt{\frac{855,106}{14,086,394}} \cdot \hat{x}_V + \sqrt{\frac{1,310,535}{14,086,394}} \cdot \hat{x}_L + \cdots + \sqrt{\frac{363,120}{14,086,394}} \cdot \hat{x}_H$$

How far are the base structures apart? The inverse cosine formula applies in the usual way:

$$\theta_{\text{orca,shark}} = \cos^{-1}\left\{ \sqrt{\frac{990,791}{13,985,706}} \cdot \sqrt{\frac{882,782}{14,086,394}} \hat{x}_A \cdot \hat{x}_A + \sqrt{\frac{842,744}{13,985,706}} \cdot \sqrt{\frac{855,106}{14,086,394}} \cdot \hat{x}_V \cdot \hat{x}_V + \cdots \right\}$$

$$\approx 0.0469 \text{ radians} \approx 2.69°$$

That the $\mathbf{\Pi}$-sets are so close is testament to their highly mutual information. This tells us that if we take the trouble to learn about the base structure of one encoding system, we acquire close knowledge of another as a bonus.

Distance measures apply to all *AB*-pairs of sets. For **Π**-sets, which ones are close and which ones are far apart? To get a feel for this, Figure Fifteen shows the total distribution $F(\theta)$ obtained from measuring the distances between *AB*-pairs involving 14 organisms (cf. Appendix Six). There are $\frac{14 \times 13}{2} = 91$ distance measurements so presented. The large filled circles are placed by measures in which one pair member is the human proteome; the labels specify the "other" organism. The small open circles are placed by the distance measures *not* involving the human proteome. There are $\frac{13 \times 12}{2} = 78$ points in this category. The properties to note are the alignments: organisms of similar evolutionary stripe evince the closest alignments of **Π**-set vectors. The separations are limited to a few degrees in spite of marked differences in the proteome sizes and environments. Topological analyses such as by information vectors complement the ones provided by BLAST, MSA, and other bioinformatics software.

The lesson to close the section is that vectors present a method to discriminate proteome base structures. This is an amenity of the mutual information across species. With only modest inspection, we can discern that the ones below are not bona fide. The 20D vectors so constructed point in the wrong directions:

$A_{106,842} V_{800,442} L_{37,125} I_{400,335} P_{392,302} F_{934,603} W_{565,711} M_{305,097} G_{122,330} S_{868,941} T_{934,250} C_{31,662} Y_{770,656} N_{955,926}$
$Q_{63,482} D_{84,965} E_{550,418} K_{605,395} R_{54,712} H_{300,933}$

$A_{88,284} V_{555,386} L_{358,502} I_{730,292} P_{49,866} F_{695,103} W_{653,822} M_{830,982} G_{27,944} S_{210,064} T_{732,204} C_{794,117} Y_{698,930} N_{813,730}$
$Q_{998,484} D_{72,394} E_{602,337} K_{804,482} R_{46,393} H_{663,112}$

Then where mutual information rules the day, new messages can be composed quickly and grammatically. With only a little work, we can assemble proteome bases that are plausible. For example, we fix N_{total} arbitrarily, say, at 10^6. A base structure obtains methodically viz.

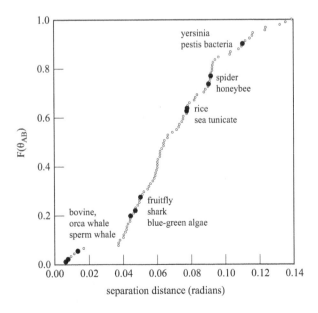

FIGURE FIFTEEN   Distribution $F(\theta)$ obtained from measuring the distances between *AB*-pairs involving 14 organisms (cf. Appendix Six). The large filled circles are placed by distances in which one pair member is the human proteome. The small open circles are placed by distance measures *not* involving the human proteome.

$$A: 0.0737 \times 10^6 = 73,700 \leftrightarrow A_{73,700}$$

$$V: 0.0600 \times 10^6 = 60,000 \leftrightarrow V_{60,000}$$

$$L: 0.0926 \times 10^6 = 92,600 \leftrightarrow V_{92,600}$$

Such a composition is close to those of natural proteomes, as reflected by K_I and vector angles.

D   SETS AND INTERNAL STRUCTURES

Protein sets present external topologies by their volumes and diameters. Their *internal* topologies deserve the same attention. Things work as follows. The sequence and base for an individual protein A underpin an information vector that points in a 20D space. For example, from the human proteome, consider:

MDPKYFILILFCGHLNNTFFSKTETITTEKQSQPTLFTSSMSQVLANSQNTTGNP
LGQPTQFSDTFSGQSISPAKVTAGQPTPAVYTSSEKPEAHTSAGQPLAYNT
KQPTPIANTSSQQAVFTSARQLPSARTSTTQPPKSFVYTFTQQSSSVQIPSRKQIT
VHNPSTQPTSTVKNSPRSTPGFILDTTSNKQTPQKNNYNSIAAILIGVLLTSM
LVAIIIIVLWKCLRKPVLNDQNWAGRSPFADGETPDICMDNIRENEISTKRT
SIISLTPWKPSKSTLLADDLEIKLFESSENIEDSNNPKTEKIKDQVNGTSEDSAD
GSTVGTAVSSSDDADLPPPPPLLDLEGQESNQSDKPTMTIVSPLPNDSTSLPP
SLDCLNQDCGDHKSEIIQSFPPLDSLNLPLPPVDFMKNQEDSNLEIQCQEFSIPPN
SDQDLNESLPPPPAELL

$$A_{21}V_{17}L_{38}I_{28}P_{47}F_{16}W_3M_6G_{15}S_{48}T_{45}C_6Y_5N_{29}Q_{31}D_{27}E_{21}K_{23}R_8H_4 \quad \longrightarrow$$

As the vector points at zero degrees with respect to itself, it defines an origin for measuring distances of other proteins in the system. Then consider fellow proteome members B and C, viz.

MKSEAKDGEEESLQTAFKKLRVDASGSVASLSVGEGTGVRAPVRTATDDTKP
KTTCASKDSWHGSTRKSSRGAVRTQRRRSKSPVLHPPKFIHCSTIASSSSSQ
LKHKSQTDSPDGSSGLGISSPKEFSAGESSTSLDANHTGAVVEPLRTSVPRL
PSESKKEDSSDATQVPQASLKASDLSDFQSVSKLNQGKPCTCIGKECQCK
RWHDMEVYSFSGLQSVPPLAPERRSTLEDYSQSLHARTLSGSPRSCSEQARVFV
DDVTIEDLSGYMEYYLYIPKKMSHMAEMMYT

$$A_{19}V_{17}L_{19}I_6P_{17}F_6W_2M_7G_{17}S_{51}T_{20}C_7Y_7N_2Q_{12}D_{17}E_{18}K_{22}R_{18}H_8 \quad \longrightarrow$$

MVFPAKRFCLVPSMEGVRWAFSCGTWLPSRAEWLLAVRSIQPEEKERIGQFVFA
RDAKAAMAGRLMIRKLVAEKLNIPWNHIRLQRTAKGKPVLAKDSSNPYPN
FNFNISHQGDYAVLAAEPELQVGIDIMKTSFPGRGSIPEFFHIMKRKFT
NKEWETIRSFKDEWTQLDMFYRNWALKESFIKAIGVGLGFELQRLEFDLSPLN
LDIGQVYKETRLFLDGEEEKEWAFEESKIDEHHFVAVALRKPDGSRHQDVPSQ
DDSKPTQRQFTILNFNDLMSSAVPMTPEDPSFWDCFCFTEEIPIRNGTKS

$$A_{21}V_{15}L_{24}I_{18}P_{19}F_{24}W_9M_9G_{16}S_{20}T_{12}C_4Y_4N_{12}Q_{12}D_{17}E_{26}K_{21}R_{20}H_6 \quad \longrightarrow$$

Their vectors point at distances from the origin measured by angles θ_{AB} and θ_{AC}. Since there is no reason to favor one member of a proteome over others, we require that every sequence and base play the role of origin. The results of measurements are thus an extended set of angles $\{\theta_{ij}\}$ characteristic of the internal topology of the proteome.

An internal topology can be encapsulated as a set of *two*-point correlations via the probability function $P_2(\theta)$. The quantity $P_2(\theta) \cdot \Delta\theta$ measures the probability of observing $\leftrightarrow$ "finding" a vector B, C, D, etc., at a distance *somewhere* between θ and $\theta + \Delta\theta$, *given A at the origin*.

As with set volumes, $P_2(\theta)$ should also ring bells. Recall how the internal and surface structures of folded proteins were explored in Chapter One via the radial distribution function $G(r)$. The structures originated from X-ray diffraction of protein single crystals. In constructing $G(r)$, we viewed each α-carbon site as an origin and looked at the placement of neighbors, near and far. More fundamentally, $G(r)$ and $P_2(\theta)$ connect with the pair distribution functions underpinning the structure thermodynamics of liquids, gases, and solids [14]. We call further attention to the pair correlation functions for membrane proteins recorded by freeze-fracture techniques and electron microscopy [15,16].

But the topology of proteomes, as with liquids and solids, is not restricted to pair correlations. When exploring proteome information, we enjoy the luxury of *three*-point correlations via the *joint* probability $P_3(\theta_1, \theta_2)$: $P_3(\theta_1, \theta_2) \cdot \Delta\theta_1 \cdot \Delta\theta_2$ measures the probability of observing a base vector C at distance between θ_2 and $\theta_2 + \Delta\theta_2$ from A, *given* that there is a B *already* at distance between θ_1 and $\theta_1 + \Delta\theta_1$. In other words, we are able to track the *joint* probability of placing neighbors with respect to all available origins. We witnessed in Chapter One how $G(r)$ plots are closely correspondent for folded systems with otherwise different biochemical functions. It should then not surprise that the correspondences extend to the base topologies of Π-sets.

Figure Sixteen shows $P_2(\theta)$ measured for the human and Yersinia Pestis proteomes. The human proteome data obtain from random sampling of sequences such as A, B, C above, forming 20D component vectors, and applying the inverse cosine formula. The error bars mark ± 1 SD about the average based on multiple random samples.

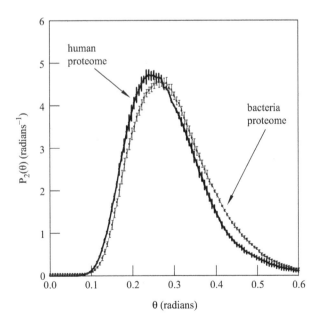

FIGURE SIXTEEN $P_2(\theta)$ for the human and Yersinia Pestis proteomes. The probability distribution of pair distances is obtained from representative sets of sequences of the proteomes. The error bars mark ± 1 SD about the average based on sampling multiple populations.

The feature to note is the shell structure expressed within Π-sets. The placement and width reflect the diversity of bases encoded by the genome. Pick a protein at random encoded by the system. There is an abstract shell around its base formed by neighbor proteins peaked at distance near 0.25 radians. Within the shell, there is an exclusion zone of width ca. 0.10 radians—the neighbors get only so close. The probability density falls off sharply beyond 0.60 radians—neighbors never stray too far way! Both the exclusion and outer zones reflect information design and selection across the Π-set. The pair correlations are the antithesis of flat and featureless, for such would signal an absence of selection and design.

Observe also how the probability density for the two systems is strikingly correspondent. There are only minor differences in the peak placement and width; ditto for the exclusion and outer zones. The number of set elements Y for human-encoded Π far exceeds Y for bacteria, and the component measures are somewhat divergent as in Figure Two. Even so, the sets are closely matched in their internal topology. The correspondence can be framed another way: the penalty for assuming $P_2(\theta)$ for human-encoded protein sets to be approximated by $P_2(\theta)$ for bacterial sets is but slight. Similar correspondences are observed for the two-point correlations across other Π-systems.

Figure Seventeen looks to higher orders by showing $P_3(\theta_1, \theta_2)$ for the human proteome. Here the joint probability structure is represented by a contour plot. The density is greatest in the center contour as labeled and falls off markedly for the larger-radii contours. The density is near zero outside the contour region as shown. Three-point correlations provide more information about the topology than two-point correlations, for we observe the preferred and exclusion zones in more intricate detail.

$P_3(\theta_1, \theta_2)$ for bacterial and other organism proteomes exhibit only slight differences from Figure Seventeen. All told, we encounter new correspondences not evident from the casual inspection of sequences and folded structures, much less extended sets. There is more to take home, however, namely that the probability functions take us closer to what Π-sets are *not*, and, most importantly, what they *are* in terms of sequence and base information.

To appreciate what they are not, let us imagine assembling sequences and bases consonant with some reasonable probability set. For example, we can take the component probabilities to

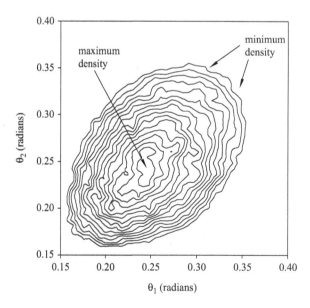

FIGURE SEVENTEEN $P_3(\theta_1, \theta_2)$ for proteins encoded by humans. Error bars are omitted for clarity without loss of structure generality.

match those of the bacteria proteome: A ↔ 0.0889, V ↔ 0.0697, L ↔ 0.105, etc., as in Figure Two. A sequence and base undergoing construction might look like:

$$AWGVTGDARTVVT... \leftrightarrow A_2V_3WG_2T_3DR...$$

There are obviously many more units to add and we can halt the procedure at any point and begin another construction.

Here is where matters turn critical. Multiple sequences and bases render a Π-set whose internal topology can be established by vector and distance analysis. Imagine that we build sets consonant with the N-distributions and component fractions of bacteria Π-sets. These are experiments that result in the tall trace toward the left in Figure Eighteen. The probability structure reflects base sets where a given element presents zero internal dependence. In other words, components have been tacked onto an element with *zero regard* to what is already there. The component probabilities are unconditioned by history: that of adding, say, A to take $A_2V_3WG_2T_3DR$ to $A_{2+1}V_3WG_2T_3DR$ is 0.0889, no more or less.

Contrast the left-side trace with $P_2(\theta)$ for the human and bacteria proteomes in Figure Eighteen. The divergence is considerable and comes with a lesson. We learn that when sequences and bases are constructed absent reflexive information, and grouped to make a set, the set elements pack too close to each other. The base structures are much too alike and the diversity falls short of what is needed by a biological system. Clearly the sets conferred by evolution are not like this. We appreciated in Chapter Three how the base of one section of lysozyme teaches significantly about others—we incur only small Kullback–Leibler penalties for assuming as much. A worthy exercise travels further by examining the penalty for wrong assumptions about reflexive information. It is the maximum possible at infinite bits!

Then consider a follow-up experiment. We assemble sequences and bases in accordance with the same probability set as before. However, we allow the probability of adding components to be

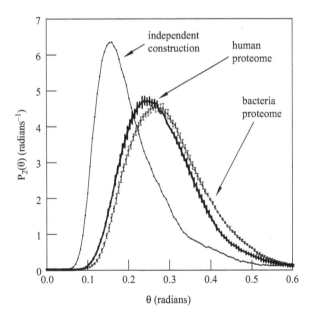

FIGURE EIGHTEEN $P_2(\theta)$ for random proteins (left-most trace) and those encoded by the proteomes for humans and Yersinia Pestis. The N-distributions track those of Chapter Four and the amino-acid fractions follow Figure Two in this chapter. The left-most trace applies to sequences and bases having no internal dependence. The error bars mark ±1 SD about the average based on sampling multiple populations.

enhanced or diminished, depending on the components in place. In other words, we give history a voice in the assembly process. In solution thermodynamics, "like" carries affinity for "like", for example, benzene combined with toluene as opposed to water. Then let the probability of adding component α (A, V, L, I, …) to a sequence and base with k units in place be governed by the following proportionality:

$$prob^{(\alpha)}(k) \propto prob_o^{(\alpha)} + \frac{\lambda N_\alpha(k)}{k}$$

$N_\alpha(k)$ tracks the α-units in the base structure having k units total; $prob_o^{(\alpha)}$ and λ represent initial probabilities and a best-fit parameter, respectively. The idea is straightforward: more α at a given stage of construction increases the chances of α joining down the road. History bears consequences whereby "like" in place shows bias toward "like" in the future. There are details in that the $prob^{(\alpha)}(k)$ changes with each unit-addition. Even so, experiments of this type are easy to carry out and illustrate. Let $prob_o^{(\alpha)}$ correspond to the component fractions of the bacteria proteome: A $\leftrightarrow$ 0.0889, V $\leftrightarrow$ 0.0697, L $\leftrightarrow$ 0.105, etc., as in Figure Two. If a sequence and base are initiated by, say, A with $prob_o^{(A)} \approx 0.0889$ then the "new" A-probability at $k = 1$ adjusts to:

$$prob^{(\alpha=A)}(k=1) = \frac{prob_o^{(A=\alpha)} + \dfrac{\lambda N_\alpha}{k}}{\sum_\alpha prob_o^{(\alpha)} + \dfrac{\lambda N_\alpha}{k}}$$

$$\approx \frac{0.0889 + \dfrac{\lambda \cdot 1}{1}}{1 + \dfrac{\lambda \cdot 1}{1}}$$

If the next addition is, say, V to give sequence and base AV, then the A-probability at $k=2$ becomes:

$$prob^{(\alpha=A)}(k=2) = \frac{prob_o^{(\alpha)} + \dfrac{\lambda \cdot N_A}{2}}{\sum_\alpha prob_o^{(\alpha)} + \dfrac{\lambda \cdot N_A}{2} + \dfrac{\lambda \cdot N_V}{2}}$$

$$\approx \frac{0.0889 + \dfrac{\lambda \cdot 1}{2}}{1 + \dfrac{\lambda \cdot 1}{2} + \dfrac{\lambda \cdot 1}{2}} = \frac{0.0889 + \dfrac{\lambda}{2}}{1 + \lambda}$$

It can be shown that the following $N = 100$ sequence and base

AVSGFQDQLLQAYLDAFLIQSDRTGGTGQATHGNNRSGASILADGTGIPSFVLS
TIAAGRIEFLSAVGLYSKAILDNLTLHITMSGTLWNTNDGNLGLFG

$$\leftrightarrow \qquad A_{10}V_3L_{14}I_7P_1F_5W_1M_1G_{14}S_9T_9C_0Y_2N_6Q_5D_6E_1K_1R_3H_2$$

underpin the A-probability shown in Figure Nineteen, λ having been taken to be 0.50.

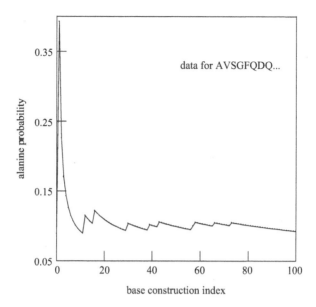

FIGURE NINETEEN   Alanine probability over the course of constructing AVSGFQDQLL…. The probability is conditioned with each unit-addition. The fitting parameter λ was taken to be 0.50.

We observe that while the probability endures wild swings, the average stays close to the component fraction for the proteome. Most importantly, if we construct bases according to such *conditioned* probabilities, we are taken to Figure Twenty. There are three traces plotted, two of them revisiting $P_2(\theta)$ for the human and Yersinia Pestis proteomes in Figures Sixteen and Nineteen. The third trace reflects a proteome topology where every element of the set is self-referential. The probability of each unit-addition to a base has taken prior history into account. The alignment is striking and yet unsurprising. The *intra-dependence* of sequences and bases in biological systems is the takeaway.

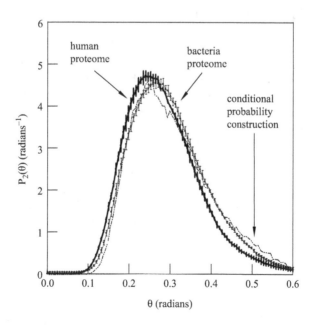

FIGURE TWENTY $P_2(\theta)$ for conditioned proteins (dash line) and those encoded by humans and Yersinia Pestis bacteria as in Figure Sixteen.

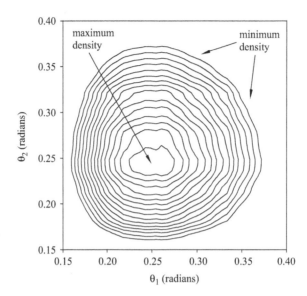

FIGURE TWENTY-ONE $P_2(\theta_1)$, $P_2(\theta_2)$ for proteins encoded by humans. This is to be compared with Figure Seventeen showing observed $P_3(\theta_1, \theta_2)$. The contour plot shows that for third-order correlations under conditions of base structure independence.

The information in three-point correlations takes us to the end of the chapter. $P_3(\theta_1, \theta_2)$ represents a *joint* probability function and offers a finer point on the base structures in proteomes. If the probability of observing a base vector C at distance between θ_2 and $\theta_2 + \Delta\theta_2$ from an origin were *independent* of the vector B placement, then we would have a case where (cf. Chapter Two)

$$P_3(\theta_1, \theta_2) = P_2(\theta_1) \cdot P_2(\theta_2)$$

The three-point correlations we observe for proteomes would equate with a product of two-point correlations. Figure Twenty-One then shows the set topology that *would* result if the above equality applied to the human proteome. The corresponding plots for bacteria and other proteomes are almost identical. The contour differences from Figure Seventeen are striking, especially toward the outer regions. This drives home that the bases and sequences in **Π**-sets express information *not* independently of one another. The base structure information of set members A *and* B provides information about C. The proteins are the antithesis of free agents, even at the ground floors of design and construction.

The major points of Chapter Five are as follows:

1. All protein sets, natural and otherwise, present base structures by their component amounts. Just as for individual proteins, set bases are conferred by evolution and underpin all the biochemical functions of an organism or virus. The amino-acid fractions for radically different systems are closely correspondent. This carries biochemical, evolutionary, and thermodynamic currency: systems share functions, ancestors, and are subject to extensive mixing and gradient-control over millennia.

2. The base structures for individual proteins are ill-correlated with genetic code and molecular weights of the amino acids (cf. Chapter Three). The correlations strengthen, however, when it comes to the larger-than-not **Π**-sets. This follows the general pattern of correspondences between biochemical needs and apportions by nature.

3. The base structures of **Π**-sets trend strongly toward family selection weights with respect to nonpolar, polar, acidic, and basic components. This is in accordance with the family

sizes conferred by evolution. The selection weights reflect the workings of maximum entropy, steady-state (or nearly so) systems.

4. The base structures of proteomes manifest topologies externally and internally that are highly correspondent across nature.

EXERCISES

1. Review the base structure for the HIV 1 virus. Use the multinomial equation from Chapter Three to estimate the surprisal in bits for obtaining the base from unselected random draws.

2. For the human proteome, the Shannon information per amino-acid unit is $I \approx 4.18$ bits. Use the proteome base structure to compute the information per unit in terms of internal, external, and ambivalent units. See details of this partition in Chapter Three.

3. Review the geometric mean of a set of real numbers and compare the expression to the set volume.

4. Under what circumstances is the volume of a set of base structures and angles zero? Does volume zero for a set imply diameter zero?

5. Construct a hypothetical set of $Y=3$ base structures with volume $(\pi/2)^3$ radians^3. How realistic is the set in terms of canonical selection weights?

6. Consider the following proteome base structure. Is it real or fake? Travel further by computing the angle between information vectors for this and the human proteome base.

$$A_{70,554} V_{53,342} \ L_{57,951} \ I_{28,956} P_{30,194} \ F_{77,474} \ W_{1401} \ M_{76,072} G_{81,449} \ S_{70,903} T_{4535} \ C_{41,403} Y_{86,261} \ N_{79,048}$$
$$Q_{37,353} \ D_{96,195} \ E_{87,144} \ K_{5623} \ R_{94,955} \ H_{36,401}$$

7. Refer again to the structure in Exercise 6. If we were to assume that natural proteomes expressed bases so prescribed, what would be the penalty in bits?

8. If we were to assume that the human proteome expressed component fractions prescribed by the HIV 1 base, what would be the penalty in bits? Is the answer surprising?

9. Revisit the base structure for the H1N1 influenza proteome from Section A:

$$A_{274} V_{264} L_{390} I_{311} P_{187} F_{186} W_{79} M_{189} G_{313} S_{363} T_{300} C_{87} Y_{133} N_{255} Q_{186} D_{210} E_{359} K_{296} R_{322} H_{85}$$

Estimate the probability of obtaining the base through draws weighted by the genetic code. Given knowledge of the base, how much information is represented by each amino-acid unit?

10. Revisit the base structure for the listeria phage viral proteome:

$$A_{437} V_{569} L_{651} I_{595} P_{270} F_{391} W_{107} M_{240} G_{496} S_{451} T_{548} C_{71} Y_{346} N_{471} Q_{318} D_{570} E_{555} K_{676} R_{365} H_{160}$$

Plot the fraction of components versus component molecular weight. Discuss the strength or weakness of correlations.

11. Consider again the base structure of the previous exercise. Construct a plot showing the fraction of amino acids versus the fraction of codons specific to the amino acid. What is the linear correlation coefficient R^2?

12. Estimate via Figure Sixteen the Kullback–Leibler penalty for assuming $P_2(\theta)$ for human-encoded protein sets to be approximated by $P_2(\theta)$ for bacterial sets.

13. Consult Figure Eighteen and estimate the Kullback–Leibler penalty for assuming that $P_2(\theta)$ for independently-random proteins conformed to $P_2(\theta)$ for human-encoded proteins.

NOTES, SOURCES, AND FURTHER READING

The amino-acid fractions in proteomes have sparked intense interest over decades. Research is supported internationally by sequencing robotics, genomic, and large-scale databases. Evolutionary and physiological issues lie at the forefront of research, for the component fractions speak to environmental impacts, mutation rates, and tolerances. References [17–24] provide a representative sampling of research directed at the challenges and principles presented by the components of proteomes. The perspective in this chapter has been primarily thermodynamic and computational. The component fractions of proteomes are not arbitrary but rather indicative of systems practicing material, energy, and entropy-production economy. In turn, the component fractions are correlated with the genetic programming codes with inherent biases toward the low-information components. The external and internal topologies for sets are natural outgrowths of these correlations. Additional correspondences of information researched by the author and students are discussed in references [25,26].

1. https://www.ncbi.nlm.nih.gov/home/proteins/
2. Appling, D. R., Anthony-Cahill, S. J., Mathews, C. K. 2016. *Biochemistry, Concepts and Connections*, Pearson Education, Hoboken, NJ, p. 125.
3. Jaynes, E. T. 1979. Where Do We Stand on Maximum Entropy? in *The Maximum Entropy Formulism*, M. Tribus, R. D. Levine, eds., MIT Press, Cambridge, MA.
4. Applebaum, D. 1996. *Probability and Information: An Integrated Approach*, Cambridge University Press, Cambridge, UK.
5. Graham, D. J., Malarkey, C., Schulmerich, M. V. 2004. Information content in organic molecules: Quantification and statistical structure via Brownian processing, *J. Chem. Inf. Comput. Sci.* 44, 1601.
6. Graham, D. J., Schulmerich, M. V. 2004. Information content in organic molecules: Reaction pathway analysis via Brownian processing, *J. Chem. Inf. Comput. Sci.* 44, 1612.
7. Graham, D. J. 2007. Information content in organic molecules: Brownian processing at low levels, *J. Chem. Info. Model.* 47, 376.
8. Graham, D. J. 2005. Information content and organic molecules: Aggregation states and solvent effects, *J. Chem. Info. Model.* 45, 1223.
9. Graham, D. J. 2011. *Chemical Thermodynamics and Information Theory with Applications*, Taylor and Francis Groups, CRC Press, Boca Raton, FL.
10. Graham, D. J. 2012. The Analysis of Organic reaction Pathways by Brownian Processing, in *Statistical Modeling of Molecular Descriptors in QSAR/QSPR*, M. Dehmer, K. Varmuza, D. Bonchev, eds., Wiley-Blackwell, Weinheim.
11. Bennett, C. H. 1982. Thermodynamics of computation—A review, *Intl. J. Theor. Phys.* 21, 905.
12. Hamming, R. W. 1991. *The Art of Probability for Scientists and Engineers*, Addison-Wesley, Redwood City, CA.
13. Stroock, D. W. 1994. *A Concise Introduction to the Theory of Integration*, 2nd ed., Birkäuser, Boston, MA.
14. Goodstein, D. L. 1985. *States of Matter*, Dover, New York.
15. Braun, J., Abney, J. R., Owicki, J. C. 1987. Lateral interactions among membrane proteins. Valid estimates based on freeze-fracture electron microscopy, *Biophys. J.* 52(3), 427–439.
16. Abney, J. R., Braun, J., Owicki, J. C. 1987. Lateral interactions among membrane proteins. Implications for the organization of gap functions, *Biophys. J.* 52(3), 441–454.
17. Bogatyreva, N. S., Finkelstein, A. V., Galzitskaya, O. V. 2006. Trend of amino acid composition of proteins of different taxa, *J. Bioinform. Comput. Biol.* 4(2), 597–608.
18. Tekaia, F., Yeramian, E., Dujon, B. 2002. Amino acid composition of genomes, lifestyle of organisms, and evolutionary trends: A global picture with correspondence analysis, *Gene* 4, 51–60.
19. Tekaia, F., Yeramian, E. 2006. Evolution of proteomes: Fundamental signatures and global trends in amino acid compositions, *BMC Genomics* 7, 307.
20. Brbic, M., Warnecke, T., Krisko, A., Supek, F. 2015. Global shifts in genome and proteome composition are very tightly coupled, *Genome Biol. Evol.* 7(6), 1519–1532. doi:10.1093/gbe/evv088.
21. Hormoz, S. 2013. Amino acid composition of proteins reduces deleterious impact of mutations, *Sci. Rep.* doi:10.1038/srep02919.
22. Gilis, D., Massar, S., Cerf, N. J., Rooman, M. 2001. Optimality of the genetic code with respect to protein stability and amino acid frequencies, *Genome Biol.* 2(11), ppRESEARCH0049.

23. Bowie, J. U., Reidhaar-Olson, J. F., Lim, W. A., Sauer, R. T. 1990. Decipering the message in protein sequences: Tolerance to amino acid substitutions, *Science* 247, 1306–1310.

24. Brooks, D. J., Fresco, J. R., Lesk, A. M., Singh, M. 2002. Evolution of amino acid frequencies in proteins over deep time: Inferred order of introduction of amino acids into the genetic code, *Mol. Biol. and Evol.* 19(10), 1645–1655.

25. Graham, D. J., Grzetic, S, May, D., Zumpf, J. 2012. Information properties of naturally-occurring proteins: Fourier analysis and complexity phase plots, *Protein J.* 31, 550–563. doi:10.1007/s10930-012-9432-7.

26. Graham, D. J. 2013. A new bioinformatics approach to natural protein collections: Permutation structure contrasts of viral and cellular systems, *Protein J.* doi:10.1007/s10930-013-9485-2.

Six Protein Structure Analysis of Base-plus Levels

The attention turns to protein information immediately above the base level. This is expressed in what will be termed base-plus (base+) structures. These form a large class of sequence substructures which are more complicated than the base. As with the base, the substructures are established by symbol counting and grouping, and their analysis proceeds through probability and information. Archetypal molecules and sets continue to be featured while the methods extend to all sectors of the protein universe. New correspondences are brought to light which further draw out the design and selection properties.

A PRELIMINARIES

We now consider the information in protein structures just above the base: that expressed in base-plus (base+) structures. Questions: what are base+ structures and why should we care about them?

It should be clear what base+ structures are not, namely the primary and beyond. The primary is represented by the total sequence while the composition establishes the base. Base+ structures fall in the vast middle: they present more information than the base, but less than the primary. We can think in level diagram terms.

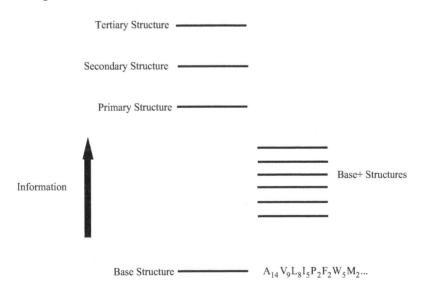

These are not to be confused with energy level diagrams. Rather they reflect that a protein presents tiers of information, all figuring into 3D folds and chemical functions. Level diagrams further emphasize that information is all we need to discern whether a structure discussed by a colleague is base+ or not. We inquire: (1) Does it provide information unavailable at the base? (2) Does it enable us to access the primary level? If the answer to both questions is negative, then the structure belongs somewhere in the base+ manifold. It turns out that base+ structures cast a wide and intricate net, even for a single protein. And whereas a single line of text does justice to the base information, multiple lines are needed for the base+ varieties.

Why explore base+ structures given that we generally obtain them from the primary—why look downward in level diagrams? The reason is that all the levels offer fresh insights into sequence design. When bases were explored in Chapter Three, guidelines emerged which were not so apparent from casual inspection of sequences and folded states. The same benefits obtain from studying base+. The biochemistry curriculum attends to these levels along multiple fronts, although with different emphases and terminology. For example, the peptide sets generated by protein hydrolysis present classes of base+ structures. Classes are further presented by folded configurations: sites such as active, binding, and recognition are base+ structures according criteria (1) and (2). In this chapter, we explore the base+ levels by symbol counting and grouping. It is easy to write sequences at whim, for example, MGDWFPTAG..., but not just any will do for a biological system. Considerations of base+ take us closer to the working architectures, and ditto for sets of sequences.

Are base+ structures confined to proteins? The answer is negative. Organic molecules express information at the base by their atom compositions, for example, $C_7H_{10}O$. Higher-level structures are represented by 2D graphs.

Base+ structures lie in the atom-bond-atom (ABA) units portrayed by a graph. The one on the left features 18 ABA units: 1 C=C, 1 C=O, 6 C–C, and 10 C–H. A base+ representation then appears as follows:

$$(C=C)_1 (C=O)_1 (C-C)_6 (C-H)_{10}$$

As with base components, the order of the units is immaterial: $C_7H_{10}O$ means the same as $H_{10}OC_7$ and $(C=O)_1(C=C)_1(C-C)_6(C-H)_{10}$ means the same as $(C=C)_1(C=O)_1(C-C)_6(C-H)_{10}$. Order is further irrelevant inside parentheses: (O=C) is synonymous with (C=O). What is important is that $(C=C)_1(C=O)_1(C-C)_6(C-H)_{10}$ cannot be inferred from $C_7H_{10}O$. Nor does $(C=C)_1(C=O)_1(C-C)_6(C-H)_{10}$ carry information to reconstruct the graph, for it applies to multiple arrangements of $C_7H_{10}O$. In effect, information in the graph relates asymmetrically to the base and intermediate levels. The graph establishes the base and base+ levels, but such levels constrain the possible graphs.

Regarding the right-side (hydroxide) compound, a base+ representation is as follows:

$$(C=C)_3 (C-C)_3 (C-O)_1 (O-H)_1 (C-H)_9$$

The same qualifiers hold regarding the unit order and information asymmetry. Note that base+ structures are not restricted to ABA-sets. We can take things up a notch by counting and grouping

ABABA components. A higher-order base+ structure for the left-side graph can be written as follows:

$$(C{=}C{-}C)_2 (C{-}C{-}C)_6 (H{-}C{=}C)_2 (H{-}C{-}C)_{16} (C{-}C{=}O)_2 (H{-}C{-}H)_5$$

In short, base+ applies to the substructures over a manifold of information levels. The substructures carry information, whether for a single compound or library. Information asymmetry holds throughout: substructures carry more information than compositions, but the latter cannot be inferred from the former. Some atoms are portrayed more than once and their identities and graph placements are generally not apparent.

Proteins are rich with base+ structures. Exploring them via ABA units is impractical to say the least, but we can grasp some key features by counting and grouping the di-peptides in a sequence. The set for lysozyme is assembled left-to-right by highlighting in bold font:

KVFERCELARTLKRLGMDGYRGISLANWMCLAKWESGYNTRATNYNAGDR STDYGIFQINSRYWCNDGKTPGAVNACHLSCSALLQDNIADAVACAKRVVRD PQGIRAWVAWRNRCQNRDVRQYVQGCGV → (**KV**)

KV**FE**RCELARTLKRLGMDGYRGISLANWMCLAKWESGYNTRATNYNAGDR STDYGIFQINSRYWCNDGKTPGAVNACHLSCSALLQDNIADAVACAKRVVRD PQGIRAWVAWRNRCQNRDVRQYVQGCGV → (**FE**)

KVFE**RC**ELARTLKRLGMDGYRGISLANWMCLAKWESGYNTRATNYNAGDR STDYGIFQINSRYWCNDGKTPGAVNACHLSCSALLQDNIADAVACAKRVVRD PQGIRAWVAWRNRCQNRDVRQYVQGCGV → (**RC**)

.
.
.
.

KVFERCELARTLKRLGMDGYRGISLANWMCLAKWESGYNTRATNYNAGDR STDYGIFQINSRYWCNDGKTPGAVNACHLSCSALLQDNIADAVACAKRVVRD PQGIRAWVAWRNRCQNRDVRQYVQGC**GV** → (**GV**)

It is straightforward to collect $N/2 = 130/2 = 65$ di-peptides. As with ABA units of a molecular graph, some di-peptides appear only once such as AV and VR. Others appear two or more times, for example, in bold underline:

KVFERCE**LA**RTLKRLGMDGYRGIS**LA**NWMC**LA**KWESGYNTRATNYNAGDRS TDYGIFQINSRYWCNDGKTPGAVNACHLSCSALLQDNIADAVACAKRVVRD PQGIRAWVAWRNRCQNRDVRQYVQGCGV

Counting and grouping symbols lead to the following set (we omit commas and brackets to make things more compact):

(AV)(AL)(AW)(AG)(AR)(VA)(VV)(VQ)(VR)(LA)$_2$(LG)(LS)(LQ)(IA)(IS)(IN)(PG)(PQ)(FQ)

(FE)(WV)(MC)(MD)(GV)(GI)$_2$(GC)(GY)$_2$(ST)(SR)(TL)(TN)(CA)(CS)(CN)(CH)(YW)(YN)

(NA)(NW)(NT)(QY)(QN)(DA)(DG)(DY)(DN)(DR)(EL)(ES)(KV)(KW)(KT)(KR)$_2$(RA)$_2$(RG)

(RC)$_2$(RN)(RD)$_2$

But this is not the only way to encapsulate the information. For example, the following list presents the equivalent:

A: V, L, W, G, R	**V**: A, V, Q, R	**L**: A(2), G, S, Q	**I**: A, S, N
P: G, Q	**F**: Q, E	**W**: V	**M**: C, D
G: V, I(2), C, Y(2)	**S**: T, R	**T**: L, N	**C**: A, S, N, H
Y: W, N	**N**: A, W, T	**Q**: Y, N	**D**: A, G, Y, N, R
E: L, S	**K**: V, W, T, R(2)	**R**: A(2), G, C(2), N, D	

The bold font marks the N-terminal-side of a di-peptide while the letters on the right identify members on the C-terminal side. The integers indicate which di-peptides appear more than once. The assembly is tedious by hand, but is expedited by programming or the find-functions of word processors. Note that there is nothing sacred about the N- to C-terminal direction. We are following convention by collecting substructures left-to-right.

Yet we immediately face a problem. If a base+ structure obtains from *di*-peptides, do structures arrive from *tri*-, *quad*-, and beyond? The answer is yes and there lies the problem. $N = 130$ for lysozyme whence 130 divided by 3 equals 43 remainder 1; 130 divided by 4 equals 32 remainder 2. How do we assemble the structures consistently while dealing with remainders?

The resolution is simple. We approach sequences *not* in chemical reaction terms, viz.

$$KV + FE + 2RC + \cdots \rightarrow \text{lysozyme}$$

Such an approach restricts the left-side units to match the right-side. Instead we imagine how a protein presents itself to molecules in solution—substrates, solvent, and other proteins. Whether folded or denatured, a protein presents a bounty of sites subject to thermal collisions. Collisions are the means of information processing as powered by Brownian motion, translational and rotational. And collisions can involve clusters of sites, not just individual amino acids [1]. Therefore, to represent a base+ structure *and* sidestep troubles with remainders, we collect units as shown:

KVFERCELARTLKRLGMDGYRGISLANWMCLAKWESGYNTRATNYNAGDRS
TDYGIFQINSRYWCNDGKTPGAVNACHLSCSALLQDNIADAVACAKRVVRD
PQGIRAWVAWRNRCQNRDVRQYVQGCGV → (**KV**)

K**VF**ERCELARTLKRLGMDGYRGISLANWMCLAKWESGYNTRATNYNAGDR
STDYGIFQINSRYWCNDGKTPGAVNACHLSCSALLQDNIADAVACAKRVVRD
PQGIRAWVAWRNRCQNRDVRQYVQGCGV → (**VF**)

KV**FE**RCELARTLKRLGMDGYRGISLANWMCLAKWESGYNTRATNYNAGDR
STDYGIFQINSRYWCNDGKTPGAVNACHLSCSALLQDNIADAVACAKRVVRD
PQGIRAWVAWRNRCQNRDVRQYVQGCGV → (**FE**)

.
.
.
.

KVFERCELARTLKRLGMDGYRGISLANWMCLAKWESGYNTRATNYNAGDR
STDYGIFQINSRYWCNDGKTPGAVNACHLSCSALLQDNIADAVACAKRVVRD
PQGIRAWVAWRNRCQNRDVRQYVQG**CG**V → (**CG**)

KVFERCELARTLKRLGMDGYRGISLANWMCLAKWESGYNTRATNYNAGDR
STDYGIFQINSRYWCNDGKTPGAVNACHLSCSALLQDNIADAVACAKRVVRD
PQGIRAWVAWRNRCQNRDVRQYVQGC**GV** → (**GV**)

We obtain (tediously) for lysozyme:

$(AV)_2(AL)(AW)_2(AG)(AT)(AC)_2(AN)(AD)(AK)_2(AR)(VA)_2(VV)(VF)(VN)(VQ)(VR)_2(LA)_3(LL)$
$(LG)(LS)(LQ)(LK)(IA)(IF)(IS)(IN)(IR)(PG)(PQ)(FQ)(FE)(WV)(WM)(WC)(WE)(WR)(MC)$
$(MD)(GA)(GV)(GI)_3(GM)(GC)(GY)_2(GD)(GK)(SA)(SL)(SG)(ST)(SC)(SR)(TL)(TP)(TN)(TD)(TR)$
$(CA)(CL)(CG)(CS)(CN)(CQ)(CE)(CH)(YV)(YW)(YG)(YN)_2(YR)(NA)_2(NI)(NW)(NS)(NT)(NY)$
$(ND)(NR)_2(QI)(QG)_2(QY)(QN)(QD)(DA)(DV)(DP)(DG)(DG)_2(DY)(DN)(DR)(EL)(ES)(ER)(KV)(KW)$
$(KT)(KR)_2(RA)_2(RV)(RL)(RG)(RS)(RT)(RC)_2(RY)(RN)(RQ)(RD)_2(HL)$

The order of the units is immaterial although order *does* matter inside parentheses, for example, (AV) is not the same as (VA). The representation is not singular as it can be condensed, viz.

A: V(2), L, W(2), G, T, C(2), N, D, K(2), R
V: A(2), V, F, N, Q, R(2)

.

.

.

H: L

A: V(2), L… indicates that AV and AL occur twice and once, respectively, in the representation.
    Two-dimensional arrays can also capture the information. The entries below note the times each row-column-labeled di-peptide appears in the sequence. The array is sparse with zero forming the majority of elements [2].

	A	V	L	I	P	F	W	M	G	S	T	C	Y	N	Q	D	E	K	R	H
A	0	2	1	0	0	0	2	0	1	0	1	2	0	1	0	1	0	2	1	0
V	2	1	0	0	0	1	0	0	0	0	0	0	0	1	1	0	0	0	2	0
L	3	0	1	0	0	0	0	0	1	1	0	0	0	0	1	0	0	1	0	0
I	1	0	0	0	0	1	0	0	0	1	0	0	0	1	0	0	0	0	1	0
P	0	0	0	0	0	0	0	0	1	0	0	0	0	0	1	0	0	0	0	0
F	0	0	0	0	0	0	0	0	0	0	0	0	0	1	0	1	0	0	0	0
W	0	1	0	0	0	0	0	1	0	0	0	1	0	0	0	0	1	0	1	0
M	0	0	0	0	0	0	0	0	0	0	0	1	0	0	0	1	0	0	0	0
G	1	1	0	3	0	0	0	1	0	0	0	1	2	0	0	1	0	1	0	0
S	1	0	1	0	0	0	0	0	1	0	1	1	0	0	0	0	0	0	1	0
T	0	0	1	0	1	0	0	0	0	0	0	0	0	1	0	1	0	0	1	0
C	1	0	1	0	0	0	0	0	1	1	0	0	0	1	1	0	1	0	0	1
Y	0	1	0	0	0	0	1	0	1	0	0	0	0	2	0	0	0	0	1	0
N	2	0	0	1	0	0	1	0	0	1	1	0	1	0	0	1	0	0	2	0
Q	0	0	0	1	0	0	0	0	2	0	0	0	1	1	0	1	0	0	0	0
D	1	1	0	0	1	0	0	0	2	0	0	0	1	1	0	0	0	0	1	0
E	0	0	1	0	0	0	0	0	0	1	0	0	0	0	0	0	0	0	1	0
K	0	1	0	0	0	0	1	0	0	0	1	0	0	0	0	0	0	0	2	0
R	2	1	1	0	0	0	0	0	1	1	1	2	1	1	1	2	0	0	0	0
H	0	0	1	0	0	0	0	0	0	0	0	0	0	0	0	0	0	0	0	0

    As in our first and false-start attempt, these and other representations carry more information than the base $A_{14}V_9L_8I_5P_2\ldots$, but considerably less than the primary KVFER….GCGV. We are able to purchase the information via the bank account in the sequence. That the amino-acid units are shared in base+ structures is déjà vu. In ABA-sets, for example, $(C=C)_3(C-C)_3(C-O)_1(O-H)_1(C-H)_9$, an atom in one unit is typically shared by another.

Base+ structures grow ponderous, even for small proteins. And if a viewer is inclined to blow past sequences at seminars, this is all the more true for substructures. For lysozyme, the base+ structure by way of overlapping tri-peptides is as follows:

(TLK)(ATN)(IRA)(TRA)(RCQ)(ESG)(RGI)(KRL)(DAV)(VVR)(RST)(VAW)(QDN)(KWE)
(CAK)(YNT)(WVA)(DVR)(PQG)(NIA)(AWR)(GDR)(DRS)(FQI)(DPQ)(NAC)(QGI)(VRQ)
(LAK)(RDP)(CLA)(RQY)(YWC)(NDG)(GMD)(KVF)(LSC)(RAT)(DNI)(VFE)(NRD)(LLQ)
(NYN)(LAN)(ADA)(GYR)(GIR)(MDG)(MCL)(AVA)(QGC)(RLG)(RNR)(WMC)(LGM)(YVQ)
(AWV)(LQD)(ISL)(NRC)(FER)(NAG)(RCE)(SGY)(IFQ)(GIF)(HLS)(IAD)(TDY)(DGY)(NSR)
(STD)(RAW)(TNY)(ERC)(LAR)(VNA)(SRY)(PGA)(AGD)(AKR)(TPG)(CND)(GYN)(ACH)
(YNA)(QYV)(ALL)(ELA)(DGK)(GAV)(CEL)(ANW)(KRV)(NTR)(CSA)(SCS)(QIN)(VAC)(RTL)
(WCN)(ART)(VQG)(SAL)(CGV)(AVN)(LKR)(AKW)(QNR)(CHL)(WES)(RDV)(KTP)(GIS)
(SLA)(YGI)(INS)(GCG)(NWM)(VRD)(RVV)(ACA)(CQN)(RYW)(WRN)(YRG)(DYG)(GKT)

But the substructures, as for all organic compounds, are not devoid of lessons. For example, the absence of subscripts in the above tells us that each unit appears only once in the primary structure. This is vital data for comparing proteins, especially via information vectors.

The base+ structures thus far derive from nearest-neighbor relations. Yet structures arrive just as significantly from long-distance relationships. For example, lysine = K is the N-terminal unit of lysozyme. A second K appears 12 units away, a third 20 after that, a fourth 36 later, and a fifth 28 units later, viz.

KVFERCELARTL**K**RLGMDGYRGISLANWMCLA**K**WESGYNTRATNYNAGDRS
TDYGIFQINSRYWCNDG**K**TPGAVNACHLSCSALLQDNIADAVACA**K**RVVRDP
QGIRAWVAWRNRCQNRDVRQYVQGCGV

Information lies in the *intervals* between consecutive K:

K: 12, 20, 36, 28

When the analysis is directed at all 20 components, we obtain:

A: 17, 6, 10, 5, 26, 3, 7, 7, 2, 2, 2, 12, 3		**V**: 72, 19, 6, 1, 10, 11, 4, 5	
L: 4, 3, 10, 6, 48, 5, 1	**I**: 33, 3, 30, 17	**P**: 32	**F**: 54
W: 6, 30, 45, 3	**M**: 12	**G**: 3, 3, 15, 11, 7, 13, 4, 33, 22, 2	
S: 12, 15, 10, 19, 2	**T**: 29, 3, 9, 18	**C**: 24, 35, 12, 4, 14, 21, 12	
Y: 18, 7, 9, 9, 61	**N**: 12, 5, 2, 14, 6, 9, 13, 26, 4	**Q**: 28, 18, 13, 6, 3	
D: 31, 4, 14, 20, 4, 11, 18	**E**: 3, 28	**K**: 12, 20, 36, 28	
R: 5, 4, 7, 20, 9, 12, 36	**H**: 0		

That zero is paired with **H** means a single histidine in the primary structure. Unlike nearest-neighbor base+, interval information *is* sufficient for constructing the base. The sequence in total, however, remains out of reach.

Partitions and families proved useful in base analysis and the ideas apply just as well to base+. For example, lysozyme base+ respelled in (a), basic (b), polar (p), and nonpolar (n) terms looks like:

$$(bb)_2(ba)_2(bp)_9(bn)_7(ab)_2(ap)_5(an)_4(pb)_7(pa)_5(pp)_{21}(pn)_{19}(nb)_8(na)_4(np)_{17}(nn)_{17}$$

The symbol string presents much-reduced information compared with the 20-letter version $(AV)_2(AL)(AW)_2(AG)\ldots$. It applies not just to lysozyme, but rather to a class of $N = 130$ proteins. Likewise, tri-peptide base+ for lysozyme in abnp-terms presents as follows:

(bbn)$_2$(ban)$_2$(bpb)(bpa)(bpp)$_3$(bpn)$_4$(bna)(bnp)$_3$(bnn)$_3$(abp)$_2$(apb)(app)$_3$(apn)(anb)(anp)(ann)$_2$

(pba)(pbp)$_4$(pbn)$_2$(pab)(pap)$_3$(pan)(ppb)$_4$(ppa)$_2$(ppp)$_9$(ppn)$_6$(pnb)$_3$(pna)(pnp)$_7$(pnn)$_7$(nbb)$_2$

(nba)(nbp)$_3$(nbn)$_2$(nab)(nap)$_2$(nan)(npb)(npa)$_2$(npp)$_6$(npn)$_8$(nnb)$_4$(nna)$_2$(nnp)$_6$(nnn)$_5$

Note that substructures need not be uniform in size. In ground-breaking research, Sanger and co-workers deduced the primary structure of insulin from sets of overlapping peptides; their data presented base+ structures of disparate sizes [3,4]. Modern day proteomics offers wet chemistry and mass spectrometry procedures for dealing with overlapping peptide collections [5,6]. To be methodical, and for simplicity, however, we will limit the discussion to equal-size substructures. These apply to select levels of the base+ manifold.

But there is a subtle (if obvious) point to make: our base+ examples thus far have been alphabetic. The representations mean the same thing irrespective of the font size and style, viz.

(AV)$_2$(AL)(AW)(AG)(AT)(AC)$_2$(AN)...

(AV)$_2$(AL) (AW)$_2$(AG) (AT) (AC)$_2$(AN)...

(AV)$_2$(AL)(AW)$_2$(AG)(AT)(AC)$_2$(AN)...

In contrast, there are base+ structures that are numerical and thus resolution-dependent. The base for lysozyme was examined in light of the amino-acid weights (cf. Chapter Three). We can view the data at a higher level by pairing the sequence units and weights in order:

	K	**V**	**F**	**E**	**R**	**C**	**E**	**L**...
Sequence Index k	1	2	3	4	5	6	7	8...
Molecular weight (g/mol)	146.2	117.2	165.2	147.2	174.2	121.2	147.2	131.2...

Numbers offer 1,000-word pictures and, in this case, noisy ones. When the weights are plotted versus index k, we obtain Figure One. It may seem that a viewer could assemble the base and

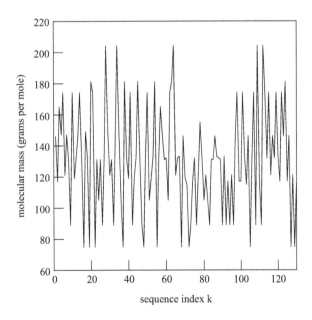

FIGURE ONE Molecular mass of lysozyme components as a function of sequence index k. The index runs from N- to C-terminal: $k = 1$ to $k = N = 130$.

primary structures via the figure and a look-up table of weights. A second thought would remind otherwise: leucine (L) and isoleucine (I) have the same formula mass. The viewer is unable to assign the 131.2 g/mol sites with certainty. The information in Figure One places in the manifold of base+ structures as discussed in Section E.

B BASE+ STRUCTURES AND INFORMATION GAPS

We have illustrated base+ structures by example. To represent the alphabetic variety, we deconstruct sequences to obtain sets of overlapping di-, tri-, etc., peptides. Alternatively, we can chart the intervals allied with components. To obtain the numerical varieties, we pair sequence units with physical properties such as molecular mass. The representations are invariably cumbersome and call for labels. We will refer to n-order base+ as base+(n): base+(2), base+(3), etc., and use base+(i) and $M(k)$ to denote interval and mass series, respectively. But now where are the lessons?

As with all aspects of protein structure, the information of base+ levels reflects both the molecules and states of knowledge on our part. Then with the level diagrams in mind, we should inquire: just how far is one state from another? The answer is important because it will demonstrate that we do not have to tend to high altitude in a diagram to nearly zero out the gap separating base+ and primary levels. This establishes that even low-lying base+ information furnishes critical guidelines about sequence design.

We measure gaps in bits. As a warm-up, let's look (again!) to the lysozyme base:

$$A_{14}V_9L_8I_5P_2F_2W_5M_2G_{11}S_6T_5C_8Y_6N_{10}Q_6D_8E_3K_5R_{14}H_1$$

We can phrase a question two ways. How much information would a neutral party—someone unfamiliar with lysozyme—require to assemble the primary structure *given* knowledge of the base? Alternatively, how surprised would *we* be if said party wrote down the primary structure after we communicated the base? We would be shocked, to put it mildly.

To answer the questions, we consider two routes, only to end up in the same place. Chapter One presented the number of distinguishable arrangements Ω of lysozyme's monomer components:

$$\Omega = \frac{N!}{N_A!N_V!N_P!\cdots N_K!N_R!N_H!}$$

$$= \frac{130!}{14!9!8!\cdots5!14!1!} \approx 10^{147}$$

This takes us at once to the gap separating levels. Taking each arrangement (structural isomer) to be equally plausible in the neutral party's eyes, we have

$$prob = \frac{1}{\Omega} \approx 10^{-147}$$

The information and our surprise become:

$$-\log_2\left(10^{-147}\right) \approx \frac{-\ln(10^{-147})}{\ln(2)} \approx 488 \text{ bits}$$

The information relation is asymmetric, as should be clear. If the neutral party is presented the primary structure, he or she could construct the base $A_{14}V_9L_8I_5\ldots$ given time, paper, and pen. Taking the acquisition of knowledge states as events, in conditional probability terms, we have:

$$prob\big(\text{sequence knowledge} \mid \text{base knowledge}\big) \approx 10^{-147}$$

This contrasts sharply with:

$$prob\big(\text{base knowledge} \mid \text{sequence knowledge}\big) = 1.$$

But we can measure gaps by a second route. Instead of aiming at the primary structure in one fell swoop, let a neutral party engage in yes/no Q & A with us. Let he or she establish the units one by one, starting with the N-terminal $\leftrightarrow k = 1$ unit. How much information is required for a correct identification? Answer: all 20 amino acids are represented in the base. The probability for a correct guess by the party and the surprise on our part become:

$$prob = \frac{1}{20} = 0.05 \qquad \text{surprisal} = -\log_2\left(\frac{1}{20}\right) \approx 4.32\,\text{bits}$$

Let us imagine that the N-terminal submission was correct, either right off the bat or after a few questions. His or her knowledge would alter irreversibly. The base for what remains of the sequence would be:

$$A_{14}V_9L_8I_5P_2F_2W_5M_2G_{11}S_6T_5C_8Y_6N_{10}Q_6D_8E_3\underline{\mathbf{K}_{5\text{-}1}}R_{14}H_1$$

The new base *still* has all 20 amino acids represented. The price for the party's knowledge of the $k = 2$ unit following Q & A would then also be $\log_2(20) \approx 4.32$ bits. Acquiring this information would alter things again. The base for the $k > 2$ units would be:

$$A_{14}\underline{\mathbf{V}_{9\text{-}1}}L_8I_5P_2F_2W_5M_2G_{11}S_6T_5C_8Y_6N_{10}Q_6D_8E_3\underline{\mathbf{K}_{5\text{-}1}}R_{14}H_1$$

And so it goes. With each succeeding unit, the state of knowledge held by the neutral party changes—information is that which changes thinking. At some point, however, there are fewer than 20 amino acids from which to choose. This lowers the information costs, but only late in the journey. In turn, the cumulative information for taking the base to the primary increases (slightly) nonlinearly with index k as plotted in Figure Two. We see that the endpoint indicated by the arrow equates with the

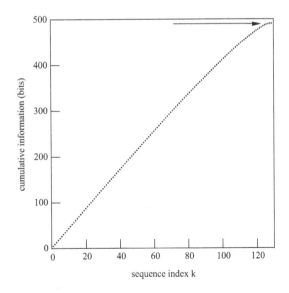

FIGURE TWO Cumulative information as a function of sequence index in the direction of N- to C-terminal. The information measures the gap between the base and primary structures. The gap is closed stepwise by strategic yes/no Q & A. The arrow points to the terminal value on the vertical axis.

information (488 bits) for the neutral party taking the base structure to primary in one swoop! The distance traveled by either route is virtually the same.

We now turn to the information gap between the base and base+(2). Recall from Section A that it is *not* zero bits. To appreciate this, imagine that a neutral party is presented with base+(2), viz.

$(AV)_2(AL)(AW)_2(AG)(AT)(AC)_2(AN)(AD)(AK)_2(AR)(VA)_2(VV)(VF)(VN)(VQ)(VR)_2(LA)_3(LL)$
$(LG)(LS)(LQ)(LK)(IA)(IF)(IS)(IN)(IR)(PG)(PQ)(FQ)(FE)(WV)(WM)(WC)(WE)(WR)(MC)$
$(MD)(GA)(GV)(GI)_3(GM)(GC)(GY)_2(GD)(GK)(SA)(SL)(SG)(ST)(SC)(SR)(TL)(TP)(TN)(TD)(TR)$
$(CA)(CL)(CG)(CS)(CN)(CQ)(CE)(CH)(YV)(YW)(YG)(YN)_2(YR)(NA)_2(NI)(NW)(NS)(NT)(NY)$
$(ND)(NR)_2(QI)(QG)_2(QY)(QN)(QD)(DA)(DV)(DP)(DG)_2(DY)(DN)(DR)(EL)(ES)(ER)(\underline{\textbf{KV}})(KW)$
$(KT)(KR)_2(RA)_2(RV)(RL)(RG)(RS)(RT)(RC)_2(RY)(RN)(RQ)(RD)_2(HL)$

Things *look* complicated. But in reality, his or her uncertainty about the base stems *solely* from ignorance of the $k = (1, 2)$ units. If we were to share this information, namely **KV**, the base construction would be a breeze. The neutral party would eliminate the first symbol of every remaining constituent (below left). The right-side symbol would be retained and added to the base as a work in progress (below right):

KV	K_1
~~K~~V	K_1V_1
~~V~~F	$K_1V_1F_1$
~~V~~N	$K_1V_1F_1N_1$
~~V~~Q…..	$K_1V_1F_1N_1Q_1$

KV is information that gets the ball rolling. The question is then how likely is the party to arrive at the keyword by lucky guess? Since $N = 130$ for lysozyme, there are $N - 1 = 129$ overlapping di-peptides. Refer to the base+(2) specifics to verify the following list:

Number of distinguishable di-peptides	Number of appearances in base+(2)
91	1
16	2
2	3

The total distinguishable di-peptides then equals $91 + 16 + 2 = 109$. We check this by noting:

$$91 \times 1 + 16 \times 2 + 3 \times 2 = 129 = N - 1$$

The modifier "distinguishable" reminds us that AV is not the same as VA. The counting tells us that the neutral party's guess at *one* of the eligible di-peptides would have 1/109 chance of being correct. This leads to the information gap between the base and base+(2). Taking the distinguishable di-peptides to be equally probable in the neutral party's mind, we have for the probability and surprisal:

$$prob = \frac{1}{109} \approx 9.17 \times 10^{-3} \qquad -\log_2\left(9.17 \times 10^{-3}\right) \approx 6.77 \text{ bits}$$

And so it goes. It should be clear how to calculate the gap separating the base and base+($n > 2$). In lysozyme we identify $N - 2 = 130 - 2 = 128$ distinguishable *tri*-peptides. The neutral party's ability to construct the base from base+(3) is limited only by ignorance of the $k = (1, 2, 3)$ unit. The probability and surprisal become:

$$prob = \frac{1}{128} \approx 7.81 \times 10^{-3} \qquad -\log_2\left(7.81 \times 10^{-3}\right) \approx 7.00 \text{ bits}$$

But now there is the gap between base+(2) and the primary structure to address. The information is considerably less than the base-primary separation. This is because the base+ constituents overlap and hold vital clues about the sequence. But here is where things get a little messy. It would seem that we should consider the probability of the neutral party guessing the primary structure in one fell swoop, given knowledge of base+(2). However, such probabilities are ill-accessible. Thus, we take the piecewise route involving yes/no Q & A whence the number of bits obtains readily.

Imagine that the neutral party is presented base+(2) for lysozyme and sets out to assemble the sequence. There is the uncertainty regarding the $k = (1, 2)$ units. A correct guess on his or her part would underpin a surprisal of

$$-\log_2\left(\frac{1}{109}\right) \approx -\log_2(9.17 \times 10^{-3}) \approx 6.77 \text{ bits}$$

Yet the chances improve dramatically for the next guess. By knowing $k = (1, 2) \leftrightarrow$ keyword **KV**, the neutral party understands that the $k = (2, 3)$ constituent has to be V-something. To figure the probability, we count the number of base+(2) units that feature V on the left. We identify eight:

$$\text{(VF) \quad (VN) \quad (VA) \quad (VV) \quad (VR) \quad (VA) \quad (VR) \quad (VQ)}$$

But the chances of a correct guess are *better* than one in eight because there are only six distinguishable units. The portion of base+(2) that the party must consider is as follows:

$$\text{(VF) (VN)(VA)}_2\text{(VV)(VR)}_2\text{(VQ)}$$

The probability and surprisal become:

$$prob = \frac{1}{6} \approx 0.167 \qquad -\log_2\left(\frac{1}{6}\right) \approx 2.58 \text{ bits}$$

Y/N questions and honest answers would take the party to the first three units of the primary structure: KVF. The cost at $k = 3$ would be $6.77 + 2.58$ bits, and the party would eliminate KV and VF from further consideration. He or she would acquire a new state of knowledge regarding base+(2) and the sequence as a work in progress.

It should be apparent how to figure the cost of knowing the $k = 4$ unit. The party has a clue via F at $k = 3$. He or she scans the possibilities and sees:

$$\text{FE, FQ}$$

Knowledge of the correct di-peptide is a bargain, requiring only 1.00 bit of information! The price of knowing KVFE is then $6.77 + 2.58 + 1.00$ bits. FE is scratched from what remains of base+(2).

And so it goes for the remainder of the primary structure. Figure Three shows the cumulative (i.e., additive) information as a function of sequence index k. The terminal value is about 226 bits indicated by the arrow.

What are the gaps separating higher-order base+ structures from the primary? These prove small because each constituent carries valuable clues about the sequence. There is always the uncertainty regarding the $k = 1$ identity. Most importantly, the distinguishable constituents for base+($n \geq 4$) scale as $N - n + 1$. In turn, the corresponding surprisal scales as

$$-\log_2\left(\frac{1}{N - n + 1}\right)$$

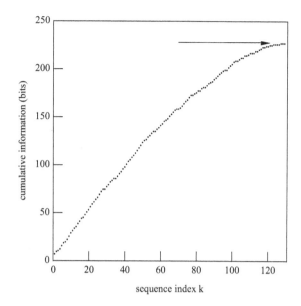

FIGURE THREE Cumulative information as a function of sequence index in the direction of the N- to C-terminal end. The information measures the gap between the base+(2) and primary structures as they grow. The arrow points to the terminal value for base+(2) on the vertical axis.

Figure Four shows the cumulative information for base+(3) and base+(4) as a function of sequence index k. The respective terminal values are about 28 and 7 bits. The gaps separating base+($n > 4$) remain at about 7 bits. This is critical: we learn that the gap separating base+ and primary structures is *nearly closed* at around the third and fourth order. What are the information gaps separating base+(i) and $M(k)$ from the primary structure? Their measures are just as worthy of exploration as encouraged in the exercises.

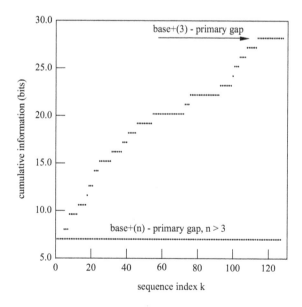

FIGURE FOUR Cumulative information of lysozyme as a function of sequence index k in the direction of the N- to C-terminal. The points mark the gap between the base+($n = 3$) and primary structures of length k. The lower-most trace marks the gap between the primary structure and base+(4). That the trace position is independent of k shows the information gap to be independent of base($n \geq 4$) structures.

C   BASE+ STRUCTURES AND INFORMATION SCALING PROPERTIES

Proteins express base+ structures with multiple takeaways. How and what to learn is suggested by ABA representations such as

$$(C{=}C)_1\,(C{=}O)_1\,(C{-}C)_6\,(C{-}H)_{10}\quad (C{=}C)_3\,(C{-}C)_3\,(C{-}O)_1\,(O{-}H)_1\,(C{-}H)_9$$

The symbol strings communicate the electronic components (alkene, carbonyl, etc.) expressed in a molecule. Just as important is the diversity: *four* distinguishable components manifest in each string as opposed to three or five. How do these ideas transfer to proteins? It is that we should begin exploring base+ structures simply by counting components. We did this instinctively for lysozyme when we measured the gaps between information levels.

Figure Five illustrates counting results for archetypal globular proteins: lysozyme, myoglobin, ribonuclease A, and phospholipase. The horizontal axis marks the order $n = 1, 2, 3, 4, 5$. The $n = 1$ case applies to the base structures assembled by monomers A, V, L, ..., H. Cases $n \geq 2$ apply to base+ representing overlapping di-peptides (KV, VF, etc.), tri-peptides (KVF, VFE, etc.), tetramers (e.g., KVFE), and pentamers (e.g., KVFER). Yet we need not look to higher order as shall become clear. The vertical axis marks the N_D, the number of *distinguishable constituents* for each structure level.

The takeaway is that base+(n) structures take a sharp turn at $n = 2$. At $n = 1$, *maximum* N_D is expressed almost exclusively, and it is the same number, namely 20, for archetypal proteins. At $n = 3$, 4, 5, maximum N_D is *still* the case, e.g., lysozyme base+(3) expresses $N_D = N - 2 = 130 - 2 = 128$, the maximum allowed. Maximum N_D for base+ at all $n \geq 3$ presents a correspondence among globular systems. This is reflected by the flattening in the right-half portion of the graph. What is N_D for base+(6) for the archetypes? We need not think twice: N_D would be $N - 5$ for these and countless other examples.

Base+ structures at $n = 2$ do not follow suit: their N_D is *not* the maximum possible, but rather variable across systems. In turn, base+(2) structures offer characteristic signatures *starting* with N_D. Figure Five shows that there are $N - 1 = 130 - 1 = 129$ components in lysozyme, 109 of which are distinguishable. In ribonuclease A, there are $N - 1 = 124 - 1 = 123$ components, 107 of which

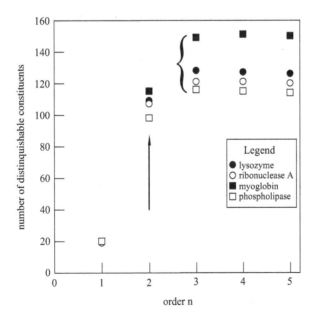

FIGURE FIVE   Number of distinguishable units as a function of order n for archetypal globular proteins. The arrow and bracket emphasize the sharp turn at $n = 2$ and flattening of N_D for $n > 2$.

are distinguishable. Revisit human myoglobin presenting $N = 154$ primary and base structures as follows:

MGLSDGEWQLVLNVWGKVEADIPGHGQEVLIRLFKGHPETLEKFDKFKHLKSEDEMK
ASEDLKKHGATVLTALGGILKKKGHHEAEIKPLAQSHATKHKIPVKYLEFISECIIQVLQ
SKHPGDFGADAQGAMNKALELFRKDMASNYKELGFQG

$$A_{12}V_7L_{17}I_8P_5F_7W_2M_4G_{15}S_7T_4C_1Y_2N_3Q_7D_8E_{14}K_{20}R_2H_9$$

The reader should verify that 115 of the $N - 1$ base+(2) components are distinguishable.

The analysis of base+ structures gets us closer to the messages embedded in a sequence. Among other things, we learn that $n \geq 2$ messages are not given to repetition. In contrast, we can think (metaphorically) of how many times we write or speak "as", "of", "in", etc. Proteins do not follow this playbook. In myoglobin, 87 di-peptides ($n = 2$) appear one time, 20 are used twice, 6 are used three times, 2 are used four times. Figure Six elaborates on this for another lesson: the *exponential* dependence for the archetypes which is apparent in the log-linear plot.

We encounter another correspondence: the *repetition* in base+(2) structures decreases *exponentially*. If $g(x)$ is the number of base+(2) components appearing x number of times, plots such as Figure Six show that

$$g(x) \approx Ae^{-ax}$$

The terms A and a represent parameters specific to a protein. But caution: the relationship is only partly appropriate for nonglobular proteins, especially large ones. For example, we show the corresponding log-linear plot in Figure Seven for an $N = 1,487$ collagen. The exponential behavior maintains through $x \approx 15$, but breaks down thereafter. Quite a few messages are popular and expressed between 20 and 54 times.

There is a follow-up experiment that is equally important. Just as a short story presents a more limited vocabulary than a 500-page novel, smaller and larger proteins manifest lesser and greater

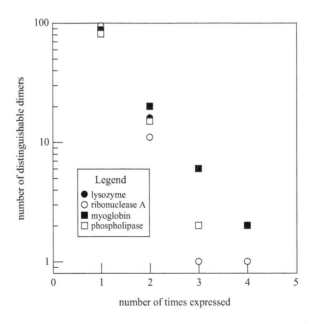

FIGURE SIX   The plot shows the number of times a given base+(2) component appears in archetypal proteins. The log-linear format brings out the exponential dependence.

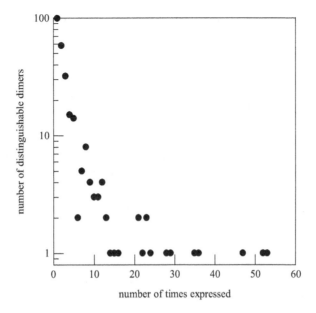

FIGURE SEVEN   The log-linear plot shows the number of times a given base+(2) component appears in a typical collagen. The exponential behavior holds until ca. 15 on the horizontal axis.

N_D. Let us explore the scaling of base+(2) N_D with the sequence unit count N. Figure Eight illustrates N_D versus N: the filled circles mark the data for archetypal proteins. The open circles with error bars mark results for *randomly-constructed* sequences of length N matching the archetypes. The data are presented as a log-log plot to draw out the scaling behavior. We observe that for both natural proteins and random constructions, the following relation applies:

$$N_D \propto N^\alpha, 0 < \alpha < 1$$

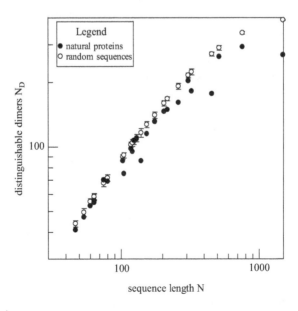

FIGURE EIGHT   Number of distinguishable dimers as a function of sequence length. The plot shows the contrasts between archetypal sequences and random strings.

The scaling coefficients for the two cases work out to be:

$$\alpha_{\text{natural proteins}} = 0.624 \pm 0.036$$

$$\alpha_{\text{random sequences}} = 0.706 \pm 0.026$$

The properties to note are that (1) the exponents do not stray far from each other, $\alpha_{\text{natural proteins}}$ being less than $\alpha_{\text{random sequences}}$ and (2) the exponents are very close to 2/3.

Figure Eight illuminates that natural proteins are *slightly* more di-peptide-vocabulary-selective than random constructions, and that the α-exponents are especially noteworthy. Globular proteins have to form secondary and tertiary structures to become biochemically active. The binding, active, and recognition sites have to position along the electronic surface. Critically, base+(2) structures foreshadow the importance of surface configurations and interactions. We see this by noting that the *volume* excluded by a polymer chain scales *linearly* with N, viz.

$$V \propto N^1$$

In turn, a chain poses an electronic surface of nominal area A upon folding whereby the scaling is nonlinear and fractional:

$$A \propto V^{2/3} \propto N^{2/3}$$

The takeaway is this: base+(2) structures drive home that larger proteins require and deliver a richer vocabulary compared with smaller ones. Moreover, the vocabulary list of di-peptides scales with the potential for forming surface area. This is attractive on intuitive grounds. To operate with high specificity, a protein has to maintain robust folds in a dynamic liquid environment. The folds have to be recognizable along multiple trajectories by substrates, inhibitors, and other proteins. But the discriminating features are effective only if positioned at the electronic surface. In turn, the recognition features are constrained by the capacity of a protein to form surfaces. It is this capacity which underpins the $N^{2/3}$ scaling.

The scaling further suggests why there are 20 amino-acid types encoded by nature. Why not 12, 16, or 24? The answer is that, in addition to the genetic code restrictions, the scaling exponent α would no longer stay so close to 2/3. If α exceeded 2/3, multiple di-peptide messages would be squandered or inappropriate at the interior core. If α were less than 2/3, a surface may present too few characteristics for recognition. The scaling behavior breaks down at about $N > 1,000$ and circles back to the size limits of proteins discussed in Chapter Four. Sequences are virtually unlimited in terms of message possibilities. This is not the case, however, at the base+(2) levels.

D   INTERVAL STRUCTURES AND INFORMATION SCALING

Base+(n) structures illuminate sequence information in pieces: AV, GC, LD, etc. Interval structures $\leftrightarrow$ base+(i) offer wider-lens perspectives, reflecting long-distance relationships. Here we find another correspondence. Revisit the sequence for lysozyme. Choose an amino acid, say, alanine (A) and inquire: where are the A-sites? Answer: The sites are noted in bold underline below. What's the big deal?

KVFERCEL**A**RTLKRLGMDGYRGISL**A**NWMCL**A**KWESGYNTR**A**TNYN**A**GDRSTDYGI
FQINSRYWCNDGKTPG**A**VN**A**CHLSCS**A**LLQDNI**A**D**A**V**A**C**A**KRVVRDPQGIR**A**WV**A**WRN
RCQNRDVRQYVQGCGV

There is a lesson in the making. To obtain it, we imagine sitting back and letting an electronic helper do the site-reading for us. Think of it as a comparator or operational amplifier sensitive to the local electronic structure, viz.

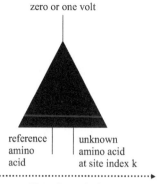

zero or one volt

reference amino acid | unknown amino acid at site index k

N- to C-terminal steps

There are two inputs as shown, the left-side one adjustable to any of the 20 amino acids. The right-side input derives from any site of a sequence placed in contact. Other circuitry guides the comparator to move stepwise from the N- to C-terminal sites. Let the device sense all sites labeled by $k = 1, 2, 3, ... N = 130$. Let it produce an output signal of 1.00 volt if the amino acid so registered at a site matches the reference and 0.00 volts otherwise. A memory chip enables storage of the signal record. Sidebar: this is not a far-fetched fantasy. There *are* devices that read biopolymers and respond in electronic terms. They are called proteins! The polymerases are Brownian computers as discussed in Chapter One: they register the identity of sites in a DNA or RNA chain with near perfect fidelity. The output is not a voltage, but rather a chemical reaction product. The memory storage obtains by synthesis of another protein or polynucleotide chain [1].

When the detector is tuned to **A**-sites of lysozyme, the output leads to the graph in Figure Nine. The plot looks like a series of needles or maybe a picket fence with missing pieces. There is no pattern that jumps out at us. Thus, if the detector erred by omitting a signal or adding a spurious one, we would unlikely to notice.

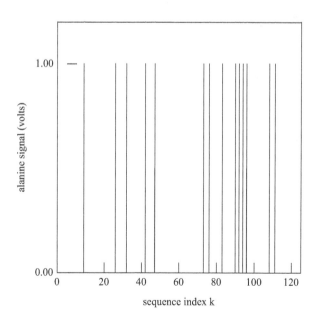

FIGURE NINE Output of digital electronic device that is tuned for registering alanines in the primary structure of lysozyme.

The lack of a pattern for any single component offers clues about base+(i). The absence of patterns holds for other amino acids as can be shown. Yet the intervals *collectively* raise a probability structure to the surface. This presents a different genre of pattern and indeed a vital correspondence among sequences. Things work as follows. Any single component A, V, D, etc., is but a minor contribution to a sequence. So as our electronic detector moves stepwise, there is only a small chance that it will happen upon *the* reference amino acid that yields a 1.00 volt signal. If the detector does happen upon a voltage-generating site, there is but a small chance that the next-adjacent site will also trigger 1.00 volt.

Let p represent a *small average probability* that the detector *will* encounter the site for which it is tuned. Then $1 - p$ is the much greater (average) probability that the detector *will not* encounter the site. This is the probability of keeping the output at 0.00 volts.

Now let's say that the detector happens upon a site matching the reference. Then when it takes the next step, the probability that the new site *does not* match the reference is as follows:

$$(1-p)^1$$

The detector advances another step. The probability that *neither* of the two sites matches the reference is as follows:

$$(1-p)^1 \cdot (1-p)^1 = (1-p)^2$$

We are implying that the probabilities of matching the reference are independent of each other. This seems justified because we discern no pattern in the component placements. Hence we can think of no reason for tying one probability term to another. Then what is the *total* probability that after three steps, the detector has still *not* matched the reference? Why it would be:

$$(1-p)^1 \cdot (1-p)^1 \cdot (1-p)^1 = (1-p)^3$$

We recognize no pattern in the A-, V- etc., signals, but their collective behavior begins to take shape. It is that the probability of the detector matching the reference site is small. But sooner or later, a match *is* made and 1.00 volt appears at the output. If $p \ll 1$, then from the Taylor Series in Chapter Four, we have

$$(1-p) \approx e^{-p}$$

This tells us that after n steps, to good approximation:

$$(1-p)^n \approx e^{-np}$$

This is the probability that the detector *will not* encounter a site matching the reference amino acid. Such probability falls *exponentially* with n as should be familiar. The takeaway is this: base+(i) structures take us to a probability rule in the family of gamma functions [7]. The gamma family was used in Chapter Four in modeling the N-distributions of protein sets. We meet it again in the design of base+(i) structures.

To see how the exponential rule applies, revisit lysozyme with the alanine sites noted in bold underline:

KVFERCELARTLKRLGMDGYRGISLANWMCLAKWESGYNTRATNYNAGDRSTDY
GIFQINSRYWCNDGKTPGAVNACHLSCSALLQDNIADAVACAKRVVRDPQGIRAWVAWR
NRCQNRDVRQYVQGCGV

An experiment entails recording all the *sequential gaps* between *identical* amino acids. For example, our digital detector registered 1.00 volt signals for **A** at sites $k = 9, 26, 32, 42, 47, 73, 76, 83, 90, 92, 94, 96, 108,$ and 111. This establishes a *set* of sequential gaps:

$$\{26 - 9 = 15, 32 - 26 = 6, 42 - 32 = 10, \ldots, 108 - 96 = 12, 111 - 108 = 3\}$$

We are ignoring all nonsequential gaps such as $42 - 9$, $96 - 9$, and so forth.

The probability rule takes shape when we construct sets and tabulate the gaps for *all* 20 amino acids, *not* just A. It presents that the fraction $f(n)$ of n number of steps taken by the detector *without* encountering a reference amino-acid scales, on average, as follows:

$$f(n) \approx Ae^{-pn}$$

The parameters A and p are readily determined. We know that the fractions $\leftrightarrow$ probability measures have to sum to 1. Let us approximate the summation as an integral:

$$\int_0^{n_{max}} f(n)dn = \int_0^{n_{max}} Ae^{-pn}dn = 1$$

We learn:

$$1 = \int_0^{n_{max}} Ae^{-pn}dn = \frac{-A}{p} \cdot e^{-pn}\Big|_0^{n_{max}} = \frac{-A}{p} \times \left[e^{-pn_{max}} - 1\right]$$

And if we take n_{max} as infinitely large, this takes us to

$$1 = \frac{-A}{p} \times \left[e^{-pn_{max}} - 1\right] \approx \frac{-A}{p} \times [0 - 1]$$

The landing is concise since:

$$1 \approx \frac{+A}{p}$$

The parameters A and p thus relate to each other in the simplest way: they are approximately equal! Returning to the probability function, we have:

$$f(n) \approx pe^{-pn}$$

The *sum* of probabilities with n then scales as follows:

$$F(n) = \int_0^n f(n')dn' \approx 1 - e^{-pn}$$

$F(n)$ models the *total distribution* for the component intervals across a protein sequence. In words, $F(n)$ approximates the *total fraction* of steps less than or equal to n in which a component fails to match the reference. Again, we have to think in terms of digital detectors moving across and reading sequences site by site.

Figure Ten illustrates a plot of base+(i) data for lysozyme along with total distribution $F(n)$. The data and function present a close match in spite of the approximations. $F(n)$ indeed offers a signature for the interval information expressed by a protein.

The signature for lysozyme is not an isolated case. Figure Eleven shows the results of applying the interval probability function to ribonuclease A. The agreement between the function and base+(i) is just as good. The agreement can be so-so in some cases, e.g., the reader is encouraged

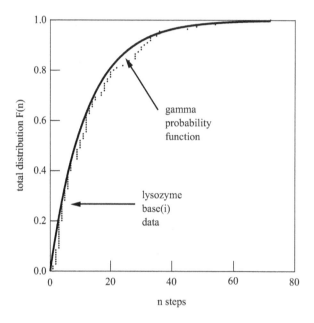

FIGURE TEN   The plot shows the gamma probability function $F(n)$ versus n, based on the gaps between identical components in lysozyme.

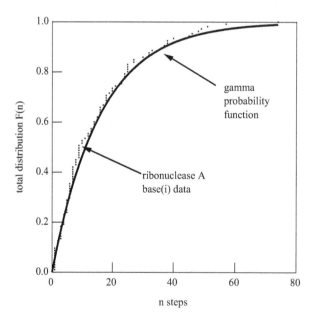

FIGURE ELEVEN   The plot shows the gamma probability function $F(n)$ versus n, based on the gaps between identical components in ribonuclease A.

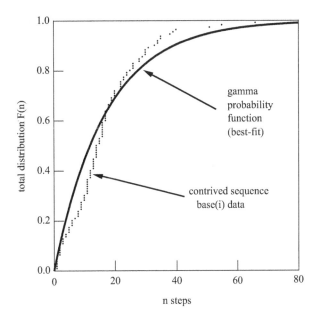

FIGURE TWELVE   The plot shows the total distribution $F(n)$ versus n, based on the gaps between identical components in a typical contrived sequence.

to apply the probability model to myoglobin. But the so-so cases are the minority party and do not negate the information correspondence. The placement of any single component in a sequence appears without chemical rhyme or reason. Thus a protein is designed to express near-maximum uncertainty (from our point of view) in its site-identities. But for a base+(i) structure viewed as a whole, there *is* rhyme and reason as tracked by a probability function.

Interval information provides not just characteristic signatures, but also red flags when things are amiss. The signatures for true-blue proteins are not easy to forge, especially if we just write primary structures at whim. If we generate sequences, say, by typing units randomly, and then carry out base+(i) analysis, we obtain distributions skewed from exponential. Here is a sample of an $N = 130$ sequence contrived by the author by entering letters off the top of his head:

MNYRWILPLDSERSACVNKLEQYIADGGSRKPCNKIQWTEIPSDAYIEWQNCSGEGGKLR
TQQNIPAWSGGVAARIPSGKHWKYWVCLIETYYKDWQSAFGSKLYIPSDYIWMVCALA
LAHIPYTWGQKN

Is this a protein conferred by nature? Of course not, and we can discern this by inspecting the component gaps and best fit $F(n)$. The results appear in Figure Twelve. Note the divergence between the probability function and the contrived sequence at almost all n. The best fit is not a very good fit at all! There is a takeaway: the probability function applies to proteins selected through evolution over the long haul, *not* to momentary contrivances on our part.

E   BASE+ STRUCTURES AND NUMERICAL INFORMATION SCALING

In Section A, we saw how base+ structures express numerical information. Most notably, the mass series $M(k)$ for lysozyme tells a noisy story as in Figure One and countless other proteins follow suit. But then what takeaways lie in a bunch of noise? Noise is something to throw away, right?

There are takeaways, but we have to re-package things to draw them out. We will take two steps with no compromise of sequence information; we will always be able to backtrack to recover the original. First, we will examine not $M(k)$ per se, but rather its accumulation across a sequence. At root, $M(k)$ tracks local density fluctuations that are critical to thermodynamic behavior as discussed in Chapter One. To delineate the fluctuations in a protein more clearly, we will inspect how they add, subtract, and cancel in places. Second, we will insist that the fluctuations *not* involve molecular weights, but rather distances measured by reduced variables. We look to the 20 amino acids for guidance.

With help from a calculator, we find the 20 amino-acid weights to have the following average and standard deviation:

<molecular weight>	136.9 g/mol
Standard deviation	30.1 g/mol

These values enable us to associate a *distance variable Z* with each component in a sequence, based on the library of 20. Z measures the *distance* of any single mass from the 20-element *set average*—not the sequence average—in units of the *set* standard deviation.

Why go through all this trouble? The reason is that while $M(k)$ tells a noisy story, it is not without biases and amenability to alternative lens-viewing. Recall from Chapters Three and Five that there is a favoring of the lightweight amino acids in natural proteins. By concentrating on the distances, we get closer to the biases and their positioning across a sequence. In turn, we more easily identify domains presenting definitive trends. Concomitantly, Z-variables are dimensionless and liberate us from having to attach grams per mole in places.

Revisit the sequence and mass series for lysozyme:

	K	**V**	**F**	**E**	**R**	**C**	**E**	**L**...
Index k	1	2	3	4	5	6	7	8...
Molecular weight (g/mol)	146.2	117.2	165.2	147.2	174.2	121.2	147.2	131.2...

The series accordingly presents Z-terms:

$$Z_K = \frac{MW_K - \langle MW \rangle}{\sigma_{MW}} \approx \frac{146.2 - 136.9}{30.1} \approx +0.308$$

$$Z_V = \frac{MW_V - \langle MW \rangle}{\sigma_{MW}} \approx \frac{117.2 - 136.9}{30.1} \approx -0.656$$

$$Z_F = \frac{MW_F - \langle MW \rangle}{\sigma_{MW}} \approx \frac{165.2 - 136.9}{30.1} \approx +0.940$$

We list all 20 Z-values in no particular order as follows:

$Z_A = -1.589$	$Z_R = +1.239$	$Z_N = -0.1588$	$Z_D = -0.1263$
$Z_C = -0.5232$	$Z_Q = +0.3071$	$Z_E = +0.3401$	$Z_G = -2.055$
$Z_H = +0.6066$	$Z_I = -0.1905$	$Z_L = -0.1905$	$Z_V = -0.6565$
$Z_K = +0.3088$	$Z_M = +0.4092$	$Z_F = +0.9404$	$Z_P = -0.7237$
$Z_S = -1.057$	$Z_T = -0.5910$	$Z_W = +2.238$	$Z_Y = +1.472$

Let the *cumulative Z* be represented as follows:

$$\Phi(k) = \sum_{i=1}^{k} Z_\alpha^{(i)}$$

The subscript α refers to the particular component at site i. With few exceptions, the summation will trend negative by virtue of the bias toward low-weight components. For lysozyme, we have the *sum* as we move along the sequence is

$$\Phi(k = 130) = \sum_{i=1}^{i=k=130} Z_\alpha^{(i)} = Z_K + Z_V + Z_F + Z_E + \cdots + Z_C + Z_G + Z_V$$

$$= 0.308 - 0.656 + 0.940 + \cdots - 0.5232 - 2.055 - 0.6565$$

Numerical information offers 1,000-word pictures. If we plot $\Phi(k)$, the result is the jagged trace in Figure Thirteen. Observe how the story tracks smoothly and negatively on account of the mass biases. Observe further how linear domains of the density function are illuminated by $\Phi(k)$. These are indicated in places via the trend lines superimposed.

Lysozyme is not an isolated case as we show the plot for ribonuclease A in Figure Fourteen. For this protein, two principal domains are brought to light. Yet also note the critical correspondences: $\Phi(k)$ scales *linearly* with k with similar slopes and fluctuation amplitudes: globular proteins follow suit by and large. Least squares analysis applied to the summations yields correlation coefficients R^2 typically 0.85 or better. Therein lies a takeaway: base+(M) presents information *equivalently* via $M(k)$ and $\Phi(k)$. But while $M(k)$ presents obtusely on the screen or printed page, its cousin $\Phi(k)$ is generous with patterns. The latter function further demonstrates how proteins express cumulative distances in highly correspondent ways.

But $\Phi(k)$ plots contribute further as marked by the trend lines in the figures. These highlight the density scaling domains and shifts which are not so obvious in source $M(k)$. In lysozyme, for example, we identify three domains over the $5 \le k \le 60$, $70 \le k \le 95$, and $105 \le k \le 130$ intervals. The patterns point to correlations between the sequence numerical information and the secondary and tertiary structures. Trend lines have been included in the figure for ribonuclease A for additional connections with higher-level structures. Note that the correlations are considerably more obtuse if approached solely via the base+ information in strictly alphabetic terms.

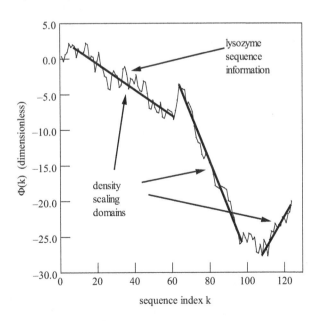

FIGURE THIRTEEN   The plot shows the cumulative mass series for lysozyme in dimensionless terms Z. Trend lines mark scaling domains.

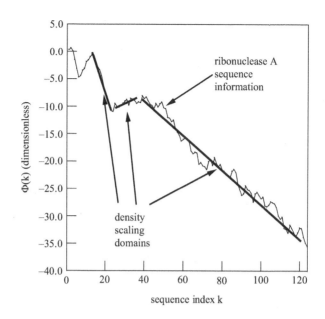

FIGURE FOURTEEN   The plot shows the cumulative mass series for ribonuclease A in dimensionless terms Z. Trend lines mark scaling domains.

F   BASE+ STRUCTURES AND NATURAL SETS

Chapter Five focused on the base structures of sets of sequences, the most illustrious being proteomes. We saw that the structures of radically different proteomes (humans, bacteria, drosophila, etc.) and bacteria are closely correspondent at the base level; ditto for their external and internal topologies.

Sets can be tracked at the higher information levels just as readily. To ascertain a base+(2) structure, we count all the overlapping di-peptides AA, AV, AL…encoded by a genome. For base+(3), we similarly count tri-peptides AAA, AAV, AAL….. As with base analysis, all counting proceeds independent of a protein's biochemical role and population in the source.

The set structures demonstrate that the correspondences maintain at the higher levels—they are robust across nature and not just incidentals at the base. We illustrate data for the principal sets of Chapter Four via Figure Fifteen with the data for other organism sets near equivalent. We observe how the base+(2) fractions ↔ probability structures for human- and bacteria-encoded proteins exhibit close correspondence. The slope of the best-fit trend line is about 0.766 with linear correlation R^2 of 0.705. The information is taken up a level by the base+(3) plot in Figure Sixteen.

Figures Fifteen and Sixteen contain another lesson via the dotted lines (cf. Figure Two of Chapter Five). These track the dispersion of the base+(2) and base+(3) components that appear when comparing proteomes. We observe how the deviations in the component fractions of one encoder (e.g., human) from another encoder (e.g., bacteria) scale with the fractions themselves. In other words, the probability of a component being enhanced or diminished in system B is proportional to its usage in A. If A expresses a component with high frequency, we observe that B modestly increases or decreases the frequency. Informally, B finds new applications for the component or discovers work-arounds via substitutions. In contrast, if A expresses a component at low frequency, B does little to alter the frequency. Informally, B stays close to the way that A communicates at the base+ levels.

Base+ structures in abnp-terms amplify these points. Figures Seventeen and Eighteen show the respective base+(2) and base+(3) fractions for human- and bacteria-encoded proteins. But note how the dispersion collapses for low- and high-frequency components alike. It is as if different sources of proteins practice different dialects regarding vocabulary, hence the dispersion in Figures Fifteen

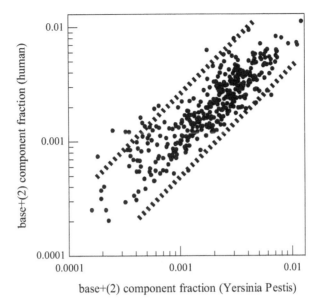

FIGURE FIFTEEN Base+(2) component fractions for bacteria- versus human-encoded proteins. This compares with Figure Two of Chapter Five. The dotted lines follow the dispersion of fractions left-to-right ascendant.

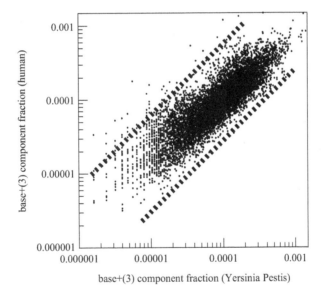

FIGURE SIXTEEN Base+(3) component fractions for bacteria- versus human-encoded proteins. The dotted lines follow the dispersion of fractions left-to-right ascendant.

and Sixteen. The sources, however, practice the same rules of grammar witnessed by the minimal dispersion in Figures Seventeen and Eighteen.

The major points of Chapter Six are as follows:

1. All proteins present base+ structures expressing information *between* the base and primary levels. The nearest-neighbor varieties are expressed by di-, tri-, etc., peptides contained in a sequence. Just as important are the base+ structures conferred by long-distance

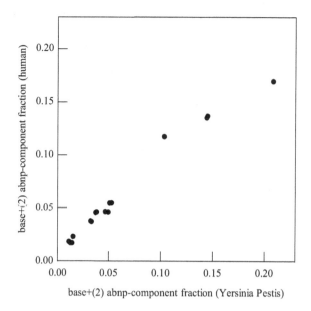

FIGURE SEVENTEEN   Fraction of each type of base+(2) unit in abnp-terms in bacteria versus human-encoded proteins. Note the lack of dispersion compared with Figure Fifteen.

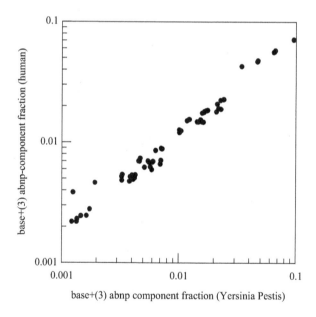

FIGURE EIGHTEEN   Fraction of each type of base+(3) unit in abnp-terms in bacteria versus human-encoded proteins. Note the lack of dispersion compared with Figure Sixteen.

    relationships. Base+ structures are further underpinned by numerical information such as mass placements across a sequence.

2. As with base-level structures, probability and information illuminate the design and selection of base+ properties. Also brought to light are scaling relationships that offer characteristic signatures for proteins.

3. Base structures express correspondences across proteomes. The correspondences maintain at the base+ levels.

EXERCISES

1. Consider the organic molecules represented by the following graphs.

  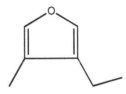

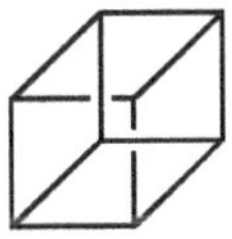

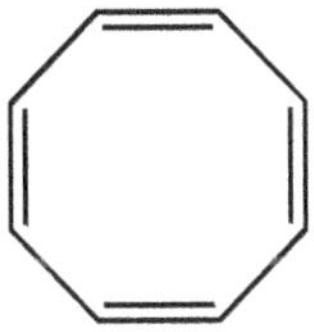

What are base+ structures in ABA terms?

2. Consider the information vectors for the base+ structures of lysozyme and ribonuclease A sequences. Estimate the projection angle for vectors based on the overlapping di-peptides, (base+(2)). Does this provide clues about the angle for tri-peptides?

3. The following represents the base+(i) structure for a much-researched protein.

A: 38, 9, 5, 13, 6, 4, 31, 2, 3, 4, 9
V: 3, 4, 11, 40, 33, 13
L: 7, 2, 18, 3, 8, 9, 12, 8, 3, 4, 13, 15, 11, 20, 2, 12
I: 9, 45, 11, 13, 8, 4, 1
P: 15, 51, 12, 20
F: 10, 3, 60, 17, 15, 13
W: 7
M: 55, 76, 11
G: 4, 10, 8, 2, 10, 30, 8, 1, 6, 41, 3, 5, 21, 3
S: 48, 7, 34, 16, 9, 27
T: 28, 3, 25
Y: 43
N: 120, 13
Q: 18, 65, 22, 3, 12, 24
D: 16, 24, 9, 7, 62, 4, 15
E: 12, 9, 11, 3, 11, 2, 5, 24, 2, 20, 4, 27, 12
K: 18, 8, 3, 2, 3, 6, 6, 1, 14, 1, 1, 8, 9, 2, 4, 16, 15, 7, 7
R: 108
H: 12, 12, 16, 17, 1, 11, 4, 22

What is the much-researched protein? What component is conspicuous by its absence?

4. Consider the base+(M) information for lysozyme. To how many proteins does the information apply?

5. Show that the number of distinguishable constituents for base+($n \geq 4$) scales as $N - n + 1$.

6. Revisit the sequence for ribonuclease A. Show that it contains 123 base+(2) components, 107 of which are distinguishable.

7. Revisit the sequence for human myoglobin. Show that it contains 153 base+(2) components, 115 of which are distinguishable. What is the expected number of distinguishable base+(3) components?

8. Consider a protein with $N \approx 1,000$. Estimate the number of distinguishable base+(2) components.

9. Does the scaling propensity of distinguishable base+(2) components place a limit on N? Estimate such a limit.

10. Construct the leucine (L) interval graph for lysozyme analogous to Figure Nine. Discuss the presence or absence of patterns in the graph?

11. Review the masses of the amino acids. Which ones are farthest from the average? Which ones are closest? Construct the Z-sum series for a sequence of choice. Is the scaling approximately linear?

12. Revisit the sequence for alpha synuclein, viz.

MDVFMKGLSKAKEGVVAAAEKTKQGVAEAAGKTKEGVLYVGSKTKEGV
VHGVATVAEKTKEQVTNVGGAVVTGVTAVAQKTVEGAGSIAAATGFVK
KDQLGKNEEGAPQEGILEDMPVDPDNEAYEMPSEEGYQDYEPEA

Construct a plot of the numerical information $M(k)$. Describe the plot in words.

13. Now construct numerical information $\Phi(k)$ for alpha synuclein. Is the plot nominally linear? What is the linear correlation coefficient?

14. An experiment was conducted in which the standard deviation of $Z(k)$ was measured for a set of $N = 154$ myoglobins across species. The results appear in Figure Nineteen. The upper dotted trace shows the results for random $N = 154$ strings of amino acids. Discuss the significance of the wild swings in the lower solid line trace observed for myoglobins. At which amino-acid sites are the deviations greatest?

15. Another experiment was conducted in which the standard deviation of $Z(k)$ was measured for a set of $N = 154$ proteins encoded by the human genome. The results are traced by the bold solid line in Figure Twenty. The dotted-line trace marks the results for random strings of amino acids. Discuss the significance of the similarities and differences of the two traces.

NOTES, SOURCES, AND FURTHER READING

The cumulative mass distributions can be taken further using Fourier transform and spectral analysis [8]. For molecular structure analysis, there exists no better source of lessons and inspiration than from the organic synthesis literature. Corey expounds on the importance of rigorous analysis of natural products structure [9]. The literature is indeed rich with development of deconstruction

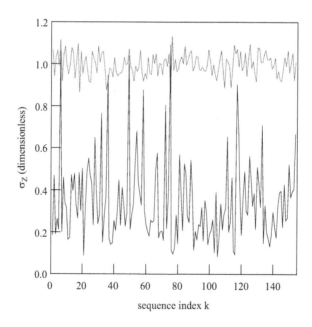

FIGURE NINETEEN   Standard deviation of $Z(k)$ observed for $N = 154$ myoglobins across species (solid line). The dotted trace above shows the results for $N = 154$ random strings.

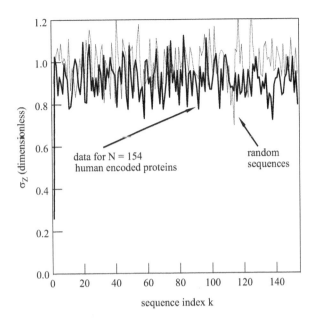

FIGURE TWENTY Standard deviation of $Z(k)$ observed for $N = 154$ proteins encoded by the human proteome (solid line). The dotted trace shows the results for $N = 154$ random strings.

techniques over decades: paying attention to nuances of organic structure without losing sight of the big picture. We recommend to the reader the landmark text by Fleming as well as more contemporary treatises [10–13]. Structure analysis, especially of the nearest-neighbor variety, is at center-stage in each. Proteins are as natural as any natural products. Curiously, they are the only class of natural products whose total synthesis recipe is a bonus of the knowledge of the primary structure. We also point out the structure analysis dedicated to the synthesis of enzyme mimics [14]. The bioinformatics literature is rich with computational approaches to protein structure analysis [15,16].

1. Bennett, C. H. 1982. Thermodynamics of computation—A review, *Int. J. Theor. Phys.* 21, 905.
2. Tewarson, R. P. 1973. *Sparse Matrices*, Academic Press, New York.
3. Sanger, F. 1945. The free amino acid groups of insulin, *Biochem. J.* 39(5), 507–515.
4. Appling, D. R., Anthony-Cahill, S. J., Mathews, C. K. 2016. *Biochemistry, Concepts and Connections*, Pearson Education, Hoboken, NJ.
5. Lundblad, R. L. 2006. *The Evolution from Protein Chemistry to Proteomics: Basic Science to Clinical Application*, CRC/Taylor & Francis, Boca Raton, FL.
6. Palzkill, T. 2002. *Proteomics*, Kluwer Academic Publishers, Boston, MA.
7. Balakrishna, N. 2003. *A Primer on Statistical Distributions*, Wiley, Hoboken, NJ.
8. Bracewell, R. 1965, *The Fourier Transform and Its Applications*, McGraw-Hill, New York.
9. Corey, E. J. 1967. General methods for the construction of complex molecules, *Pure Appl. Chem.* 14, 19.
10. Fleming, I. 1973. *Selected Organic Syntheses: A Guidebook for Organic Chemists*, John Wiley & Sons, London.
11. Starkey, L. S. 2012. *Introduction to Strategies for Organic Synthesis*, Wiley, Hoboken, NJ.
12. Serratosa, F. 1990. *Organic Chemistry in Action: The Design of Organic Synthesis*, Elsevier, Amsterdam.
13. Hudlicky, T., Reed, J. W. 2007. *The Way of Synthesis: Evolution of Design and Methods for Natural Products*, Wiley-VCH, Weinheim.
14. Gutsche, C. D. 1989. *Calixarenes*, Royal Society of Chemistry, Cambridge, UK.
15. Edwards, Y. J., Cottage, A. 2003. Bioinformatics methods to predict protein structure and function: A practical approach, *Mol. Biotechnol.* 23, 139.
16. Fiser, A. 2009. Comparative Protein Structure Modeling, in *From Protein Structure to Function with Bioinformatics*, D. J. Rigden, ed., Springer, Berlin, p. 57.

Seven Analysis of Sequences Internal to Proteins

The discussion of base+ information turns to the sequences internal to proteins. These obtain through the disulfide bonds between select cysteine sites in favorable environments. The bonds make for auxiliary covalent paths from N- to C-terminal as introduced in Chapter One. We examine sets of paths for archetypal proteins using information vectors and the base+(2) ideas of the previous chapter. These enable construction of pair distribution functions for true and false sets alike. The functions capture the sets and structures in visual terms and provide signatures for comparison. All of this brings principles and correspondences to light.

A PRELIMINARIES

Molecular structure is famously represented using graphs, such as for the lightweight amino acids:

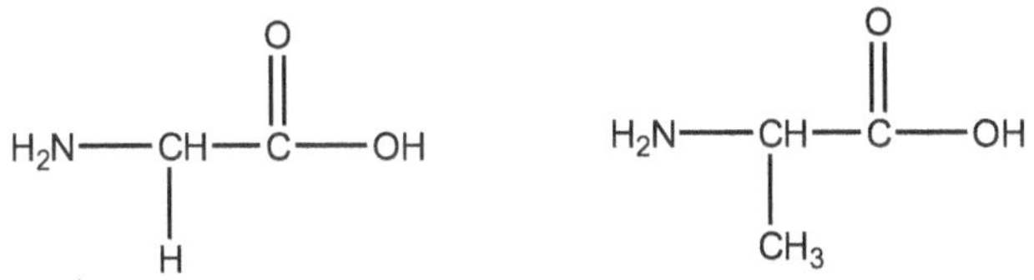

A graph communicates to the viewer the atomic components, functional groups, and points of stability [1]. It further emphasizes the electronic network and path structures. Revisit the representation for squalene from Chapter One:

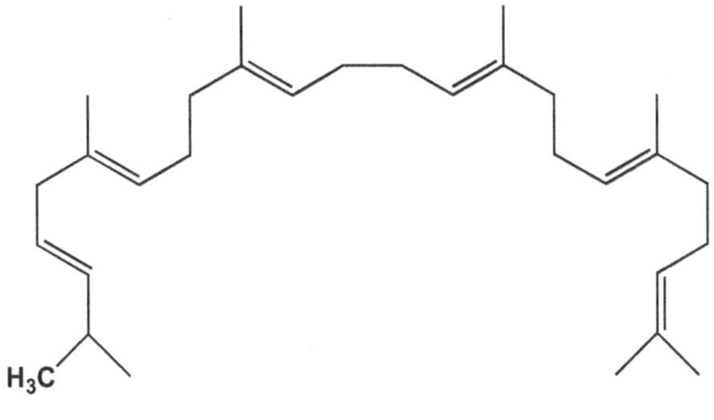

Then imagine parking on the methyl group marked in bold. As things stand, the graph offers a single route to the opposite end. The situation changes if squalene undergoes cyclization [2]:

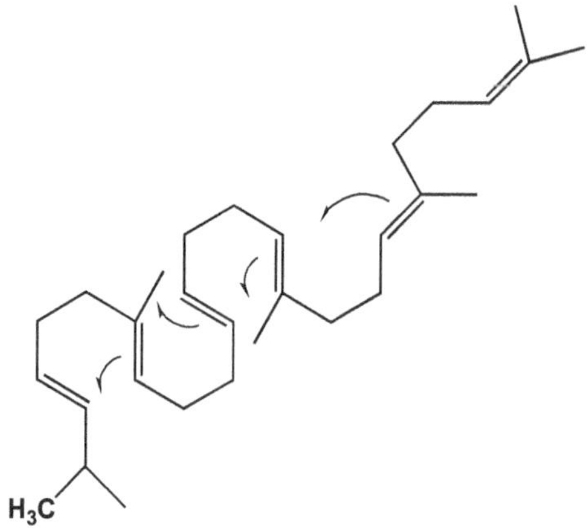

The steroid offers new travel options such as in bold:

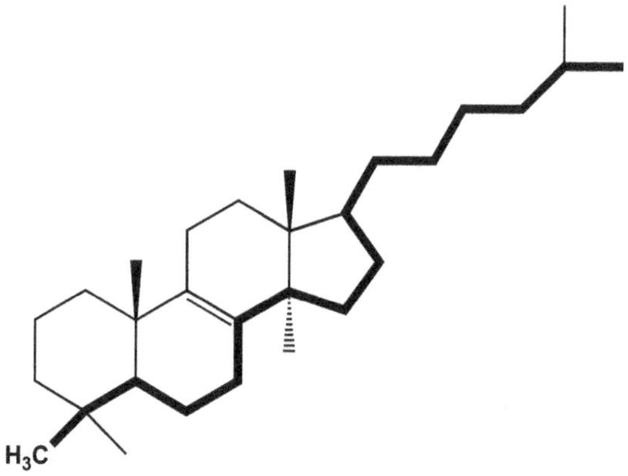

It is interesting to compare the path structures in the various classes of organic compounds: aromatics, steroids, and so forth. Graph theory formally addresses the mathematics and topology of molecular representations and has contributed a wealth of insights over the years [3–5].

Proteins are rich with path structures. A sequence represents the amino acids linked by peptide bonds, for example, for lysozyme:

KVFERCELARTLKRLGMDGYRGISLANWMCLAKWESGYNTRATNYNAGDRSTDY
GIFQINSRYWCNDGKTPGAVNACHLSCSALLQDNIADAVACAKRVVRDPQGIR
AWVAWRNRCQNRDVRQYVQGCGV

This tracks what can be termed the *principal* covalent path from the N- to C-terminal units. But there is much more to the story in oxidative environments [6]. The cysteines (**C**) indicated in bold allow disulfide bridges:

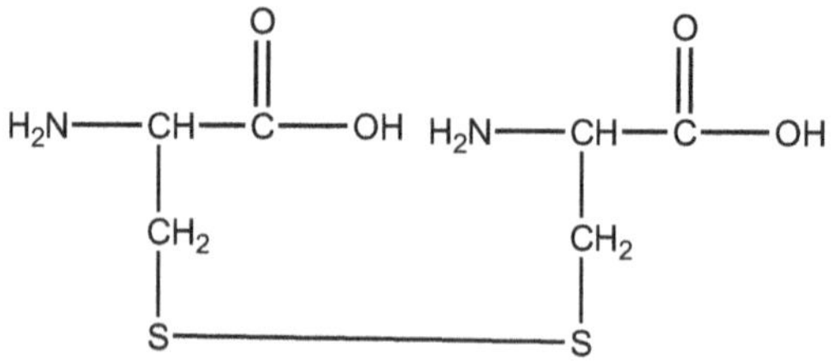

In lysozyme, the bridges link sites as follows, the k-indices noted in the superscripts:

$$^6C-^{128}C \quad ^{30}C-^{116}C \quad ^{65}C-^{81}C \quad ^{77}C-^{95}C$$

In turn, lysozyme carries *auxiliary* sequences—*alternative* paths for end-to-end travel with no switchbacks:

KVFER^{6}C^{128}CGV

KVFERCELARTLKRLGMDGYRGISLANWMCLAKWESGYNTRATNYNAGDRSTDYGI
FQINSRYW^{65}C^{81}CSALLQDNIADAVACAKRVVRDPQGIRAWVAWRNRCQNRDV
RQYVQGCGV

KVFERCELARTLKRLGMDGYRGISLANWM^{30}C^{116}CQNRDVRQYVQGCGV

KVFERCELARTLKRLGMDGYRGISLANWMCLAKWESGYNTRATNYNAGDRSTDYGI
FQINSRYWCNDGKTPGAVNA^{77}C^{95}CAKRVVRDPQGIRAWVAWRNRCQNRDVRQ
YVQGCGV

If we imagine parking on K at the N-terminal site, we have *five* choices for the journey to V at C-terminal. Each route presents base+ information according to criteria discussed in the preceding chapter. An auxiliary sequence carries more information than the base structure $A_{14}V_9L_8I_5\ldots$ $R_{14}H_1$, but less than the principal path $\leftrightarrow$ primary structure. Auxiliary sequences can be parsed and compared for alignments using bioinformatics software. Most interesting is what makes their information so special. Lysozyme expresses one of 105 possible path sets—ways to bridge the eight cysteines. Each set hosts the principal path plus auxiliaries. If we regard all the sets as equally plausible, then our knowledge of *the* set represents $\log_2(105) \approx 6.71$ bits of information. In contemplating the possibilities, we would observe that false sets, for example, generated by

$$^{77}C-^{116}C \quad ^6C-^{81}C \quad ^{65}C-^{128}C \quad ^{30}C-^{95}C$$

also feature four auxiliaries, namely:

KVFERCELARTLKRLGMDGYRGISLANWMCCAKRVVRDPQGIRAWVAWRNRCQNRDV
RQYVQGCGV

KVFERCCSALLQDNIADAVACAKRVVRDPQGIRAWVAWRNRCQNRDVRQYVQGCGV

KVFERCELARTLKRLGMDGYRGISLANWMCLAKWESGYNTRATNYNAGDRSTDYGIF
QINSRYWCCGV

KVFERCELARTLKRLGMDGYRGISLANWMCLAKWESGYNTRATNYNAGDRSTDYGIFQ
INSRYWCNDGKTPGAVNACCQNRDVRQYVQGCGV

The sets generated by

$$^{77}C-^{116}C \quad ^{65}C-^{81}C \quad ^{95}C-^{128}C \quad ^6C-^{30}C$$

$$^6C-^{30}C \quad ^{81}C-^{116}C \quad ^{95}C-^{128}C \quad ^{65}C-^{77}C$$

are just as false and confer nine and eleven auxiliaries, respectively. That leaves us 101 more sets to ponder, just for one enzyme! Clearly the sets are rich with information, but only one registers as true-blue. Question: could we have spotted it in advance or at least narrowed the possibilities?

Lysozyme is not an isolated case. Revisit the sequence for ribonuclease A [7]:

KETAAAKFERQHMDSSTSAASSSNYCNQMMKSRNLTKDRCKPVNTFVHESLADVQAVC
SQKNVACKNGQTNCYQSYSTMSITDCRETGSSKYPNCAYKTTQANKHIIVACE
GNPYVPVHFDASV

Disulfide bonds join the cysteines as follows:

$$^{26}C-^{84}C \qquad ^{40}C-^{95}C \qquad ^{58}C-^{110}C \qquad ^{65}C-^{72}C$$

Four auxiliary sequences are born in the process:

KETAAAKFERQHMDSSTSAASSSNYCCRETGSSKYPNCAYKTTQANKHIIVACEGNPYVP
VHFDASV

KETAAAKFERQHMDSSTSAASSSNYCNQMMKSRNLTKDRCKPVNTFVHESLADVQAV
CCEGNPYVPVHFDASV

KETAAAKFERQHMDSSTSAASSSNYCNQMMKSRNLTKDRCCAYKTTQANKHIIVACE
GNPYVPVHFDASV

KETAAAKFERQHMDSSTSAASSSNYCNQMMKSRNLTKDRCKPVNTFVHESLADVQAV
CSQKNVACCYQSYSTMSITDCRETGSSKYPNCAYKTTQANKHIIVACEGNPYVP
VHFDASV

Just as in lysozyme, there is a true-set of five paths—plus 104 false sets. For example, let the cysteines bridge as follows:

$$^{26}C-^{58}C \qquad ^{65}C-^{72}C \qquad ^{84}C-^{95}C \qquad ^{40}C-^{110}C$$

These bring about eight auxiliaries:

KETAAAKFERQHMDSSTSAASSSNYCCSQKNVACCYQSYSTMSITDCRETGSSKY
PNCAYKTTQANKHIIVACEGNPYVPVHFDASV

KETAAAKFERQHMDSSTSAASSSNYCCSQKNVACKNGQTNCYQSYSTMSITDCCAY
KTTQANKHIIVACEGNPYVPVHFDASV

KETAAAKFERQHMDSSTSAASSSNYCNQMMKSRNLTKDRCKPVNTFVHESLA
DVQAVCSQKNVACKNGQTNCYQSYSTMSITDCCAYKTTQANKHIIVACEGN
PYVPVHFDASV

KETAAAKFERQHMDSSTSAASSSNYCCSQKNVACKNGQTNCYQSYSTMSITDCRET
GSSKYPNCAYKTTQANKHIIVACEGNPYVPVHFDASV

KETAAAKFERQHMDSSTSAASSSNYCNQMMKSRNLTKDRCKPVNTFVHESL
ADVQAVCSQKNVACCYQSYSTMSITDCRETGSSKYPNCAYKTTQANKHIIVACEGN
PYVPVHFDASV

KETAAAKFERQHMDSSTSAASSSNYCNQMMKSRNLTKDRCCEGNPYVPVHFDASV

KETAAAKFERQHMDSSTSAASSSNYCNQMMKSRNLTKDRCKPVNTFVHESLADV
QAVCSQKNVACCYQSYSTMSITDCCAYKTTQANKHIIVACEGNPYVPVHFDASV

KETAAAKFERQHMDSSTSAASSSNY**CC**SQKNVA**CC**YQSYSTMSITD**CC**AYKTTQANKHI
IVA**C**EGNPYVPVHFDASV

These and other archetypes raise questions: (1) what is so special about *true*-path sets, as opposed to *false*? (2) How, if at all, is their information correspondent? Or does each protein present its own peculiar case? Cross-links and auxiliary paths help stabilize the folded configurations. Another question is then: what links and auxiliaries are antithetical to folds and functions?

The issues are freighted because multiple variants of a protein would favor the same path sets. The ones appropriate to

KVFERCELARTLKRLGMDGYRGISL<u>V</u>NWMCLAKWESGYNTRATNYNA<u>A</u>DRSTD
YGIFQINSRYWCNDGKTPGAVNACHLSCSALLQDNIA<u>E</u>AVACAKRVVRDP
QGIRAWVAWRNRCQNRDVRQYVQGCGV

would be identical to wild type lysozyme, the **V**, **A**, **E**-substitutions noted in bold underline. Thus we should approach this facet of proteins less via component identities (A, V, L, …, R, H) and more in family terms. The questions should be rephrased: what is special about true-path sets for given *classes* of proteins? Are there correspondences among the classes?

Auxiliary sequences and sets and their information form the subject of this chapter. Principles and correspondences will come to light by associating each path with a base+(2) information vector. As we will see, the vectors furnish abstract structures for true and false sets alike. They pave the way for visual signatures which are useful for comparing sets. This follows the spirit of Chapter Five where we explored proteomes using information vectors and pair distribution functions.

Can we anticipate correspondences in advance? The answer is affirmative. First, recall that proteins can present more than one sequence regardless of cysteines. For example, phospholipase is a homologous protein with identical chains *A* and *B* (cf. Chapter One) [8]:

NLYQFKNMIQCTVPSRSWWDFADYGCYCGRGGSGTPVDDLDRCCQVHDNCYNEAE
KISGCWPYFKTYSYECSQGTLTCKGGNNACAAAVCDCDRLAAICFAGAPYNDND
YNINLKARC

NLYQFKNMIQCTVPSRSWWDFADYGCYCGRGGSGTPVDDLDRCCQVHDNCYNEA
EKISGCWPYFKTYSYECSQGTLTCKGGNNACAAAVCDCDRLAAICFAGAPYN
DNDYNINLKARC

Because *A* and *B* carry the same message, their information vectors are collinear at all orders. They are positioned by evolution to be zero radians apart—as close as possible!

In contrast, proteins like insulin host different chains [9]:

GIVEQCCTSICSLYQLENYCN
FVNQHLCGSHLVEALYLVCGERGFFYTPKT

The *A*, *B*-messages diverge and the vectors reflect as much. The messages are positioned to be far apart.

Same- and different-chain systems foreshadow the correspondences of internal sequence sets. For homologous systems, the sequences reinforce each other by chemical affinity. Just as for an ideal solution, *A* presents a welcoming environment for identical twin *B*, and vice-versa. In contrast, mixed systems carry complementary messages. *A* cannot handle all the chemistry by itself; it needs *B*, and vice-versa. *A* combined with *B* resembles a non-ideal solution where the collective properties are removed from any single component.

These observations point us to the following: where proteins pose principal and auxiliary sequences, the information in each is either closely positioned or far removed. Evolution does not aim messages toward the middle grounds. The two chains in phospholipase do not look like:

NLYQFKNM<u>A</u>QCTVPSRS<u>Y</u>WDFADYGCYCGR<u>G</u>GSGTPVDDLDRCCQVHDNCYNEAEKIS
GCWPYFKTYS<u>Y</u>ECSQGTLTCKGGNNACAAAVCDCDRLAA<u>I</u>CFAGAPYNDNDYNIN
LKARC

NLYQFKNM<u>I</u>QCTVPSRS<u>W</u>WDFADYGCYCGR<u>S</u>GSGTPVDDLDRCCQVHDNCYNEAEKIS
GCWPYFKTYS<u>R</u>ECSQGTLTCKGGNNACAAAVCDCDRLAA<u>V</u>CFAGAPYNDN
DYNINLKARC

If the differences noted in bold underline were the case, *A* and *B* would be close information-wise, but not as close as they could be. Insulin does not look like:

GIVEQCCTSICSLYQLENYCN

GIVEQCCTSICSLYQLENYCNERGFFYTPKT

If so, *A* and *B* would not be as far removed as they could be. These ideas anticipate auxiliary sequences with information properties that are *either* highly coalescent *or* divergent. The middle grounds are avoided.

There is a third characteristic to consider, namely that we encounter multiple-sequence proteins infrequently. Phospholipase and insulin carry two chains each, not ten. This foreshadows auxiliaries to be minimalist in number: why should a biological system support ten when two can do the job? The rest of the chapter elaborates with archetypes providing the visuals and takeaways.

B   COVALENT PATHS, INFORMATION, AND SETS

A protein with disulfide bridges presents a set of paths, principal, and auxiliary. If we were to park at the N-terminal site, we would have a choice of routes to the opposite end. The paths are directed in that left-to-right travel is not the same as the reverse. Both the choices and mileage depend on the bridges, their number, and placement in the sequence. This applies to all bridge configurations, true and false.

Let us address the information by first learning how to count sets and paths. Let the number of bridge sets, true *plus* false, be denoted by Ω. An example will take us to the counting formula. We will use boldface to distinguish cysteine sites $\leftrightarrow$ **C** from ordinary font-C as in C-terminal.

We imagine a sequence with, say, ten C-sites. These allow $X = 10/2 = 5$ disulfide bridges. Let the sites be scattered throughout. We ignore their precise placement and index them as they occur in the N- to C-terminal direction, viz.

^{1}**C** ^{2}**C** ^{3}**C** ^{4}**C** ^{5}**C** ^{6}**C** ^{7}**C** ^{8}**C** ^{9}**C** ^{10}**C**

We then contemplate the information needed to identify the *true*-set of bridges. Obviously, the greater the possibilities Ω, the greater information we require, for the number of bits equates with $\log_2\Omega$. Note that we are describing a classic problem in protein structure analysis [10]. We address it by considering the options piecewise, beginning with ^{1}**C**–^{2}**C**. As things stand, ^{1}**C** has nine candidates for opposite-side $^?$**C**. Then knowledge of where the side is placed means information in quantity $\log_2\Omega = \log_2(9) \approx 3.17$ bits.

Let's say that experiments show ^{1}**C** to bond to ^{5}**C**. This would eliminate two sites from further consideration, viz.

$${}^{1}\mathcal{C} \quad {}^{2}C \quad {}^{3}C \quad {}^{4}C \quad {}^{5}\mathcal{C} \quad {}^{6}C \quad {}^{7}C \quad {}^{8}C \quad {}^{9}C \quad {}^{10}C$$

But then how much information is needed to learn about the ${}^{2}C$–${}^{?}C$ bridge? As we have $9 - 2 = 7$ candidates for ${}^{?}C$, the bits compute as $\log_2 \Omega = \log_2(7) \approx 2.81$ bits. Let experiments show ${}^{2}C$ to link with ${}^{6}C$. This would leave us with

$$\ {}^{1}\mathcal{C} \quad {}^{2}\mathcal{C} \quad {}^{3}C \quad {}^{4}C \quad {}^{5}\mathcal{C} \quad {}^{6}\mathcal{C} \quad {}^{7}C \quad {}^{8}C \quad {}^{9}C \quad {}^{10}C$$

The pattern should be apparent. With each bridge accounted for, the cost of learning new ones decreases. The price of ascertaining the true-set becomes:

$$\log_2(9) + \log_2(7) + \log_2(5) + \log_2(3) + \log_2(1)$$

$$\approx 9.88 \text{ bits}$$

This method of counting is simple and transferable. If a sequence presents 12 **C**-sites, knowledge of the true-set is information in amount:

$$\log_2 \Omega = \log_2(11 \cdot 9 \cdot 7 \cdot 5 \cdot 3 \cdot 1) \approx 13.3 \text{ bits}$$

It should then be clear why lysozyme and ribonuclease A present $\Omega = 105$: with eight **C**-sites, $\Omega = (8-1) \cdot (6-1) \cdot (4-1) \cdot (2-1)$. And caution: it may appear that sequences presenting adjacent **C**s require an adjustment to the counting formula. Further consideration tells us otherwise: just because two cysteines are adjacent means that they automatically host a disulfide bond. Look to phospholipase (1POB) from Chapter One for an example.

Disulfide bridges allow extra paths from N- to C-terminal which can be duly counted. It is confusing, however, to count via the letter strings for paths. It is easier if we let diagrams do the heavy lifting.

We imagine a protein devoid of bridges whereby there is *only* the principal path. Let us denote the bridge and path numbers by X and p, respectively. Then the *state* labeled by $X = 0$, $p = 1$ applies to *all* proteins in non-oxidative environments, *or* where the **C**-sites number one or zero. For example, collagens are typically devoid of **C**-sites while molecules like myoglobin can present a single **C**.

Let the $X = 0$, $p = 1$ state be represented by a horizontal solid line while a dotted arrow tracks *the* N- to C-terminal directed path:

Matters change with the addition of **C**-sites and cross-links. A pair of sites A and B takes the state from $X = 0$, $p = 1$ to $X = 1$, $p = 1 + 1 = 2$. To represent this, we modify the diagram to show the principal path (lower dotted arrow) as before and a shortcut tracked by the upper dotted arrows:

There is only one way to bridge the **C**-sites. This diagram applies to all proteins where $X = 1$, $\Omega = 1$, $p = 2$.

Now consider the case of four **C**-sites and two bridges. This applies to $X = 2$ systems where the path sets number $\Omega = 3 \cdot 1$. For diagrams, we have:

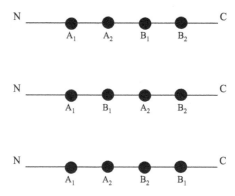

The subscripts indicate how the **C**-sites are linked: A_1 to B_1, and A_2 to B_2. The diagrams show principal paths in common, but pose different shortcuts. The top-most case offers three shortcuts, for example, whence $p = 3 + 1$, viz.

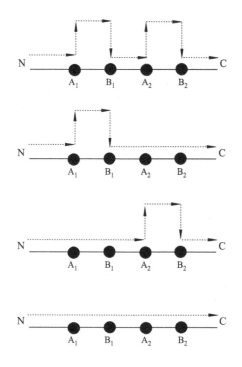

The other two graphs describe $X = 2$, $p = 3$ states. All total, with two **C**-sites present, we have $X = 2$, $\Omega = 3$, $p = 3, 3,$ *or* 4. For simplicity, we are confining attention to examples where the **C**-sites are not nearest neighbors.

The analysis is more tedious for $X > 2$. It proceeds by counting *all* the n-bridge paths with $0 \le n \le X$. Consider a diagram appropriate to $X = 3$:

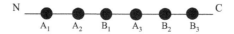

Disulfide bridges link A_i and B_i sites (note the identical subscripts). The principal path has *zero*-bridge-crossings: N-$A_1A_2B_1A_3B_2B_3$-C. The *single*-bridge paths are three in number: N-A_1B_1-C, N-A_2B_2-C, and N-A_3B_3-C. There is a single *two*-bridge path N-$A_1B_1A_3B_3$-C and *none* at higher order. The N- to C-travel options sum as follows:

$$p = 1 + 3 + 1 = 5.$$

An example of $X = 4$ is represented by

N ● ● ● ● ● ● ● ● C
    A_1 A_2 A_3 B_1 A_4 B_2 B_3 B_4

There is a *zero*-bridge principal while the *one*-bridge paths are N-A_1B_1-C, N-A_2B_2-C, N-$A_3B_3A_4B_4$-C. There is a *two*-bridge path N-$A_1B_1A_4B_4$-C and zero paths crossing three or more bridges. For this case,

$$p = 1 + 4 + 1 = 6.$$

This all seems tedious, but we should notice a pattern in the making. For a sequence presenting X bridges, minimum and maximum p equate with $1 + X$ and 2^X, respectively. The number of *two*-bridge paths arrives by counting B_iA_j pairs with $i \leq j$. The number of *three*-bridge paths equates with the $B_iA_j A_k$ triples with $i < j < k$. The number of *four*-bridge paths is obtained by counting *quadruples*: $B_iA_j A_k A_l$, $i < j < k < l$. And so forth. Total p is the sum of *all n*-bridge crossings where n ranges from zero to X. The details for $X = 3, 4$ are summarized as follows:

$$X = 3,\ \Omega = 5 \cdot 3 \cdot 1 = 15,\ p = 4,4,4,\ldots,6,\text{ or } 8$$

$$X = 4,\ \Omega = 7 \cdot 5 \cdot 3 \cdot 1 = 105,\ p = 5,5,5,\ldots,12,\text{ or } 16$$

Diagrams are doing the heavy lifting. But they are burdening us with too many integers. Fortunately, X, Ω, and p can be presented visually. Figure One illustrates critical information about $X = 3$ sets by plotting Ω versus p. Note how six of the possible $X = 3$ sets feature four N- to C-terminal paths, four sets present five paths, and so on. Just as important are the absences, for example, none of the possible sets provide seven transit routes. Note how the heights of the needles sum to the *total* number of sets $\Omega = 15$. Further, observe how Ω decreases with p. This is a clue and reflects that the greater the travel options, the more restrictions are placed on the routes. There is a tradeoff between the workable routes and the fine print. For example, *one* of the $X = 3$ sets allows eight choices for getting from one end of the protein to the other. But there is *only* one set so generous. In contrast, six sets allow four travel options. It is easy to think of analogies using airlines and airports.

Figure Two summarizes the properties for $X = 4$ proteins. The interpretation follows that of Figure One: twenty-four of the possible sets feature five paths—one principal and four auxiliaries. Eighteen sets present six paths, and so on. Zero sets allow 11, 13, and 14 travel options; one offers sixteen options. Note how the needle heights sum to total $\Omega = 105$. We again witness how an increase in transit routes arrives with fewer sets supplying the routes. If we want 16 ways to get from N- to C-terminal, there is only one set that fits the bill.

The ideas find purchase in archetypal proteins. We saw in Section A how the disulfide bridges in lysozyme and ribonuclease allow four auxiliary paths. For these enzymes, $p = 4 + 1$ by including the principal. Observe that the paths are the *minimum* allowed by $X = 4$. There are shortcuts from N- to C-terminal, but evolution is reserved in conferring them. Put another way, evolution's conferral of minimum p forces Ω to be the maximum possible. This is reflected in the figures: the tallest needles are allied with minimum p.

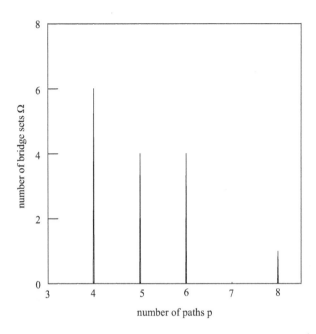

FIGURE ONE   Path information for $X = 3$ disulfide bridge sets in visual terms. The heights of the needles sum to *total* $\Omega = 15$.

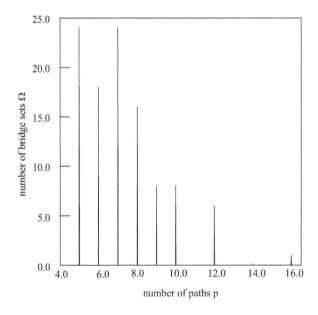

FIGURE TWO   Path information for $X = 4$ disulfide bridge sets in visual terms. The heights of the needles sum to *total* $\Omega = 105$.

An $X = 3$ case is presented by the anti-coagulent protein (PDB 1TAP), the cysteines noted in bold as usual [11]:

YNRL⁵**C**IKPRDWIDE¹⁵**C**DSNEGGERAYFRNGKGG³³**C**DSFWI³⁹**C**PEDHTGADYYSSYRD⁵⁵
CFNA⁵⁹**C**I

The bridges place as follows:

$$^5C-{}^{59}C \qquad {}^{15}C-{}^{39}C \qquad {}^{33}C-{}^{55}C$$

It can be shown then that the bridge placements confer $p = 3 + 1$: three auxiliary and one principal. Sequence YNRLC… and its folded structure diverge radically from lysozyme and ribonuclease A. Even so, all parties elect minimum p and maximum Ω. These are correspondences among small proteins with $X = 3$ or 4.

C   PATHS, INFORMATION VECTORS, AND PAIR DISTRIBUTIONS

Revisit the representations of organic compounds, for example, amino acids:

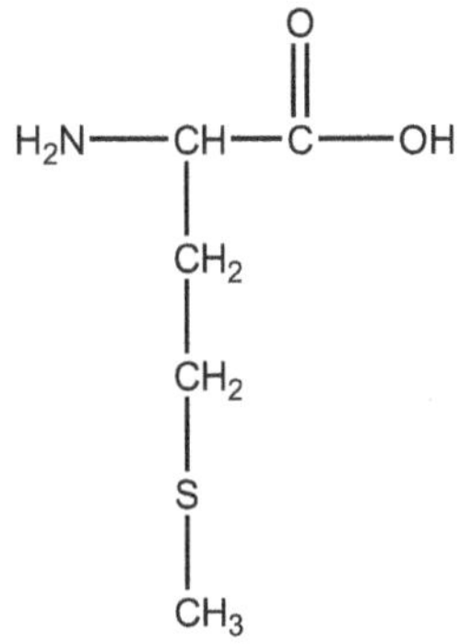

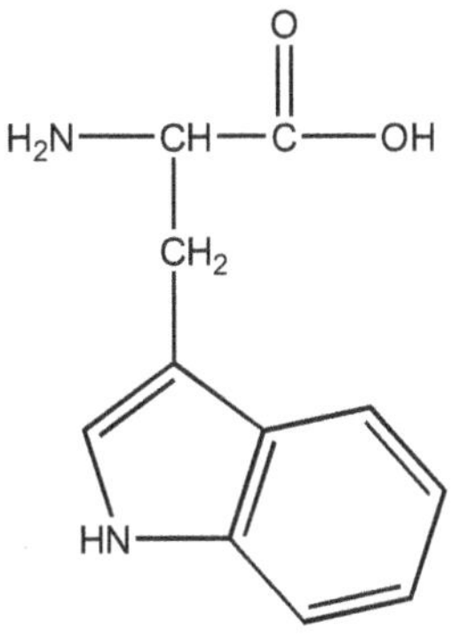

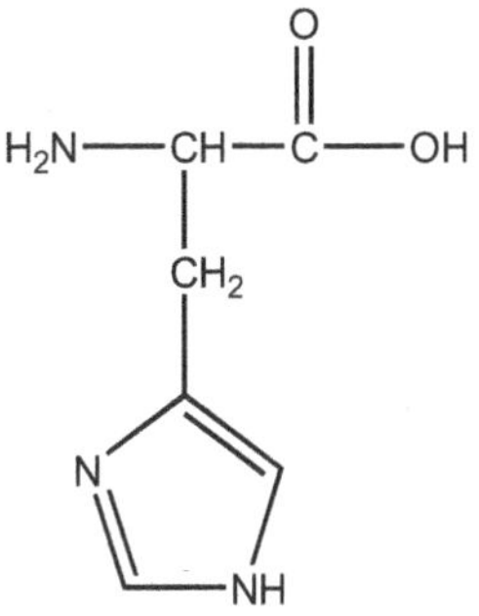

Besides paths and networks, the graphs emphasize the role of nearest-neighbor interactions: C=O, C–C, C–H, etc., which are overriding in determining configurations and chemical behavior. We saw in Chapter Six how the base+(2) structures (di-peptides) of proteins do likewise: KV, VF, FE, etc. The structures make for tedious lists, especially if we include ones for auxiliary sequences. To simplify the analysis, we appeal to multi-component information vectors. We further leverage the information of partitions and families and view the amino acids by their type: acidic (a), basic (b), polar (p), and non-polar (n). In so doing, our analysis of cross-linked proteins encompasses whole classes, not just individuals.

Then consider lysozyme in abnp-terms (cf. Chapter Three):

**bnnabpannbpnbbnpnappbpnpnnpnnpnnbnapppppbnppppnpabppappnnpnppbpnppapbpnp
nnpnpbnpppnnnpapnnannnpnbbnnbanppnbnnnnnbpbpppbanbppnppppn**

Further consider the auxiliaries from Section A:

<u>bnnabpppn</u>

bnnabpannbpnbbnpnappbpnpnnpnnpnnbnapppppbnppppnpabppappnnpnppbpnpppnnnpapnna
nnnpnbbnnbanppnbnnnnnbpbpppbanbppnppppn

bnnabpannbpnbbnpnappbpnpnnpnnppppbanbppnppppn

bnnabpannbpnbbnpnappbpnpnnpnnpnnbnapppppbnppppnpabppappnnpnppbpnppapbpnpnnpn
ppnbbnnbanppnbnnnnnbpbpppbanbppnppppn

The sequences carry information much reduced from 20-letter representations. Their base+(2) vectors point with strengths f_i lie along axes labeled aa, ab, an, ap, ba, bb, bn, bp, na, nb, np, nn, pa, pb, pn, and pp. Following Chapter Six, the strengths obtain by parsing each sequence for overlapping di-peptides. Five base+(2) assemblies arrive as follows:

$$\mathbf{(bb)_2(ba)_2(bp)_9(bn)_7(ab)_2(aa)_0(ap)_5(an)_4(pb)_7(pa)_5(pp)_{21}(pn)_{19}(nb)_8(na)_4(np)_{17}(nn)_{17}}$$

$$\underline{(bb)_0(ba)_0(bp)_1(bn)_1(ab)_1(aa)_0(ap)_0(an)_0(pb)_0(pa)_0(pp)_2(pn)_1(nb)_0(na)_1(np)_0(nn)_1}$$

$$(bb)_2(ba)_2(bp)_8(bn)_6(ab)_2(aa)_0(ap)_4(an)_4(pb)_5(pa)_4(pp)_{20}(pn)_{16}(nb)_8(na)_4(np)_{13}(nn)_{16}$$

$$(bb)_1(ba)_1(bp)_4(bn)_2(ab)_1(aa)_0(ap)_1(an)_2(pb)_2(pa)_1(pp)_8(pn)_7(nb)_3(na)_2(np)_5(nn)_4$$

$$(bb)_2(ba)_2(bp)_9(bn)_6(ab)_2(aa)_0(ap)_4(an)_3(pb)_6(pa)_4(pp)_{20}(pn)_{17}(nb)_8(na)_3(np)_{14}(nn)_{12}$$

The subscripts show how the base+(2) terms diverge, both significantly and obtusely. What do we take from the information?

Figures can make the important points. Figure Three shows the di-peptide fractions for the principal path in lysozyme, so highlighted above in bold. The most substantive fractions lie with pp and pn while np and nn lag not far behind. The remaining ones are scattered along bb, ba, etc., with zero strength in aa. Note the asymmetries: the pn-strengths do not match those of np; bp-strengths $\neq$ pb, etc. There is an absence as zero strength is expressed along the aa axis. The vector space is formally 16D. But the principal path vector is silent along one of the dimensions.

Compare this with Figure Four showing the component fractions for the shortest auxiliary path: KVFER^{6}C^{128}CGV $\leftrightarrow$ bnnabpppn. Note how the strengths register zero along several axes. The figure emphasizes how the principle and auxiliaries carry markedly different messages: their information vectors point not just a little differently. From this perspective, we see how lysozyme harbors traits of a heterologous protein.

It is straightforward to proceed quantitatively. The caveat is that the unit number N becomes a new variable, depending on the covalent path. For example, bnnabpppn presents $N = 9$ with $N - 1 = 8$ di-peptides in base+(2). Several vector components have zero strength as shown in the figure, for example,

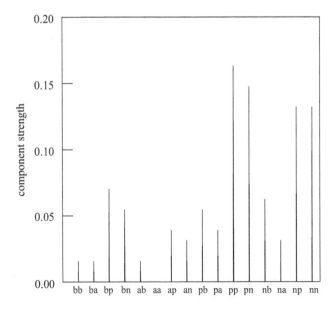

FIGURE THREE Component strengths of the base+(2) information vector in abnp-terms for the principal path of lysozyme.

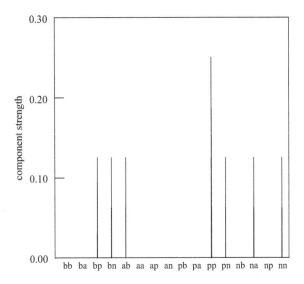

FIGURE FOUR Component strengths of the base+(2) information vector for the shortest auxiliary path in lysozyme.

$$bb : \sqrt{\frac{0}{9-1}} \qquad ba : \sqrt{\frac{0}{9-1}} \qquad ap : \sqrt{\frac{0}{9-1}} \qquad an : \sqrt{\frac{0}{9-1}}$$

Some of the non-zero components look like:

$$bp : \sqrt{\frac{1}{9-1}} \qquad bn : \sqrt{\frac{1}{9-1}} \qquad ab : \sqrt{\frac{1}{9-1}} \qquad pp : \sqrt{\frac{2}{9-1}}$$

Dot products establish the vector projections as usual while angles follow from inverse cosine functions. Such calculations can be directed to all the path-pairs presented by a set. The result is an *abstract* structure and we can fashion structures for true and false sets alike. If a set presents p number of paths, the structure features $p(p-1)/2$ non-zero angles. The *true* bridges in lysozyme manifest $p = 1 + 4$. The abstract structure features five vectors and angles that number:

$$\frac{p \cdot (p-1)}{2} = \frac{5 \cdot (5-1)}{2} = 10$$

The same number applies to *all* $p = 5$ sets. For $X = 4$ systems, maximum $p = 2^X = 2^4 = 16$ whence 120 angles are required to describe the abstract structure.

Abstract structures provide 1,000-word pictures. This is because the angles can be packaged as a pair distribution function $g(\theta_{\alpha\beta})$. Every vector α in a set points at angle zero radians with respect to itself *and* can serve as an origin. In turn, every neighbor vector β projects at some angle $\theta_{\alpha\beta} > 0$. In effect, β lies at a precise, measureable distance from the origin. Then to assemble abstract structures, we take $g(\theta_{\alpha\beta}) = 1$ for $\alpha \neq \beta$ and zero otherwise. We should think of structures in probability terms: $g(\theta_{\alpha\beta})$ represents the *conditional* probability of finding a neighbor at $\theta_{\alpha\beta}$, given a vector fixed at the origin. And this should ring bells: pair distributions $G(r)$ were used to represent the internal structures of proteins in Chapter One. We called them for duty again in Chapter Five to explore proteome topology. Suffice to say that pair distributions apply to any set of internal sequences, whether true or false. They are especially inviting in the present application because $g(\theta_{\alpha\beta})$ is restricted to values zero and one. We either find a vector at angle $\theta_{\alpha\beta}$ or we don't!

Figure Five illustrates $g(\theta_{\alpha\beta})$ for the principal and auxiliary sequences in lysozyme. We observe how the vectors for sequences aggregate in clusters. There is a far-end crowd near $\theta_{\alpha\beta} \approx 0.70$ radians and two near-crowds at $\theta_{\alpha\beta} < 0.14$ radians. This tells us that lysozyme carries both homologous and heterologous traits on account of its cross-links. It is as if a few paths are selectively attracted to each other by their messages. At the same time, paths are repelled selectively by messages.

Let us then view the angle distances through a second lens. This obtains in Figure Six which presents the $g(\theta_{\alpha\beta})$-information as a checkerboard. The pixel density in each square reflects the distance $\theta_{\alpha\beta}$ between sequence vectors. The row and column labels refer to principal and auxiliaries in the order listed in Section A. That the board is blank along the left-to-right ascendant diagonal reflects $\theta_{\alpha\beta} = 0$ for $\alpha = \beta$; the symmetry is consistent with $\theta_{\alpha\beta} = \theta_{\beta\alpha}$. What do we take home? It is that the shortest path KVFER^{6}C^{128}CGV $\leftrightarrow$ bnnabppp is something of an outlier. Its vector points not just a little differently from the neighbors. Further, the sibling paths sponsor vectors that point more or less uniformly. Needle-plots for $g(\theta_{\alpha\beta})$ and their corresponding checkerboards and provide visual signatures. In Figures Five and Six we have the signatures for *true*-path lysozyme.

Then what do false signatures look like? Figures Seven and Eight show examples derived from the lysozyme false set presented in Section A:

$$^{77}C\text{--}^{116}C \qquad ^{6}C\text{--}^{81}C \qquad ^{65}C\text{--}^{128}C \qquad ^{30}C\text{--}^{95}C$$

The set features one principal and four auxiliaries with sequences specified in Section A. We observe in Figure Seven how the $g(\theta_{\alpha\beta})$-needles cluster together instead of sequestering. There is no near- or far-crowd, but rather one mixed crowd with no outliers. Figure Eight follows with the second-lens view. We notice here how the pixel densities are more evenly distributed: there is conformity in the path messages and how their vectors point.

We can illustrate the signatures for the 103 remaining false sets. To save paper and screens, however, we simply look to the average $\langle\theta_{\alpha\beta}\rangle$ and variance $\sigma^2 = \langle\theta_{\alpha\beta}^2\rangle - \langle\theta_{\alpha\beta}\rangle^2$ for quick summaries. These are the probability distribution moments for the angles marked by $g(\theta_{\alpha\beta})$ and paint the signatures in broad-brush terms. Figure Nine subsequently shows $\langle\theta_{\alpha\beta}\rangle$ for $\Omega = 105$ sets for lysozyme. Each point is placed by (1) constructing the N- to C-terminal paths and vectors, and (2) computing

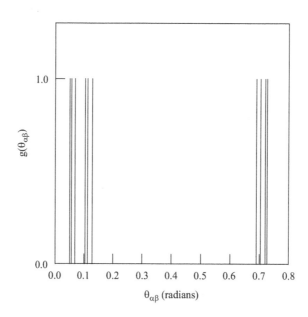

FIGURE FIVE Pair distribution function for the *true*-set of base+(2) information vectors for lysozyme.

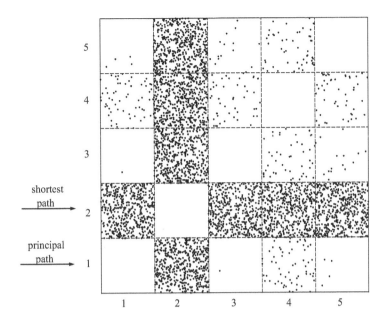

FIGURE SIX   Pair distribution for the *true*-set of lysozyme viewed through a second lens. The pixel density in each square is proportional to the angle between base+(2) information vectors. The rows and columns are numbered according to the lysozyme paths presented in Section A.

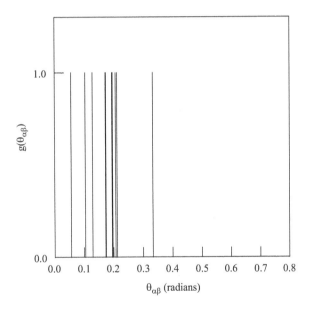

FIGURE SEVEN   Pair distribution function for the *false* set of base+(2) information vectors for lysozyme. The false set corresponds to that presented in Section A.

all the angles followed by moments. The large filled circle is placed by the *true*-path set while the small open circles are placed by false sets.

There are two features of note. First, *true*-set $\langle \theta_{\alpha\beta} \rangle$ does not fall just anywhere. Rather it places in the uppermost tier of a multi-modal distribution: it places fourth highest of the 24 $p = 5$ sets—and not too far from the leaders!

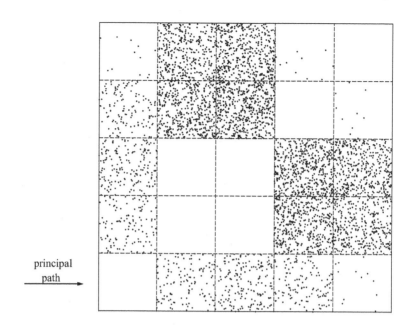

FIGURE EIGHT   Second-lens view of pair distribution for a false set of lysozyme. The pixel density is proportional to the angle between information vectors. The false set corresponds to that presented in Section A and Figure Seven.

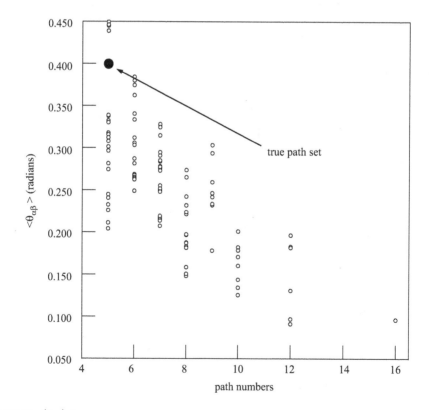

FIGURE NINE $\langle \theta_{\alpha\beta} \rangle$ for the pair distribution functions for all $\Omega = 105$ path sets for lysozyme. The point for the true-set is marked by the large filled circle. The small open circles are placed by false sets.

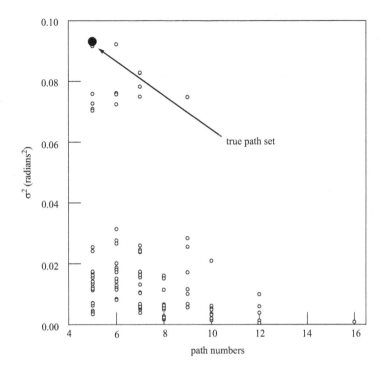

FIGURE TEN $\sigma^2 = \left\langle \theta_{\alpha\beta}^2 \right\rangle - \left\langle \theta_{\alpha\beta} \right\rangle^2$ for the pair distribution functions for all $\Omega = 105$ path sets for lysozyme. The point for the true-set is marked by the large filled circle. The open circles are placed by false sets.

The second feature involves the variances σ^2 in Figure Ten. Note how the point for true-set lysozyme places the *highest* of all the possible sets. This shows the impact of the outlier path as it dramatically expands the probability and information structure. The path structure is not unlike mixed systems like insulin, where the sequences are far removed message-wise. The extreme placement of σ^2 further echoes proteome traits. Recall that proteome volumes are significantly expanded compared with independent random sequences. The expansion acutely reflects natural selection aimed at maximum diversity.

The signature of true-path lysozyme is emulated by multiple proteins. This illuminates new correspondences of sequence information. For example, look to two more $X = 4$ systems: the sweet protein brazzein (1BRZ) and a plant toxin (1GPS).

The principal path (total sequence) of brazzein is represented by

EDKCKKVYENYPVSKCQLANQCNYDCKLDKHARSGECFYDEKRNLQCICDYCEY

The true-path set derives from the following disulfide bonds [12]:

$$^4\text{C}-^{52}\text{C} \quad ^{16}\text{C}-^{37}\text{C} \quad ^{22}\text{C}-^{47}\text{C} \quad ^{26}\text{C}-^{49}\text{C}$$

The plant toxin has principal path [13]:

KICRRRSAGFKGPCMSNKNCAQVCQQEGWGGGNCDGPFRRCKCIRQC

The bridges place as follows:

$$^3\text{C}-^{47}\text{C} \quad ^{14}\text{C}-^{34}\text{C} \quad ^{20}\text{C}-^{41}\text{C} \quad ^{24}\text{C}-^{43}\text{C}$$

Pictures take us the rest of the way. Figure Eleven shows $\langle \theta_{\alpha\beta} \rangle$ versus p for the $\Omega = 105$ path sets for brazzein. The point for the true-set is placed by the large filled circle while false sets place the open circles. Just as with lysozyme, the minimum p is expressed along with near-maximum expansion of the angle set. The messages carried by the sequences are few in number, but selected to be far apart information-wise. The expansion places $\langle \theta_{\alpha\beta} \rangle$ at the apex of the $p = 5$ possible sets.

The checkerboard signature for true-set brazzein appears in Figure Twelve. Note its resemblance to true-set lysozyme in Figure Six. Both signatures reflect systems containing one outlier and four conformant paths.

Figure Thirteen illustrates $\langle \theta_{\alpha\beta} \rangle$ versus p for the plant toxin sets. The point for the true-set is placed by the large filled circle while open circles apply to false sets. Again, minimum p is expressed by the protein along with near-maximum expansion of the angle set. The bridges and auxiliary paths place $\langle \theta_{\alpha\beta} \rangle$ (again!) at the apex of the $p = 5$ possibilities.

Figure Fourteen completes the story by showing the checkerboard. It should not surprise that the signature reminds us of lysozyme and brazzein.

But lysozyme's signature is not universal. Contrasting cases are found when the principal path vector strongly favors one component axis. For example, revisit the sequence for $X = 4$ ribonuclease A [7]:

KETAAAKFERQHMDSSTSAASSSNYCNQMMKSRNLTKDRCKPVNTFVHESLAD
VQAVCSQKNVACKNGQTNCYQSYSTMSITDCRETGSSKYPNCAYKTTQANKHIIVACEG
NPYVPVHFDASV

Figure Fifteen shows the strengths for the information vector. The arrow draws our attention to the pp-needle. Contrast this with Figure Three for lysozyme. The pp-strength leaves the other needles shortchanged and scattered.

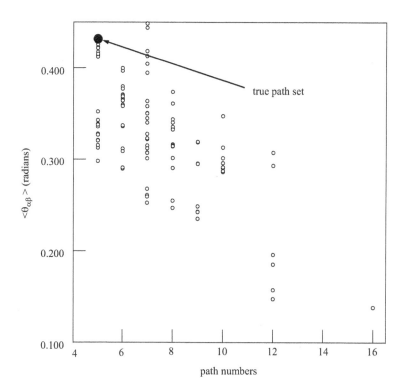

FIGURE ELEVEN $\langle \theta_{\alpha\beta} \rangle$ for the pair distribution functions for all $\Omega = 105$ path sets for brazzein (PDB 1BRZ). The point for the true-set is marked by the large filled circle. The open circles are placed by false sets.

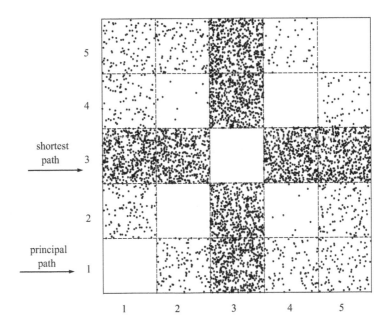

FIGURE TWELVE   Pair distribution function for the *true*-set of brazzein in checkerboard terms. The number of pixels in each square of the checkerboard is proportional to the angle between base+(2) information vectors.

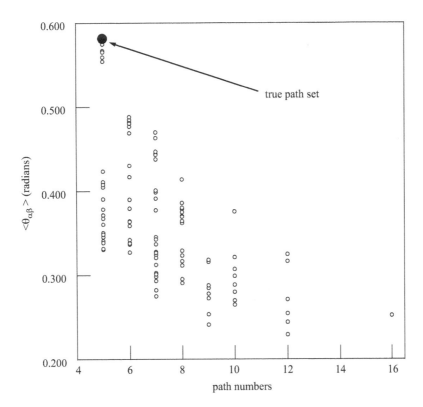

FIGURE THIRTEEN $\langle \theta_{\alpha\beta} \rangle$ for the pair distribution functions for all $\Omega = 105$ path sets for plant toxin (PDB 1GPS). The point for the true-set is marked by the large filled circle. The small open circles are placed by false sets.

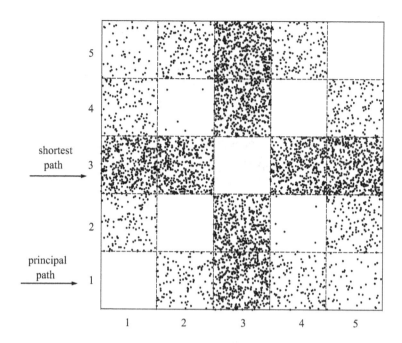

FIGURE FOURTEEN   Pair distribution function for the *true*-set of plant toxin in checkerboard terms. The number of pixels in each square of the checkerboard is proportional to the angle between base+(2) information vectors.

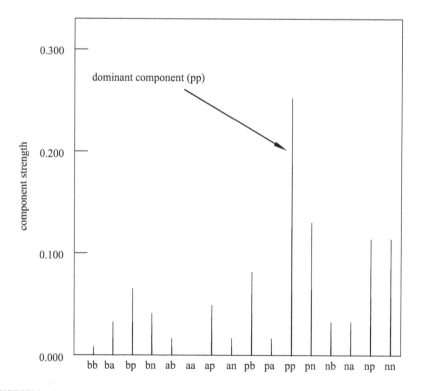

FIGURE FIFTEEN   Component strengths of the base+(2) information vector for the principal path of ribonuclease A (PDB 1FS3).

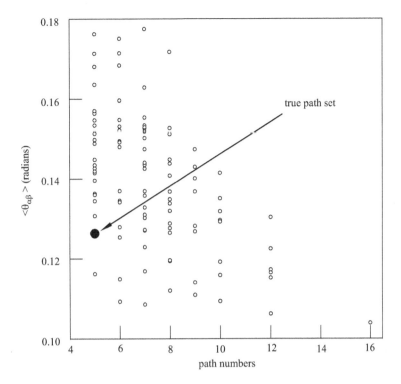

FIGURE SIXTEEN $\langle\theta_{\alpha\beta}\rangle$ for the pair distribution functions for all $\Omega = 105$ path sets for ribonuclease A. The point for the true-set is marked by the large filled circle. The small open circles are placed by false sets.

Figure Sixteen shows the impact on internal paths by way of $\langle\theta_{\alpha\beta}\rangle$. As with lysozyme, brazzein, and plant toxins, minimum $p = 5$ set is selected. But now observe how the first moment for the angles is pulled toward the bottom rungs. The strong information bias *contracts* the path structure. This makes the true-set substantially compressed compared with false sets.

Figure Seventeen shows the second-lens view. Note how the pixels are dispersed fairly evenly. This reflects the high conformity of the path messages. Ribonuclease A lacks an outlier as all five sequences are close information-wise.

The scorpion neurotoxin (1SNB) follows much the same playbook. The principal sequence is given by [14]:

GRDAYIADSENCTYFCGSNPYCNDVCTENGAKSGYCQWAGRYGNACYCIDLPASERIK
EGGRCG

The bridges place as follows:

$$^{12}C-^{63}C \qquad ^{16}C-^{36}C \qquad ^{22}C-^{46}C \qquad ^{26}C-^{48}C$$

Figure Eighteen shows the component strengths for the principal vector. As for ribonuclease A, the arrow draws attention to the pp-component.

The information bias compresses the angles marked by $g(\theta_{\alpha\beta})$ as seen via $\langle\theta_{\alpha\beta}\rangle$ in Figure Nineteen. We observe how $\langle\theta_{\alpha\beta}\rangle$ falls at the bottom of the $p = 5$ menu. Figure Twenty shows the checkerboard signature. Here things are a little mixed. Whereas the needle-plot for $g(\theta_{\alpha\beta})$ is highly compressed, the allied checkerboard indicates one of the paths to be an outlier.

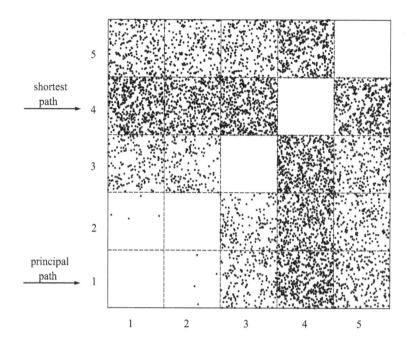

FIGURE SEVENTEEN Pair distribution function for the *true*-set of ribonuclease A in checkerboard terms. The number of pixels in each square of the checkerboard is proportional to the angle between base+(2) information vectors.

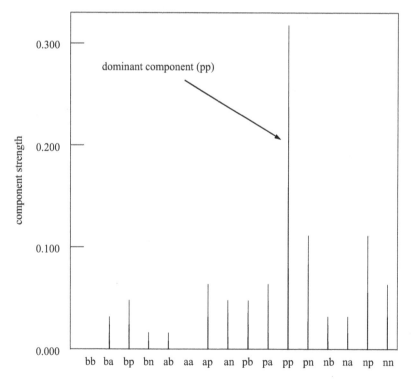

FIGURE EIGHTEEN Component strengths of the base+(2) information vector for the principal path of scorpion neurotoxin (PDB 1SNB).

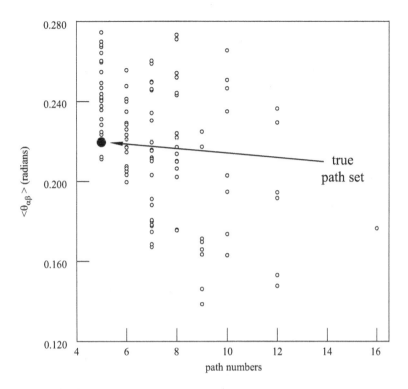

FIGURE NINETEEN $\langle \theta_{\alpha\beta} \rangle$ for the pair distribution functions for all $\Omega = 105$ path sets for scorpion neurotoxin. The point for the true-set is marked by the large filled circle. The small open circles are placed by false sets.

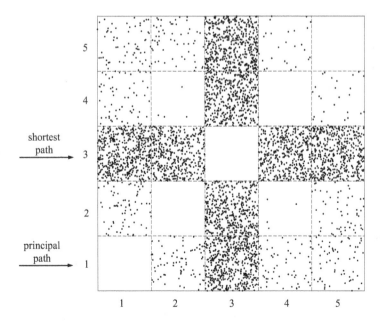

FIGURE TWENTY Pair distribution function for the *true*-set of scorpion neurotoxin in checkerboard terms. The number of pixels in each square of the checkerboard is proportional to the angle between base+(2) information vectors.

D PATHS, INFORMATION VECTORS, AND DEVIATIONS

The information of the sequences internal to proteins demonstrates multiple correspondences. Where do correspondences fray and go astray?

To answer the first question, recall that auxiliary sequences are subsidiary to the cysteine placements. A **C** at position k carries different information if switched with, say, alanine at j. Evolution confers all amino-acid positions and the information is impacted if the boat is rocked. We can appreciate this in lysozyme if we exchange the cysteines with their left- or right-side neighbors. For example, left-side (i.e., N-terminal side) exchanges give:

KVFE**CR**ELARTLKRLGMDGYRGISLANW**CM**LAKWESGYNTRATNYNAGDRSTDY
GIFQINSRY**CW**NDGKTPGAVN**CA**HL**CS**SALLQDNIADAV**CA**AKRVVRDPQGIRAWVAW
RN**CR**QNRDVRQYVQ**C**GGV

The true-set of disulfide bridges is rendered false:

$$^{6-1}C-^{128-1}C \qquad ^{30-1}C-^{116-1}C \qquad ^{65-1}C-^{81-1}C \qquad ^{77-1}C-^{95-1}C$$

Right-side (C-terminal side) exchanges yield:

KVFER**EC**LARTLKRLGMDGYRGISLANWM**L**CAKWESGYNTRATNYNAGDRSTDYG
IFQINSRYW**NC**DGKTPGAVNA**HC**LSS**C**ALLQDNIADAVA**AC**KRVVRDPQGIRAW
VAWRNR**QC**NRDVRQYVQGG**C**V

The disulfide bridges become:

$$^{6+1}C-^{128+1}C \qquad ^{30+1}C-^{116+1}C \qquad ^{65+1}C-^{81+1}C \qquad ^{77+1}C-^{95+1}C$$

Figure Twenty-One illustrates the impact of left-shifted **C**-sites. The small open circles are placed by $\Omega - 1 = 104$ bridge sets for the corrupt sequence. The half-filled large circle placed by the true-made-false set while the 100% filled circle corresponds to pristine $\leftrightarrow$ true-blue lysozyme. Note how the variance of angles marked by $g(\theta_{\alpha\beta})$ is diminished slightly when the principal sequence is perturbed. In other words, when the sequence carries slightly deviant information, the natural bias toward expanded $g(\theta_{\alpha\beta})$ is weakened.

Figure Twenty-Two shows the results for right-shifted **C**-sites. As in the previous figure, the open circles are placed by $\Omega - 1 = 104$ bridge sets for the corrupted sequence. The half-filled circle is placed by true-made-false set while the 100% filled circle corresponds to pristine lysozyme. Again, the variance of angles marked by $g(\theta_{\alpha\beta})$ decreases when **C**-sites are steered from their natural position. If the sequence does not follow the correct grammar, the information is not as message-diverse.

Minimum p is the overriding rule for small $X = 4$ proteins with $N \leq 150$. Further, the information structures are highly expanded *or* highly compressed. Things become mixed, however, when it comes to larger proteins. For example, consider the sequence for $X = 4$ cathepsin A (4CIA) with $N = 455$ [15]:

SRAPDQDEIQRLPGLAKQPSFRQYSGYLKGSGSKHLHYWFVESQKDPENSPV
VLWLNGGPGCSSLDGLLTEHGPFLVQPDGVTLEYNPYSWNLIANVLYLESPAGVGF
SYSDDKFYATNDTEVAQSNFEALQDFFRLFPEYKNNKLFLTGESYAGIYIPTLAVLV
MQDPSMNLQGLAVGNGLSSYEQNDNSLVYFAYYHGLLGNRLWSSLQTH
CCSQNK**C**NFYDNKDLE**C**VTNLQEVARIVGNSGLNIYNLYAPCAGGVPSHFRYEKD
TVVVQDLGNIFTRLPLKRMWHQALLRSGDKVRMDPP**C**TNTTAASTYLNNPYVRKAL
NIPEQLPQWDM**C**NFLVNLQYRRLYRSMNSQYLKLLSSQKYQILLYNGDVDMA**C**NFM

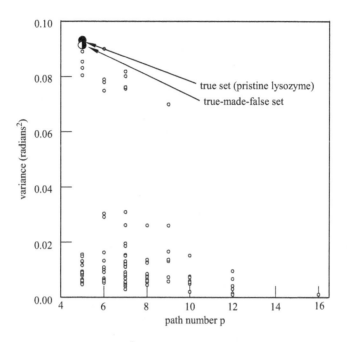

FIGURE TWENTY-ONE $\sigma^2 = \left\langle \theta_{\alpha\beta}^2 \right\rangle - \left\langle \theta_{\alpha\beta} \right\rangle^2$ for the pair distribution functions for all $\Omega = 105$ path sets for corrupt lysozyme. The point for the uncorrupted sequence is placed by the large filled circle. The small open circles are placed by false bridge sets. The half-filled large circle is placed by the true-made-false set from left-shifting **C**-sites.

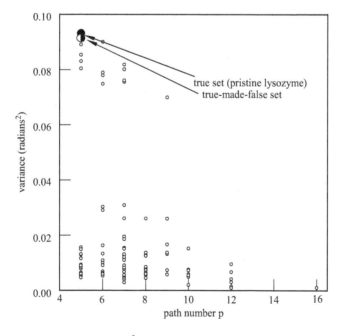

FIGURE TWENTY-TWO $\sigma^2 = \left\langle \theta_{\alpha\beta}^2 \right\rangle - \left\langle \theta_{\alpha\beta} \right\rangle^2$ for the pair distribution functions for all $\Omega = 105$ path sets for corrupt lysozyme. The point for the uncorrupted sequence is marked by the large filled circle. The open circles are placed by false bridge sets. The half-filled circle is placed by the true-made-false set from right-shifting **C**-sites.

GDEWFVDSLNQKMEVQRRPWLVKYGDSGEQIAGFVKEFSHIAFLTIKGAGHMVPT
DKPLAAFTMFSRFLNKQPYE

The disulfide bridges places as follows:

$$^{60}C-^{334}C \qquad ^{212}C-^{228}C \qquad ^{213}C-^{218}C \qquad ^{253}C-^{303}C$$

Figure Twenty-Three shows the variance of angles marked by $g(\theta_{\alpha\beta})$ for all possible path sets. The true-set variance is marked by the large filled circle while small open circles mark false sets. Observe how true marks the apex for $p = 7$. Minimum $p = 5$ is *not* the case as for smaller proteins like lysozyme and ribonuclease A. But the $g(\theta_{\alpha\beta})$ distribution *is* the most expanded of the $p = 7$ choices. Figure Twenty-Four is the follow-up showing the corresponding checkerboard. The signature reminds us somewhat of lysozyme as one stripe is much more prominent than the others. One of the auxiliaries is something of an outlier, while its siblings are conformist.

The sequence for $X = 5$ chymotrypsinogen (1EX3, $N = 245$, $\Omega = 945$) is [16]:

CGVPAIQPVLSGLSRIVNGEEAVPGSWPWQVSLQDKTGFHFCGGSLINENWVVTAA
HCGVTTSDVVVAGEFDQGSSSEKIQKLKIAKVFKNSKYNSLTINNDITLLKLSTAASFSQ
TVSAVCLPSASDDFAAGTTCVTTGWGLTRYTNANTPDRLQQASLPLLSNTNCK
KYWGTKIKDAMICAGASGVSSCMGDSGGPLVCKKNGAWTLVGIVSWGSSTCS
TSTPGVYARVTALVNWVQQTLAAN

The disulfide bridges place as follows:

$$^{1}C-^{122}C \qquad ^{42}C-^{58}C \qquad ^{136}C-^{201}C \qquad ^{168}C-^{182}C \qquad ^{191}C-^{220}C$$

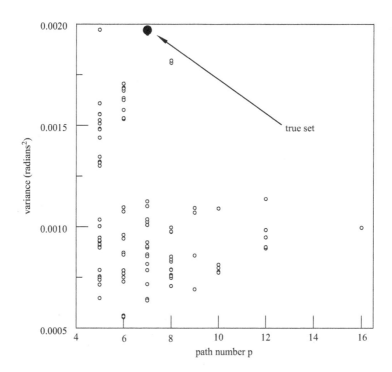

FIGURE TWENTY-THREE $\sigma^2 = \langle \theta_{\alpha\beta}^2 \rangle - \langle \theta_{\alpha\beta} \rangle^2$ for the pair distribution functions for all $\Omega = 105$ path sets for cathepsin A (PDB 4CIA). The point for the true bridge set is marked by the large filled circle. The small open circles are placed by false sets.

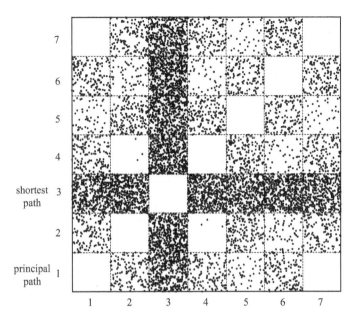

FIGURE TWENTY-FOUR   Pair distribution function for the *true*-set of cathepsin A in checkerboard terms. The number of pixels in each square of the checkerboard is proportional to the angle between base+(2) information vectors.

Figure Twenty-Five shows the variance of angles marked by $g(\theta_{\alpha\beta})$. There are $\Omega = 945$ points in the figure with the true-set marked by the large filled circle. Note how true places toward the bottom—much as in ribonuclease A. Minimum $p = 6$ is not the case. However, the internal sequences demonstrate the most compressed information structure of the $p = 10$ possibilities.

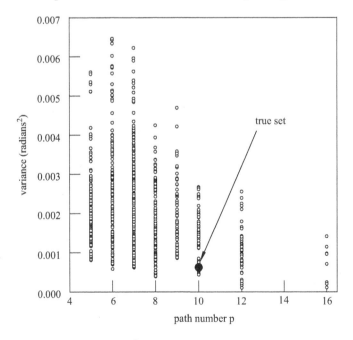

FIGURE TWENTY-FIVE $\sigma^2 = \left\langle \theta_{\alpha\beta}^2 \right\rangle - \left\langle \theta_{\alpha\beta} \right\rangle^2$ for the pair distribution functions for all $\Omega = 945$ path sets for chymotrypsinogen A (PDB 1EX3). The point for the true bridge set is marked by the large filled circle. The small open circles are placed by false sets.

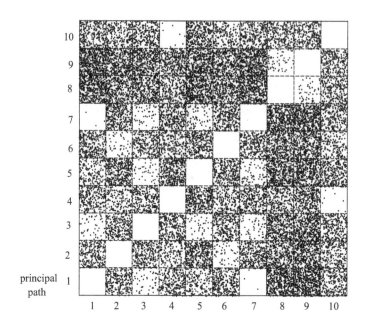

FIGURE TWENTY-SIX   Pair distribution function for the *true*-set of chymotrypsinogen A in checkerboard terms. The number of pixels in each square of the checkerboard is proportional to the angle between base+(2) information vectors.

Figure Twenty-Six shows the corresponding checkerboard. The pixels are spread out over the board in a manner evocative of ribonuclease A. There is no outlier path as the ten paths are conformist in their information expression.

The major points of this chapter are as follows:

1. Proteins express information by their paths, principal and auxiliary. The probability and information structures are drawn out by vectors, angles, and pair distributions $g(\theta_{\alpha\beta})$. These emphasize the auxiliaries to be held to the minimum for small proteins, and substantial bias toward either highly expanded or compressed information structures.
2. The larger proteins elect more than the minimum paths for folding and function. But they retain the acute biases toward expanded or compressed information structures.
3. What are abiotic proteins? At the least, they are ones in opposition to the biases and correspondences of natural proteins. Among other properties, they over-express the N- to C-terminal paths and steer toward the middle-range of $g(\theta_{\alpha\beta})$, $\langle\theta_{\alpha\beta}\rangle$, and σ^2 possibilities.

EXERCISES

1. Consider opposite carbon sites of anthracene indicated by arrows, viz.

How many covalent paths separate the sites?

2. Consider the yeast steroid, ergosterol:

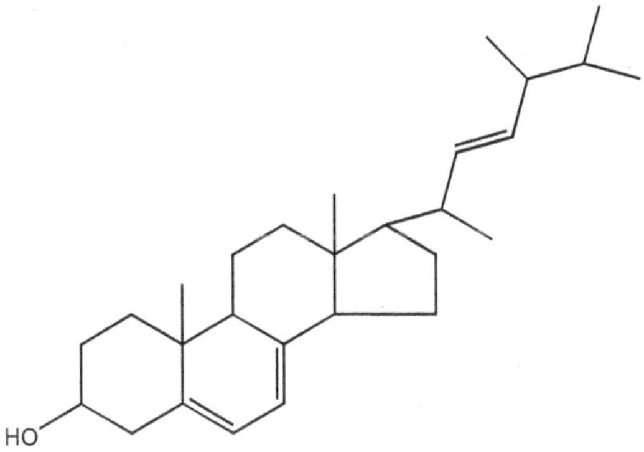

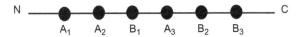

There are northern and southern paths from the hydroxyl to the farthest methyl group? Which is shorter?

3. A false set of disulfide bridges for lysozyme is as follows.

$$^6C-^{128}C \qquad ^{30}C-^{116}C \qquad ^{65}C-^{81}C \qquad ^{77}C-^{95}C$$

How many covalent paths separate the N- and C-terminal sites? What are the internal sequences? How many angles would we need to represent the distance measures?

4. Revisit the sequences in insulin.

GIVEQCCTSICSLYQLENYCN
FVNQHLCGSHLVEALYLVCGERGFFYTPKT

Construct 20D information vectors and compute the intervening angle in radians.

5. Revisit a sequence graph showing six cysteine sites:

N ●——●——●——●——●——● C
   A_1 A_2 B_1 A_3 B_2 B_3

Can we cross three bridges to travel from N- to C-terminal?

6. Revisit a sequence graph showing eight cysteine sites:

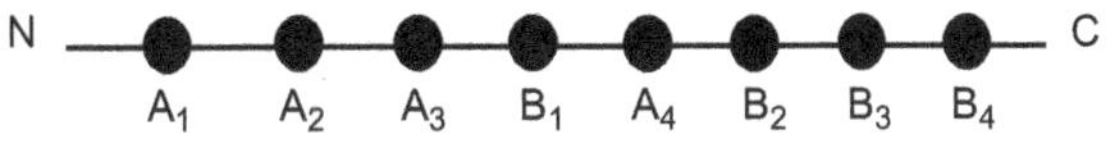

N ●——●——●——●——●——●——●——● C
   A_1 A_2 A_3 B_1 A_4 B_2 B_3 B_4

Can we cross three bridges to travel from N- to C-terminal?

7. Revisit the sequence and bridge placements for the anti-coagulent protein (PDB 1TAP):

YNRL^{5}CIKPRDWIDE^{15}CDSNEGGERAYFRNGKGG^{33}CDSFWI^{39}CPEDH
TGADYYSSYRD^{55}CFNA^{59}CI

$$\leftrightarrow \qquad ^5C-^{59}C \qquad ^{15}C-^{39}C \qquad ^{33}C-^{55}C$$

How many paths separate the N- and C-terminal sites?

8. Consider Exercise One as applied to naphthalene. How many paths separate opposite points of the molecule?

9. Consider Exercise One as applied to tetracene. How many paths separate opposite points of the molecule? What is the maximum number of covalent bonds crossed in travel between opposite points?

NOTES, SOURCES, AND FURTHER READING

Disulfide bridges enhance the stability of a protein and limit the incorrect folding possibilities. They are a relatively infrequent practice of nature as disulfide bridges manifest primarily in the proteins exported from cells [17]. Disulfide bridges have been studied for decades and the reader is referred to Flory for the fundamentals of cross-linked polymers [18]. In modern day, Abkevich and Shakhnovich have explored the multiple facets of disulfide bonds using molecular mechanics and bioinformatics [19]. Scheraga and co-workers have investigated the impact of disulfide bonds on folding mechanisms and trajectories [20,21]. A number of cross-linked systems have provided valuable pharmacological leads, particularly venoms, toxins, and ribonucleases [22–26]. Cross-links confer an intrinsic chirality in proteins quite apart from the α-carbons of amino acids. This property has attracted the attention of graph and knot theorists. This is discussed in an elegant text by Flapan [27]. Regarding the pathway structures in molecules, Randić, Balaban, Bonchev, Trinajstić, and their co-workers have contributed valuable graph and topological index tools over the years [28–34]. For proteins specifically, the reader is directed to the review by Randić et al. that includes multiple applications of primary structure analysis [34]. Researchers have also explored alternative representations for amino-acid sequences [35,36]. These target the essential information in primary structures and utilize fewer symbols than the standard 20 for amino acids. Significant research has also led to prediction algorithms for cross-link configurations. The approaches include the application of evolutionary bioinformatics, neural networks, support vector machines, and electronic structure modeling [37–42].

1. le Noble, W. J. 1974. The structural theory, in *Highlights of Organic Chemistry*, Marcel Dekker, Inc., New York.

2. Roberts, J. D., Caserio, M. C. 1965. The chemistry of natural products, in *Basic Principles of Organic Chemistry*, W. A. Benjamin, Inc., New York.

3. Ray, J. 1993. *Molecular Orbital Calculations Using Chemical Graph Theory*, Springer-Verlag, New York.

4. Trinajstić, N. 1991. *Computational Chemical Graph Theory: Characterization, Enumeration, and Generation of Chemical Structures by Computer Methods*, E. Horwood, New York.

5. Balaban, A. T. 1976. Enumeration of cyclic graphs, in *Chemical Applications of Graph Theory*, A. T. Balaban, ed., Academic Press, London.

6. White, A., Handler, P., Smith, E. L. 1973. The proteins I, in *Principles of Biochemistry*, 5th ed., McGraw-Hill, New York.

7. Chatani, E., Hayashi, R., Moriyama, H., Ueki, T. 2002. Conformational strictness required for maximum activity and stability of bovine pancreatic ribonuclease A as revealed by crystallographic study of three Phe120 Mutants at 1.4 A resolution, *Protein Sci.* 11, 72–81.

8. White, S. P., Scott, D. L., Otwinowski, Z., Gelb, M. H., Sigler, P. B. 1990. Crystal structure of cobra-venom phospholipase A2 in a complex with a transition-state analogue, *Science* 250, 1560–1563.

9. Palivec, V., Viola, C. M., Kozak, M., Ganderton, T. R., Krizkova, K., Turkenburg, J. P., Haluskova, P., Zakova, L., Jiracek, J., Jungwirth, P., Brzozowski, A. M. 2017. Computational and structural evidence for neurotransmitter-mediated modulation of the oligomeric states of human insulin in storage granules, *J. Biol. Chem.* 292, 8342–8355.

10. Lehninger, A. L. 1970. *Biochemistry*, Worth Publishers, New York.

11. Antuch, W., Guntert, P., Billeter, M., Hawthorne, T., Grossenbacher, H., Wuthrich, I. 1994. NMR solution structure of the recombinant tick anticoagulant protein (rTAP), a factor Xa inhibitor from the tick *Ornithodoros moubata*, *FEBS Lett.* 352, 251–257.

12. Caldwell, J. E., Abildgaard, F., Dzakula, Z., Ming, D., Hellekant, G., Markley, J. L. 1998. Solution structure of the thermostable sweet-tasting protein brazzein, *Nat. Struct. Biol.* 5, 427–431.

13. Bruix, M., Jimenez, M. A., Santoro, J., Gonzalez, C., Colilla, F. J., Mendez, E., Rico, M. 1993. Solution structure of gamma 1-H and gamma 1-P thionins from barley and wheat endosperm determined by 1H-NMR: A structural motif common to toxic arthropod proteins, *Biochemistry* 32, 715–724.

14. Li, H. M., Wang, D. C., Zeng, Z. H., Jin, L., Hu, R. Q. 1996. Crystal structure of an acidic neurotoxin from scorpion *Buthus martensii* Karsch at 1.85 A resolution, *J. Mol. Biol.* 261, 415–431.

15. Schreuder, H. A., Liesum, A., Kroll, K., Boehnisch, B., Buning, C., Ruf, S., Sadowski, T. 2014. Crystal structure of Cathepsin A, a novel target for the treatment of cardiovascular diseases, *Biochem. Biophys. Res. Comm.* 445, 451.

16. Pjura, P. E., Lenhoff, A. M., Leonard, S. A., Gittis, A. G. 2000. Protein crystallization by design: Chymotrypsinogen without precipitants, *J. Mol. Biol.* 300, 235–239.

17. Appling, D. R., Anthony-Cahill, S. J., Mathews, C. K. 2015. The three-dimensional structure of proteins, in *Biochemistry, Concepts and Connections*, Pearson Education, Hoboken, NJ.

18. Flory, P. J. 1953. Cross-linking of polymer chains, in *Principles of Polymer Chemistry*, Cornell University Press, Ithaca, NY.

19. Abkevich, V. I., Shakhnovich, E. I. 2000. What can disulfide bonds tell us about protein energetics, function, and folding: Simulations and bioinformatics analysis, *J. Mol. Biol.* 300, 975–985. doi:10.1006/jmbi.2000.3893.

20. Wedemeyer, W. J., Welker, E., Narayan, M., Scheraga, H. A. 2000. Disulfide bonds and protein folding, *Biochemistry* 39, 4207–4216. doi:10.1021/bi992922o.

21. Welker, E., Wedemeyer, W. J., Narayan, M., Scheraga, H. A. 2001. Coupling of conformational folding and disulfide-bond reactions in oxidative folding of proteins, *Biochemistry* 40, 9059–9064. doi:10.1021/bi010409g.

22. Escoubas, P. 2006. Molecular diversification in spider venoms: A web of combinatorial peptide libraries, *Mol. Diver.* 10, 545–554. doi:10.1007/s11030-006-9050-4.

23. Norton, R. S., Pallaghy, P. K. 1998. The cystein knot structure of ion channel toxins and related polypeptides, *Toxicon* 36, 1573–1583. doi:10.1016/s0041-0101(98)00149-4.

24. Shu, Q., Lu, S. Y., Gu, X. C., Liang, S. P. 2002. The structure of spider toxin huwentoxin-II with unique disulfide linkage: Evidence for structural evolution, *Prot. Sci.* 11, 245–252. doi:10.110/ps30502.

25. Tamiya, T., Fujimi, T. K. 2006. Molecular evolution of toxin genes in Elapidae snakes, *Mol. Divers.* 10, 529–543. doi:10.1007/s11040-006-9049-x.

26. Semple, C. A., Gautier, P., Taylor, K., Dorin, J. R. 2006. The changing of the guard: Molecular diversity and rapid evolution of beta-defensins, *Mol. Divers.* 10, 575–584. doi:10.1007/s11030-006-9031-7.

27. Flapan, E. 2000. Möbius ladders and related molecular graphs, in *When Topology Meets Chemistry, A Topological Look at Molecular Chirality*, Cambridge University Press, Cambridge, UK.

28. Randić, M., Wilkins, C. L. 1979. Graph theoretical ordering of structures as a basis for systematic searches for regularities in molecular data, *J. Phys. Chem.* 83, 1525–1540. doi:10.1021/j10047ra032.

29. Randić, M., Zupan, J., Vikić-Topić, D. 2007. On representation of proteins by star-like graphs, *J. Mol. Graph. Model.* 26, 290–305.

30. Randić, M. 1975. Characterization of molecular branching, *J. Am. Chem. Soc.* 97, 6609–6615. doi:10.1021/ja0085a001.

31. Balaban, A. T. 1982. Highly discriminating distance-based topological index, *Chem. Phys. Lett.* 80, 399–404.

32. Bonchev, D., Buck, G. A. 2005. Quantitative measures of network complexity, in *Complexity in Chemistry, Biology, and Ecology*, D. Bonchev, D. H. Rouvray, eds. Springer, New York.

33. Bonchev, D., Trinajstić, N. 1977. Information theory, distance matrix, and molecular branching, *J. Chem. Phys.* 67, 4517–4533. doi:10.1063/1.434593.

34. Randić, M., Zupan, J., Balaban, A. T., Dräzen, V.-T., Plavšić, D. 2011. Graphical representation of proteins, *Chem. Revs.* 111, 790–862. doi:10.1021/cr800198j.

35. Wang, J., Wang, W. 2000. Modeling study on the validity of a possibly simplified representation of proteins. *Phys. Rev. E.* 61, 6981–6986. doi:10.1103/PhysRevE.61.6981.

36. Zhao, E., Liu, H.-L., Tsai, C.-H., Tsai, H.-K., Chan, C.-H., Kao, C.-Y. 2005. Cysteine separations profiles on protein sequences infer disulfide connectivity. *Bioinformatics* 21, 1415–1420. doi:10.1093/bioinformatics/bti179.

37. Chen, Y.-C., Lin, Y.-S., Lin, C.-J., Hwang, J.-K. 2004. Prediction of the bonding states of cysteines using the support vector machines based on multiple feature vectors and cysteine state sequences, *Proteins* 55, 1036–1042. doi:10.1002/prot.20079.

38. Chuang, C.-C., Chen, C.-Y., Yang, J.-M., Lyu, P.-C., Hwang, J.-K. 2003. Relationship between protein structures and disulfide-bonding patterns, *Proteins* 53, 1–5. doi:10.1002/prot.10492.

39. Fariselli, P., Casadio, R. 2001. Prediction of disulfide connectivity in proteins, *Bioinformatics* 17, 957–964. doi:10.1093/bioinformatics/17.10.977.

40. Fariselli, P., Riccobelli, P., Casadio, R. 1999. Role of evolutionary information in predicting the disulfide-bonding state of cysteine in proteins, *Proteins* 36, 340–346.

41. Martelli, P. L., Fariselli, P., Malaguti, L., Casadio, R. 2002. Prediction of the disulfide-bonding state of cysteines at 88% accuracy. *Prot. Sci.* 11, 2735–2739. doi:10.1110/ps.0219602.

42. Vullo, A., Frasconi, P. 2004. Disulfide connectivity prediction using recursive neural networks and evolutionary information. *Bioinformatics* 20, 653–659. doi:10.1093/bioinformatics/btg463.

Eight Writing and Refining Protein Sequences

The previous chapters focused on the analysis of sequences we encounter in labs, reading, and seminars. This (hopefully!) redirected our inclination to look past letter strings, or to examine them only in cursory terms. Information at the base and base+ levels was the most amenable to exploration absent sophisticated computation. The aims were twofold: (1) reading skills—how we can approach unfamiliar sequences through probability and information, and (2) recognizing correspondences across individuals and sets. The present chapter revisits core ideas with a view toward writing sequences—the flip side of reading. The tools and procedures of previous chapters offer guidance and invitations to explore.

A PRELIMINARIES

The spotlight of previous chapters was on analysis: ways to view amino-acid sequences, archetypal and otherwise. The base and base+ structure levels offered ready starting points along with counting and grouping experiments. These enhanced our reading skills and brought to light multiple correspondences. Nothing sophisticated was demanded in the way of theory or computer modeling.

But now we inquire: by appreciating base and base+ properties, are we able to write new and potentially interesting sequences? The chemistry curriculum emphasizes writing skills across families: aromatics, carbohydrates, and so forth. A family is characterized by one or more motifs such as the fused rings in steroids:

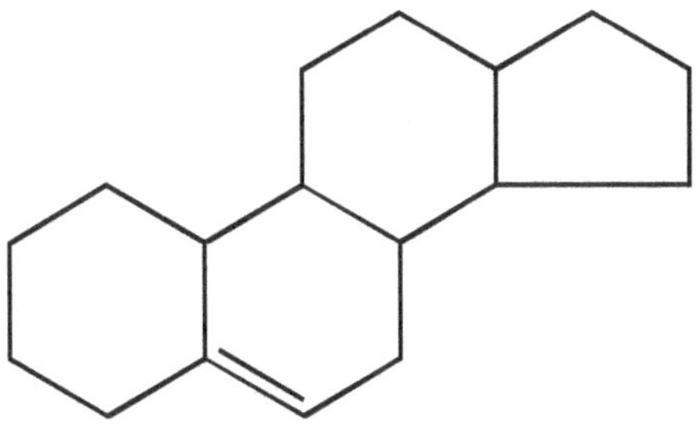

To represent new molecules, we append or modify in places, following chemical structure principles. We write, for instance:

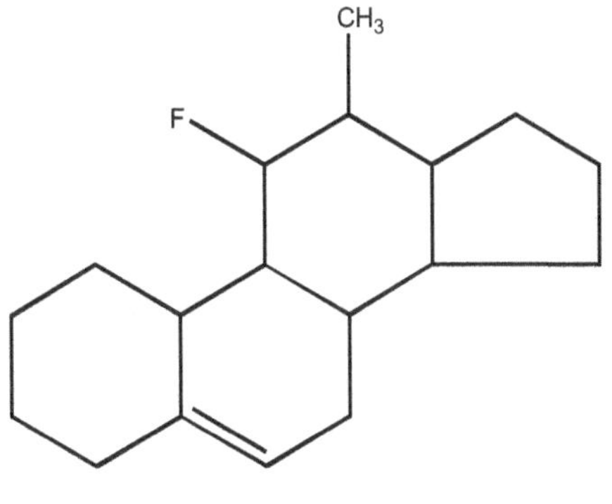

The new representation need not be contemplated in a vacuum. Textbooks and databases can be consulted for family relations, chemical, and physical properties. We can further direct computational models (e.g., density functional) to probe the charge distribution and geometry.

The building blocks of proteins present motifs and accommodate variations just as much. For example, tyrosine inspires:

Motifs are springboards throughout organic and pharmaceutical chemistry [1]. Their variations motivate new experiments that originate with diagrams such as above.

But now we come to the organic family called proteins. A primary structure is rich with motifs given its length, diversity, and aperiodicity. And the information is trivial to modify: we need only modify in places using the 20-letter alphabet—we instinctively rule out halogenations, oxidations, and other possibilities. Consider the representation for lysozyme:

KVFERCELARTLKRLGMDGYRGISLANWMCLAKWE**S**GYNTRATNYNAGDRS
TDYGIFQINSRYWCNDGKTPGAVNACHLSCSALLQDNIADAVACAKRVVRDPQ
GIRAWVAWRNRCQNRDVRQYVQGCGV

We can pick any of 130 sites, for example, **S** in bold underline, and have 19 choices for substitution—more if we put the post-genomic amino acids in play. We can modify the above as, say,

KVFERCELARTLKRLGMDGSRGISLANWMCLAKWE**A**GYNTRATNYNAGDRS
TDYGIFQINSRYWCNDGKTPGAVNACHLSCSALLQDNIADAVACAKRVVRDP
QGIRAWVAWRNRCQNRDVRQYVQGCGV

Or we can pick two sites and substitute by writing:

KVFERCELARTLKRLGMDGYR**A**ISLANWMCLAKWESGYNTRATNY**Q**AGDRS
TDYGIFQINSRYWCNDGKTPGAVNACHLSCSALLQDNIADAVACAKRVVRDPQ
GIRAWVAWRNRCQNRDVRQYVQGCGV

We can select three sites and so on. In all cases, so much of the information holds firm. We would refer to the first composition as the **S20A** variant while the second would be **G22A, N46Q.**

But substitutions are not the only route. We can trim a sequence at either end, viz.

~~KVFER~~CELARTLKRLGMDGYRGISLANWMCLAKWESGYNTRATNYNAGDRST
DYGIFQINSRYWCNDGKTPGAVNACHLSCSALLQDNIADAVACAKRVVRDPQ
GIRAWVAWRNRCQNRDVRQYVQGCGV

KVFERCELARTLKRLGMDGYRGISLANWMCLAKWESGYNTRATNYNAGDRS
TDYGIFQINSRYWCNDGKTPGAVNACHLSCSALLQDNIADAVACAKRVVRDP
QGIRAWVAWRNRCQNRDVRQYV~~QGCGV~~

We can invert a sector, for example, ….**MCLA**…. giving

KVFERCELARTLKRLGMDGYRGISLANW**ALCM**KWESGYNTRATNYNAGDRS
TDYGIFQINSRYWCNDGKTPGAVNACHLSCSALLQDNIADAVACAKRVVRDPQGIR
AWVAWRNRCQNRDVRQYVQGCGV

We can delete in places:

KVFERCELARTLKRLGMDGYRGISLA~~N~~WMCLAKWESGYNTRATNYNAGDRS
TDYGIFQINSRYWCNDG~~K~~TPGAVNACHLSCSALLQDNIADAVACAKRVVRDPQ
GIRAWVAWRNRCQNRDVRQYVQGCGV

We can insert:

KVFERCELARTLKRLGMDGYRGISLANWMCLAK**A**WESGYNTRATNYNAGDRS
TDYGIFQINSRYWCNDGKTPGAVNACHLSCSALLQDNIADAVACAKRVVRDP
QGIRAWVAWRNRCQNRDVRQYVQGCGV

Or we can trade:

KVFERCELARTLKRLGMDGY**G**GISLANWMCLAKWESGYNTRATNYNAGDRST
DYGIFQINSRYWCNDGKTPGAVNACHLSCSALLQDNIADAVACAKRVVRDPQ**R**I
RAWVAWRNRCQNRDVRQYVQGCGV

And we can think of more things to try. In every case, we are representing a new and potentially interesting protein at the sequence level. And we are following the spirit of pharmaceutical chemistry by letting established motifs or "hits" do the heavy lifting—we count on evolution for the major R & D [1]. We might well be writing primary structures that have never appeared, and may never appear in databases. And the proteins so represented may demonstrate modest or marked deviations from the hits regarding folds and activity. As with small organics, new representations do not stop at the drawing board, for they invite database sessions via BLAST, MSA, and more [2]. At the same time, alignment software will reflect that the above examples are not *completely* new; they resemble the lysozymes of multiple species. But we already knew that! Most importantly, the new sequences initiate studies at the lab bench.

We considered variant sequences in Chapter Three, and touched upon the writing process in places thereafter. Here we engage more closely and inquire how probability and information can assist. To set the stage, we remind ourselves of the Everest-height challenges. We can type:

FMSDTECSADGEIQLCLDLAVBMDM….
WISALCCVLSIWKKFLLSKCKGIESDQP…,

Ditto for thousands more and we can elaborate with chemical structure graphs given time and paper. What's the big deal? The answer is that in typing strings at whim, we are working in a dark space of size 20^N. We are throwing a dart *somewhere* in the space and hoping for the best, viz.

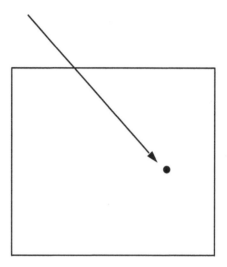

Granted, we are representing stable primary structures. But we should have zero expectations that they are biochemically relevant or even interesting. There are all manner of thermodynamic forces, solvent effects, and folding pathways that are being ignored. Thus it is crucial to have guidance and this is where motifs enter. They do not address all the thermodynamic and biochemical issues. But they are known quantities with track records. How could we do better?

Then if we start with, say, lysozyme and substitute in a few places, we are sampling points in 20^N-space that are *very* close to each other *and* to the motif:

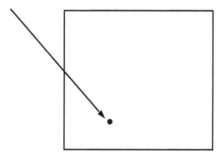

But this is not the only way to proceed. We can begin with the lysozyme information subject to component partitions and respellings, for example,

bnnabpannbpnbbnpnappbpnpnnpnnpnnbnapppppbnppppnpabppappnnpnppbpnppa
pbpnpnnpnpbnpppnnnpapnnannnpnbbnnbanppnbnnnnnbpbpppbanbppnppppn

*eiieeaeiaeaieeiaieaaeaiaiaaaiaiaeaeaaaaaeaaaaaaaaeeaaeaaiieiaaeaaaaeaeaaaa
iaaaeiaaaaiieeaiaeaiaaaeeiieeaeaieaaiaaeaeaeaeeieeaieaaai*

These representations stake out regions of the space. Through elaboration with the 20-letter alphabet, we can narrow a region to a single point:

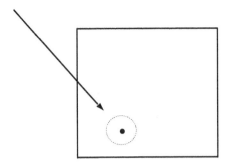

Or we can initiate writing at the base level:

$$A_{14}V_9L_8I_5P_2F_2W_5M_2G_{11}S_6T_5C_8Y_6N_{10}Q_6D_8E_3K_5R_{14}H_1 \leftrightarrow a_{11}b_{20}n_{47}p_{52} \leftrightarrow a_{67}e_{37}i_{26}$$

These are templates which give us more room to explore:

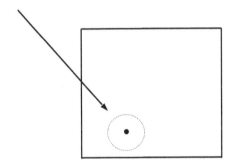

Expanding a 20-symbol base $A_{14}V_9L_8\ldots$ to a sequence collapses the neighborhood to a single point. Expanding bases written in family terms ($a_{11}b_{20}n_{47}p_{52}$, $a_{67}e_{37}i_{26}$, etc.) narrows a region to a locus of points. These are just a few of the options available to the motivated writer. What should we be thinking about as we try different options? The remaining sections address this question using lysozyme as a principal motif.

B   ALTERING MOTIFS TO OBTAIN NEW SEQUENCES

In writing sequences based on motifs, we would like to have some biochemical direction in mind, say, enhancing an enzyme's turnover rate. The reality is that the low information levels offer limited guidance, and it would be unreasonable to expect otherwise. Every motif has eons of evolution to its name whereby the sequence arrives through extensive trial-and-error. Besides, small changes can wield large-scale effects; for example, the wrong substitution can corrupt the folding or catalytic function. Proteins are not singular on the sensitivity account. Removing a mere hydrogen atom from any of the Section A molecules produces free radicals.

But all is not wasted by attention to low-level information, and for several reasons. The information is global and reflects whether modifications place a protein near or far from a motif. The low levels further illuminate scaling properties such as for base+(2) and base+(i). Lastly, compositions need not be evaluated in 20-letter terms. Through partitions, respellings, and numerical information (e.g., mass distributions), we enhance our perspective for whole classes allied with a sequence. This is low-hanging fruit of modest work and informs the writing process. It similarly shines light on sequences that are interesting on various fronts. In effect, global information provides a screen for judging what's deserving of benchtop and X-ray experiments, given limited time and resources. We illustrate examples whereby the writing and refinement methods extend far and wide.

Consider the following $N = 130$ sequences. As should be obvious, they are all inspired by motifs in lysozyme:

KVFERCELARTLKRLGMDGYRGISLANWMCLAKWESSSSSRATNYNAGDRST
DYGIFQINSRYWCNDGKTPGAVNACHLSCSALLQDNIADAVACAKRVVRDP
QGIRAAAAARNRCQNRDVRQYVSSSSV   (1)

KLFERCDLVRTLKRLGMESYKGISLANMMCLAKFDSTYQTRFTNYNAGDRSTE
YYVFQINSRYWYSDGKSPGAINAGHLSCYMLLYENIIDWVACAKRVVRDPQGI
RIWVAWHCRCQNRDVRQGVQGCGV   (2)

RVFERQELPRTLKKLNMENNHGIGLITWMCAAKLESGYTQRLQQYNAGDRTT
DYGIFQITSRYWCNDGKTPNAINACHLSCSMLLQDNIADAVACAKRVI
KDFQGIKAWVAWRNRCQTRDVRQYVQGCGL   (3)

KFFDKCEAARTLKRLGMEGSHGISLANWMCLAKWESGYGTRLTNYNAGDRST
DYGIFQIYSRYISNDGKTPTAVNACHLSYYFPLSDNIADAVACARHVFRDPQG
AKAWVFVRNRCQNRDPKQYVTSCGM   (4)

What should we think besides that they look a whole lot like the predominant sequence in this book? With close inspection, things should jump to mind for Sequence (1). The runs of S and A are improbable given what we expect of sequences, and elicit surprise as a consequence. But how should we regard Sequences (2), (3), and (4)? They present more cryptically on the page and screen and echo points discussed in Chapters One. Whereas the representations for small organics are readily interpreted given our chemistry training, things are more complicated for proteins.

Probability and information help critique and guide the writing process. Via symbol counting and grouping, we monitor all compositions from the ground up. This means paying attention to the base structures as we write, for example,

$$\text{KVFER}.... \leftrightarrow A_0V_1L_0I_0P_0F_1W_0M_0G_0S_0T_0C_0Y_0N_0Q_0D_0E_1K_1R_1H_0$$

$$\text{KVFERCELARTLKR}.... \leftrightarrow A_1V_1L_2I_0P_0F_1W_0M_0G_0S_0T_1C_1Y_0N_0Q_0D_0E_2K_2R_3H_0$$

$$\cdot$$
$$\cdot$$
$$\cdot$$

The structures take us to:

$$A_{17}V_8L_8I_5P_2F_2W_3M_2G_8S_{14}T_4C_7Y_5N_9Q_5D_8E_3K_5R_{14}H_1 \quad \leftrightarrow \quad \text{Sequence (1)}$$
$$A_8V_9L_9I_7P_2F_4W_4M_4G_{10}S_8T_5C_7Y_9N_7Q_6D_7E_4K_6R_{12}H_2 \quad \leftrightarrow \quad \text{Sequence (2)}$$
$$A_{11}V_6L_{10}I_8P_2F_3W_4M_3G_9S_4T_8C_7Y_5N_9Q_{10}D_7E_4K_7R_{11}H_2 \quad \leftrightarrow \quad \text{Sequence (3)}$$
$$A_{13}V_6L_7I_5P_4F_6W_3M_3G_{10}S_9T_7C_6Y_8N_8Q_4D_8E_3K_7R_{10}H_3 \quad \leftrightarrow \quad \text{Sequence (4)}$$

The bases compare with the motif:

$$A_{14}V_9L_8I_5P_2F_2W_5M_2G_{11}S_6T_5C_8Y_6N_{10}Q_6D_8E_3K_5R_{14}H_1 \quad \leftrightarrow \quad \text{lysozyme}$$

At once we are able to evaluate distances between compositions and the motif. As shown in Chapter Three, the base information underpins 20D vectors, for example we have for Sequence (4):

$$\vec{V}_{(4)} = \sqrt{\frac{13}{130}} \cdot \hat{x}_A + \sqrt{\frac{6}{130}} \cdot \hat{x}_V + \cdots + \sqrt{\frac{10}{130}} \cdot \hat{x}_R + \sqrt{\frac{3}{130}} \cdot \hat{x}_H$$

The angles between vectors obtain from the inverse cosine function and demonstrate (1) and (4) to be the nearest and farthest from the motif, respectively. Point being made: we have written sequences grounded on a motif, three of which present equal numbers of substitutions. The angles between vectors establish the hierarchy of distances between base structures.

In writing sequences, it is just as critical to monitor the base+(2) levels. Recall how the di-peptide number N_D scales as $N^{2/3}$ in line with the potential for surface area. We count $N_D = 109 \leftrightarrow$ lysozyme, $103 \leftrightarrow$ Sequence (1), $112 \leftrightarrow$ Sequence (2), $109 \leftrightarrow$ Sequence (3), $113 \leftrightarrow$ Sequence (4), and revisit Figure Eight of Chapter Six. The error bars in Figure One apply to random strings of amino acids. We observe how Sequences (2), (3), and (4) express N_D well within bounds for $N = 130$ proteins whereas Sequence (1) is deficient. Its grammatical diversity is better suited to a sequence having ten fewer units.

The base+(i) information offers additional grammar markers. Recall from Chapter Six that sequences present long-distance relationships. In Sequence (2), for instance, we have sequential **A** and **V** separated by the following intervals:

A: 6, 15, 26, 3, 18, 2, 15
V: 47, 37, 6, 1, 10, 11, 4, 5

It was shown that as we move from N- to C-terminal, the probability of landing on a unit identical to a reference (e.g., **A**) scales exponentially with the number of steps n. We saw that the sum of probabilities $F(n)$ is approximated by

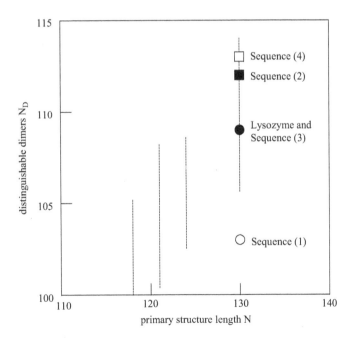

FIGURE ONE Distinguishable di-peptides for Sequences (1)–(4) based on lysozyme. The error bars pertain to random strings of amino acids.

$$F(n) \approx 1 - e^{-pn}$$

$F(n)$ represents the fraction of steps taken less than or equal to n in which a site does *not* match a reference.

Figure Two presents $F(n)$ for the compositions as labeled; the $F(n)$ plot for lysozyme is included. We learn that the compositions reflect exponential $F(n)$ as with natural proteins. By the same token, Sequences (1) and (2) offer the closest alignment with lysozyme. In (3) and (4), we observe substantially divergent information.

In critiquing sequences, we can always go further and monitor information via partitions and respellings. For example, it can be shown that all four examples present as $a_{11}b_{20}n_{47}p_{52}$, identical to lysozyme! The base+(2) information is then also identical to lysozyme, namely:

$$(bb)_2(ba)_2(bp)_9(bn)_7(ab)_2(ap)_5(an)_4(pb)_7(pa)_5(pp)_{21}(pn)_{19}(nb)_8(na)_4(np)_{17}(nn)_{17}$$

The partitions and respellings underpin information vectors:

$$\vec{V}_{base} = \sqrt{\frac{11}{130}} \cdot \hat{x}_a + \sqrt{\frac{20}{130}} \cdot \hat{x}_b + \sqrt{\frac{47}{130}} \cdot \hat{x}_n + \sqrt{\frac{52}{130}} \cdot \hat{x}_p$$

$$\vec{V}_{base+(2)} = \sqrt{\frac{2}{129}} \cdot \hat{x}_{bb} + \sqrt{\frac{2}{129}} \cdot \hat{x}_{ba} + \cdots + \sqrt{\frac{17}{129}} \cdot \hat{x}_{np} + \sqrt{\frac{17}{129}} \cdot \hat{x}_{nn}$$

We ascertain that the vectors point in the same direction as the motif *and* that the abnp-weights are near-canonical (cf. Chapter Three). The amino acids admit multiple partitions which provide perspectives in sequence writing.

In writing sequences, numerical information is just as attention-worthy. Recall from Chapter Six how mass distributions are illuminated. We focus not on a molecular weight series $M(k)$ per se, but rather the accumulation of distances. The information is repackaged in the function $\Phi(k)$:

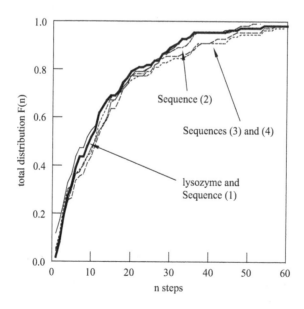

FIGURE TWO Total distribution function $F(n)$ for Sequences (1)–(4) based on lysozyme. The bold curve corresponds to $F(n)$ for lysozyme.

$$\Phi(k) = \sum_{i=1}^{k} Z_{\alpha}^{(i)}$$

The terms $Z_{\alpha}^{(i)}$ refer to the distance, positive or negative, of the amino acid of type α (e.g., alanine) at position i in the sequence up to index k. It was shown in Chapter Six how natural proteins express $\Phi(k)$ that scales (more or less) linearly with negative slope. Just as important, $\Phi(k)$ marks density scaling domains across the primary structure.

We illustrate $\Phi(k)$ for the compositions in Figure Three, the bold line allied with the motif $\leftrightarrow$ lysozyme. We observe how the trace for Sequence (4) comes closest to the motif and it especially tracks the large domain $60 \leq k \leq 100$. In contrast, Sequence (2) deviates significantly and the non-linear scaling is glaring in places. Sequence (1) follows the motif information more or less until a sharp turn near $k = 90$.

And so it goes. We are using lysozyme just as a hit in pharmaceutical chemistry and are writing variations on the themes. The compositions place close to or far from the motif at the low levels of information. We need only track a few properties to light the way.

But how can we tell if a composition is interesting? This is a subjective question, but not out of bounds. One way is to view it through a maximum-indifference lens [3,4]. Interesting compositions should have low probability and high surprisals compared with pedestrian ones, right? It was shown in Chapter Three how protein sequences are highly individualistic. Even their low structure levels stand out from crowds.

Recall that the multinomial distribution (cf. Appendix Four) assesses the probability of a base structure given draw criteria. If we practice maximum indifference, the base for, say, Sequence (2) expresses probability:

$$prob(k=8, l=9, m=9,...) = \frac{130!}{8!9!9!7!2!4!4!4!10!8!5!7!9!7!6!7!4!6!12!2!} \times \left(\frac{1}{20}\right)^{8} \times \left(\frac{1}{20}\right)^{9} \times \left(\frac{1}{20}\right)^{9} \times \cdots$$

$$= \frac{130!}{8!9!9!7!2!4!4!4!10!8!5!7!9!7!6!7!4!6!12!2!} \times \left(\frac{1}{20}\right)^{8+9+9+\cdots}$$

$$\approx 4.82 \times 10^{-21}$$

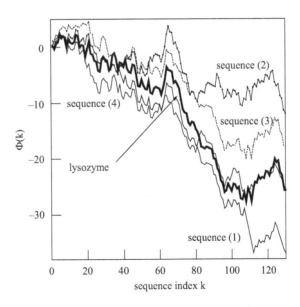

FIGURE THREE Cumulative mass distribution for Sequences (1)–(4) based on lysozyme. The bold curve corresponds to the mass distribution for lysozyme.

This carries surprisal:

$$-\log_2\left(4.82\times10^{-21}\right)\approx 63.7 \text{ bits}$$

The base further asserts a sum of squares of deviations from expected:

$$\sigma^2 = \sum_{j=1}^{20}\left(k_j - \langle k\rangle\right)^2$$

The integers k_j follow the base subscripts while

$$\langle k\rangle = N\cdot p = 130\cdot(1/20) = 6.50$$

For example, Sequence (2) presents $\sigma^2\approx 135$.

Figure Four shows the placement of compositions place in relation to lysozyme. The black circles block out the point locus for $N=130$ random bases. We learn that Sequences (2), (3), and (4) place squarely in the crowd. Their global information is more commonplace than lysozyme and renders them less interesting, or at least less individualistic. In contrast, Sequence (1) places as an outlier—even more so than lysozyme. It is quite interesting by its base information alone.

The takeaway is that we are able to write new sequences and use the low-level properties for critiques—we take stock of what is put to paper or word processor. Where compositions can be evaluated, they can be adjusted to bring them closer to or farther from the motif. With practice, we are even able to internalize the information changes as we write.

A second scan for interest involves reflexive properties. Recall that proteins express information that is internally mutual. The abnp-fractions in one part of a sequence are prone to match other parts. It was shown in Chapter Three that assumptions of this sort incur only slight penalties of the Kullback–Leibler information. As Sequences (1)–(4) express the same base and base+ information in family terms, their abnp-reflexive properties align with lysozyme. They are all attention-worthy in this sense.

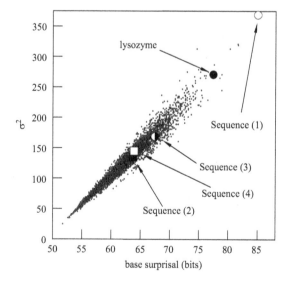

FIGURE FOUR Base surprisals and deviations from expectations for Sequences (1)–(4) viewed through zero-bias lenses. The small circles mark the distribution of base structures according to the multinomial distribution.

Compositions take more radical turns if we direct non-conservative substitutions to a motif. For example, we can start with lysozyme and write:

KVFERCELARTLKRLGMDGYRGISLANWMCLAKWEEEEEERATNYNAGDRS
TDYGIFQINSRYWCNDGKTPGAVNACHLSCSALLQDNIADAVACAKRVVRDPQ
GIRRRRRRRNRCQNRDVRQYVVVVVV (5)

KVFERCTLARTQKRHHMDHYAHNSYKNWMCLAKWSSGYNTRRTNDPTGDR
STDKGIFQINSRIIWCNDGKTPEEVVALHSSCSHLLQDNIADAVACAKQVVQDP
QGIRAWVAWRNRCQERDCRQYSQDDGV (6)

IVFERCELARTTKALGMCPLRGISLANWMCRAKWQLGYNAPATNPWAGDRF
TDYGIFEQNIRYWCRDGKTPGAENACDLSCSALLQDRIKDAVACASRVVRDPQ
GIDAWVHQRNRMQVRDVNQKVQGCGV (7)

KVFECVELARTLPRLGHDGKRGIVKRNWMALAKWESFYNTRATNYNAGDRS
TDYVIFQINSRYWCNDGMFSGTRNAWCTMWSALLQDNIAGAVACAKR
KVRSPIGGRAWVSWRNRCQNRSCRQYVQGCGS (8)

But we confront all the same issues as before. Sequence (5) draws attention by the runs of E, R, and V. Sequences (6)–(8) are more understated, although a few substitutions stand out in places.

What should we register during the writing process? We follow the same tracks as for conservative substitutions. These commence with assembling the base structures:

$$A_{12}V_{12}L_8I_5P_2F_2W_3M_2G_8S_5T_4C_7Y_5N_9Q_5D_8E_8K_5R_{19}H_1 \quad \leftrightarrow \quad \text{Sequence (5)}$$
$$A_{10}V_8L_5I_4P_3F_2W_5M_2G_6S_9T_7C_7Y_4N_8Q_9D_{11}E_4K_7R_{12}H_7 \quad \leftrightarrow \quad \text{Sequence (6)}$$
$$A_{14}V_9L_8I_6P_5F_3W_5M_3G_{10}S_4T_5C_8Y_3N_7Q_8D_9E_4K_5R_{13}H_1 \quad \leftrightarrow \quad \text{Sequence (7)}$$
$$A_{12}V_8L_6I_5P_2F_4W_7M_3G_{11}S_9T_6C_7Y_5N_{10}Q_5D_5E_3K_6R_{15}H_1 \quad \leftrightarrow \quad \text{Sequence (8)}$$

We compare them with the motif:

$$A_{14}V_9L_8I_5P_2F_2W_5M_2G_{11}S_6T_5C_8Y_6N_{10}Q_6D_8E_3K_5R_{14}H_1 \quad \leftrightarrow \quad \text{lysozyme}$$

We then look to information vectors to establish distances. If you guessed that the base for Sequence (6) is farthest from lysozyme, take a bow. Then which base stands closest?

We turn to the base+(2) levels and count $N_D = 109 \leftrightarrow$ lysozyme, $100 \leftrightarrow$ Sequence (5), $111 \leftrightarrow$ Sequence (6), $113 \leftrightarrow$ Sequence (7), $109 \leftrightarrow$ Sequence (8). Figure Five shows how the word counts place. We see how Sequence (5) is vocabulary-deficient while the others meet the bar.

We examine base+(i) information via $F(n)$ in Figure Six. Sequence (8) closely follows lysozyme while Sequence (6) veers the farthest.

Partitions and respellings offer additional guidance. The bases express as:

$a_{16}b_{25}n_{46}p_{43} \leftrightarrow$ Sequence (5)	$a_{15}b_{26}n_{39}p_{50} \leftrightarrow$ Sequence (6)
$a_{13}b_{19}n_{53}p_{45} \leftrightarrow$ Sequence (7)	$a_8b_{22}n_{47}p_{53} \leftrightarrow$ Sequence (8)

These compare with $a_{11}b_{20}n_{47}p_{52} \leftrightarrow$ lysozyme.

It is at this juncture that we can better grasp things. This is on account of expectations of the abnp-components according to canonical selection weights. Recall from Chapter Three that bases in family terms express canonical probability:

$$prob(N_a, N_b, N_n, N_p) = \frac{(N_a + N_b + N_n + N_p)!}{N_a! N_b! N_n! N_p!} \times \left(\frac{2}{20}\right)^{N_a} \cdot \left(\frac{3}{20}\right)^{N_b} \cdot \left(\frac{8}{20}\right)^{N_n} \cdot \left(\frac{7}{20}\right)^{N_p}$$

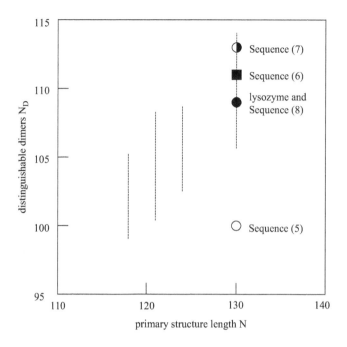

FIGURE FIVE   Distinguishable di-peptides for Sequences (5)–(8) obtained via non-conservative substitutions. The error bars pertain to random strings of amino acids.

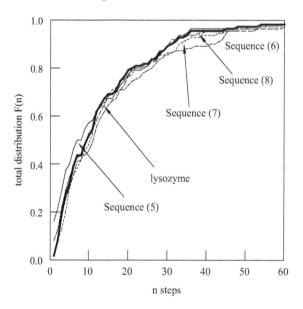

FIGURE SIX   Total distribution function $F(n)$ for Sequences (5)–(8) obtained via non-conservative substitutions. The bold curve corresponds to $F(n)$ for lysozyme.

This underpins base and sequence surprisals:

$$S(N_a, N_b, N_n, N_p) = -\log_2 prob(N_a, N_b, N_n, N_p)$$

$$S(\Omega) = \log_2 \left[\frac{(N_a + N_b + N_n + N_p)!}{N_a! N_b! N_n! N_p!} \right]$$

The surprisals place state points as in Figure Seven. Included is the locus for $N = 130$ structures with canonical selection weights.

The figure should be compared with Figure Twelve of Chapter Three. We then see how Sequences (7) and (8) land in the preferred zones for lysozyme variants. In contrast, Sequences (5) and (6) place points in uncharted territory. Their low-level information seems antithetical to lysozyme.

Information via $\Phi(k)$ appears in Figure Eight. We observe how Sequence (7) comes closest to the lysozyme trace noted in bold; the traces for Sequences (6) and (8) deviate most significantly from

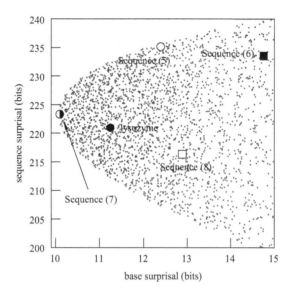

FIGURE SEVEN Sequence versus base information in abnp-terms for Sequences (5)–(8) obtained via non-conservative substitutions. The small open circles mark the $N = 130$ locus allied with canonical selection weights.

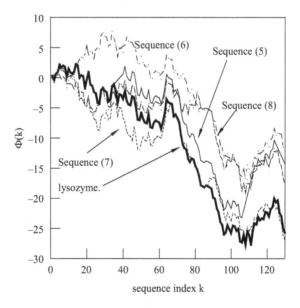

FIGURE EIGHT Cumulative mass distribution for Sequences (5)–(8) obtained via non-conservative substitutions. The bold trace corresponds to the mass distribution of lysozyme. Base surprisals and deviations from expectations viewed through zero-bias lenses.

the motif. The trace for (5) aligns with lysozyme but takes a drastic turn near $k = 105$. Thereafter it closely follows (6) and (8).

Figure Nine takes us the rest of the way with the zero-bias state points. The black circles mark the point locus for random bases. We learn that Sequence (6) places toward the edge of the crowd while the others reside in fringe or outlier territory.

The check for abnp-reflexive properties is left as an exercise. If you guessed that Sequence (7) expresses the highest mutual information internally, take a bow. Which composition expresses the lowest?

We have shown that we can craft sequences by substituting motifs. And we can monitor and critique grammatical aspects using tools and ideas of previous chapters. But we have not said enough about individual steps. How do we initiate a composition and when should we stop? These are difficult questions given the complexities of proteins.

The short if unsatisfactory answers are that we have to (1) decide the extent we wish to alter the motif information, and (2) track multiple properties simultaneously. We assess where each substitution takes the global information, and halt when we reach some tolerable limit. The properties are of equal importance and we present them in order of difficulty.

The first concerns the rarefaction. There are three ways to go with substitutions: we can hold the surprisal constant or tune it upward or downward. Natural proteins have rarefied bases as discussed in Chapter Three. In monitoring this property, we view information using the 20-letter alphabet:

KVFERCELARTLKRLGMDGYRGISLANWMCLAKWESGYNTRATNYNAGDRS
TDYGIFQINSRYWCNDGKTPGAVNACHLSCSALLQDNIADAVACAKRVVRDPQG
IRAWVAWRNRCQNRDVRQYVQGCGV

$$\leftrightarrow \qquad A_{14}V_9L_8I_5P_2F_2W_5M_2G_{11}S_6T_5C_8Y_6N_{10}Q_6D_8E_3K_5R_{14}H_1$$

At the start and throughout, we examine the component expectations according to maximum indifference. For $N = 130$, we have for the $i\underline{th}$ component:

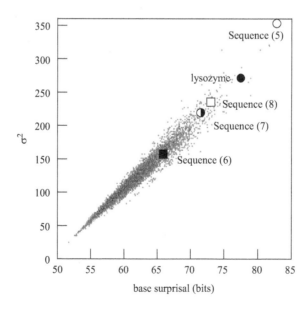

FIGURE NINE   The small circles mark the distribution of base structures according to the multinomial distribution. The large symbols mark the state points for Sequences (5)–(8).

$$\langle N_i \rangle = N \cdot p = 130 \cdot \left(\frac{1}{20} \right) = 6.50$$

Substitutions in which components trade weights are probability-neutral and do nothing to the rarefaction. For example, substituting an L in place of V gives:

KVFERCELARTLKRLGMDGYRGISLANWMCLAKWESGYNTRATNYNAGDRS
TDYGIFQINSRYWCNDGKTPGAVNACHLSCSALLQDNIADAVACAKRVVRDPQ
GIRAW<u>L</u>AWRNRCQNRDVRQYVQGCGV

$\leftrightarrow$ $\qquad$ $A_{14}\mathbf{V_{9\text{-}1}}\mathbf{L_{8+1}}I_5P_2F_2W_5M_2G_{11}S_6T_5C_8Y_6N_{10}Q_6D_8E_3K_5R_{14}H_1$

To increase the surprisal (lower the probability), we move a component weight farther from the zero-bias expectation. For example, if we put an **A** in place of **I**, we obtain:

KVFERCELARTLKRLGMDGYRG<u>A</u>SLANWMCLAKWESGYNTRATNYNAGDRS
TDYGIFQINSRYWCNDGKTPGAVNACHLSCSALLQDNIADAVACAKRVVRDPQGI
RAWVAWRNRCQNRDVRQYVQGCGV

$\leftrightarrow$ $\qquad$ $\mathbf{A_{14+1}}V_9L_8\mathbf{I_{5\text{-}1}}P_2F_2W_5M_2G_{11}S_6T_5C_8Y_6N_{10}Q_6D_8E_3K_5R_{14}H_1$

This enhances the outlier status of the base. In plots as in Figure Four, the state point is pushed deeper into right field. To make the base less rarefied, we steer in the opposite direction and move a component closer to expectation. For example, we can substitute **Y** in place of **A** to obtain:

KVFERCEL<u>Y</u>RTLKRLGMDGYRGISLANWMCLAKWESGYNTRATNYNAGDRS
TDYGIFQINSRYWCNDGKTPGAVNACHLSCSALLQDNIADAVACAKRVVRDPQ
GIRAWVAWRNRCQNRDVRQYVQGCGV

$\leftrightarrow$ $\qquad$ $\mathbf{A_{14\text{-}1}}V_9L_8I_5P_2F_2W_5M_2G_{11}S_6T_5C_8\mathbf{Y_{6+1}}N_{10}Q_6D_8E_3K_5R_{14}H_1$

This directs the state point closer to the crowd.

The second property concerns the abnp representations. From start to finish, we keep canonical expectations in mind:

$$\langle N_a \rangle = N \cdot p_a = 130 \cdot \left(\frac{2}{20} \right) = 13.0$$

$$\langle N_b \rangle = N \cdot p_b = 130 \cdot \left(\frac{3}{20} \right) = 19.5$$

$$\langle N_n \rangle = N \cdot p_n = 130 \cdot \left(\frac{8}{20} \right) = 52.0$$

$$\langle N_p \rangle = N \cdot p_p = 130 \cdot \left(\frac{7}{20} \right) = 45.5$$

In applying substitutions, we monitor whether the base becomes more or less canonical, or holds steady. For example, to move lysozyme closer to canonical, we change a polar site to acidic, viz.

KVFERCELARTLKRLGMDGYRGI<u>D</u>LANWMCLAKWESGYNTRATNYNAGDRS
TDYGIFQINSRYWCNDGKTPGAVNACHLSCSALLQDNIADAVACAKRVVRDP
QGIRAWVAWRNRCQNRDVRQYVQGCGV

$$\leftrightarrow \qquad \mathbf{a}_{11+1}b_{20}n_{47}\mathbf{p}_{52\text{-}1}$$

This pushes the state point leftward in diagrams such as Figure Seven. To make the base less canonical, we modify the opposite way, for example,

KVFERC**T**LARTLKRLGMDGYRGISLANWMCLAKWESGYNTRATNYNAGDRS
TDYGIFQINSRYWCNDGKTPGAVNACHLSCSALLQDNIADAVACAKRVVRDPQ
GIRAWVAWRNRCQNRDVRQYVQGCGV

$$\leftrightarrow \qquad \mathbf{a}_{11\text{-}1}b_{20}n_{47}\mathbf{p}_{52+1}$$

This moves the state point rightward in Figure Seven. To effect neutral changes, we keep all substitutions abnp-conservative. For example, trading N for Q gives:

KVFERCELARTLKRLGMDGYRGISLA**Q**WMCLAKWESGYNTRATNYNAGDRS
TDYGIFQINSRYWCNDGKTPGAVNACHLSCSALLQDNIADAVACAKRVVRDPQ
GIRAWVAWRNRCQNRDVRQYVQGCGV

$$\leftrightarrow \qquad a_{11}b_{20}n_{47}\mathbf{p}_{52+0\text{-}0}$$

The surprisal state point stays put.

In writing sequences, we have to track the distinguishable di-peptides. The base+(2) representation for lysozyme appeared in Chapter Six as follows:

$(AV)_2(AL)(AW)_2(AG)(AT)(AC)_2(AN)(AD)(AK)_2(AR)(VA)_2(VV)(VF)(VN)(VQ)(VR)_2(LA)_3(LL)$
$(LG)(LS)(LQ)(LK)(IA)(IF)(IS)(IN)(IR)(PG)(PQ)(FQ)(FE)(WV)(WM)(WC)(WE)(WR)(MC)$
$(MD)(GA)(GV)\underline{\mathbf{(GI)}_3}\underline{\mathbf{(GM)}}(GC)(GY)_2(GD)(GK)(SA)(SL)(SG)(ST)(SC)(SR)(TL)(TP)(TN)(TD)$
$(TR)(CA)(CL)(CG)(CS)(CN)(CQ)(CE)(CH)(YV)(YW)(YG)(YN)_2(YR)(NA)_2(NI)(NW)(NS)(NT)$
$(NY)(ND)(NR)_2(QI)(QG)_2(QY)(QN)(QD)(DA)(DV)(DP)(DG)_2(DY)(DN)(DR)(EL)(ES)(ER)(KV)$
$(KW)(KT)(KR)_2(RA)_2(RV)(RL)(RG)(RS)(RT)(RC)_2(RY)(RN)(RQ)(RD)_2(HL)$

To increase N_D, we scan the list, noting the terms with subscripts such as $\underline{\mathbf{(GI)}_3}$ in bold underline— the absence of a subscript indicates the default value of 1. We sacrifice a word, say, by substituting **R** for **I**, viz.

KVFERCELARTLKRLGMDGYRGISLANWMCLAKWESGYNTRATNYNAGDRS
TDYG**R**FQINSRYWCNDGKTPGAVNACHLSCSALLQDNIADAVACAKRVVRDP
QGIRAWVAWRNRCQNRDVRQYVQGCGV

$$\leftrightarrow \qquad A_{14}V_9L_8\mathbf{I}_{5\text{-}1}P_2F_2W_5M_2G_{11}S_6T_5C_8Y_6N_{10}Q_6D_8E_3K_5R_{14+1}H_1$$

To decrement N_D, we chose any word lacking a subscript, say, **GM** in bold. We trade M for, say, I to obtain another **GI**, viz.

KVFERCELARTLKRLG**I**DGYRGISLANWMCLAKWESGYNTRATNYNAGDRST
DYGIFQINSRYWCNDGKTPGAVNACHLSCSALLQDNIADAVACAKRVVRDPQGI
RAWVAWRNRCQNRDVRQYVQGCGV

$$\leftrightarrow \qquad A_{14}V_9L_8\mathbf{I}_{5+1}P_2F_2W_5\mathbf{M}_{2\text{-}1}G_{11}S_6T_5C_8Y_6N_{10}Q_6D_8E_3K_5R_{14}H_1$$

To hold N_D fixed, we eliminate one di-peptide and render another in the process. For example, we can lose **GA** and obtain **GL** in its place, viz.

KVFERCELARTLKRLGMDGYRGISLANWMCLAKWESGYNTRATNYNAGDRS
TDYGIFQINSRYWCNDGKTPG<u>L</u>VNACHLSCSALLQDNIADAVACAKRVVRDPQ
GIRAWVAWRNRCQNRDVRQYVQGCGV

$$\leftrightarrow \quad A_{14\text{-}1}V_9L_{8+1}I_5P_2F_2W_5M_2G_{11}S_6T_5C_8Y_6N_{10}Q_6D_8E_3K_5R_{14}H_1$$

Monitoring what we write gets a handle on the long-distance information. Natural sequences express randomness with first-order exponential distributions $F(n)$ as robust signatures. To maintain the styles of signatures, we avoid periodic patterns during substitutions. To embolden natural signatures, we iron out patterns already in place. For example, A presents in 14 places of lysozyme, including at the regular intervals noted in bold underline:

KVFERCELARTLKRLGMDGYRGISLANWMCLAKWESGYNTRATNYNAGDRS
TDYGIFQINSRYWCNDGKTPGAVNACHLSCSALLQDNI<u>A</u>DA<u>V</u>AC<u>A</u>KRVVRDP
QGIRAWVAWRNRCQNRDVRQYVQGCGV

To enhance the randomness, we scramble the intervals via trades, for example,

KVFERCELARTLKRLGMDGYRGISLANWMCLAKWESGYNTRATNYNAGDRS
TDYGIFQINSRYWCNDGKTPGAVNACHLSCSALLQD<u>A</u>NIDVCKRVVR<u>A</u>DPQ
GIR<u>A</u>WVAWRNRCQNARDVRQYVQ<u>A</u>GCGV

The base is unchanged, but the interval structure movers closer to first-order exponential $F(n)$. Intuiting this is a challenging as it reflects non-local properties. It is doable, however, with practice.

In making substitutions, we keep track of the mass distribution using $\Phi(k)$ as in Figure Ten. $\Phi(k)$ brings to light domains as emphasized by the trend lines. These help us view a sequence (new and otherwise) in sections:

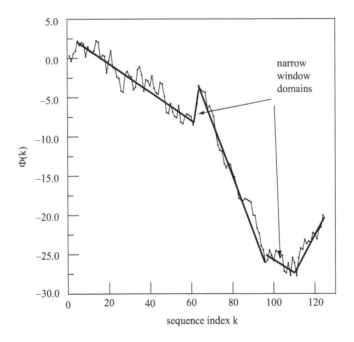

FIGURE TEN Cumulative mass distribution for wild-type lysozyme. The trend lines correspond to sectors of the sequence.

KVFERCELARTLKRLGMDGYRGISLANWMCLAKWESGYNTRATNYNAGDRS
TDYGIFQIN

SRYWC

NDGKTPGAVNACHLSCSALLQDNIADAVACA

KRVVRDPQGIRAWVAW

RNRCQNRDVRQYVQGCGV

The sections present high linear scaling as observed in the figure. By tracking the linearity, we get a feel for how a substitution disrupts or maintains domains of the mass distribution. For example, $\Phi(k)$ for the low-index domain demonstrates a linear correlation coefficient R^2 of 0.856. Trading, say, A for V raises this by a whisker, viz.

KVFERCEL<u>V</u>RTLKRLGMDGYRGISLANWMCLAKWESGYNTRATNYNAGDRST
DYGIFQIN

In contrast, if we substitute Y for A, we pull down R^2 severely to 0.662:

KVFERCEL<u>Y</u>RTLKRLGMDGYRGISLANWMCLAKWESGYNTRATNYNAGDRSTD
YGIFQIN

We also see where substitutions are most likely to jostle the distribution, in particular, the narrow-window domains marked by the lines. The smallness and high linearity of these sections arrive with enhanced sensitivity to mass changes.

The next property to track is the most challenging because it is open-ended. It is the degree of conservation for a substitution. Conservative substitutions involve trading one family member for another. But matters are not black and white because the amino acids accommodate multiple partitions and family ties as discussed in Chapter Three. An **A** → **L** substitution is conservative if we consider only the abnp-families—both components are non-polar. But **A** → **L** is non-conservative through the lens of external, internal, and ambivalent families: **L** is internal but **A** is ambivalent. How do we reconcile things?

The answer is simple qualitatively. Substitutions are conservative if the discarded and replacement components carry substantial information about each other. More information means more conservative and otherwise if the components share little in common. Each amino acid is a member of multiple families, the number depending on the partitions we prescribe. The shared information hinges on all the family ties.

This takes us to sample spaces and sets as represented schematically:

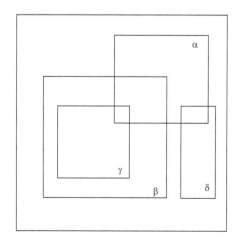

The component space is represented by the largest square while the enclosed squares and rectangles mark family spaces prescribed by various partitions. Consider the relationships among components α, β, γ, and δ. The degree of conservation depends not on the overlap of the family spaces per se, but rather the number of ties—the *number* of overlaps. Component α has ties to β, γ, and δ. But β and γ have no family members in common with δ. Then consider replacing α with β, γ, or δ. By counting ties, we have that β and γ offer the more conservative choices for substitution while δ offers conservation to a lower degree. Matters are fluid, however, because we can always apply more partitions to the space.

Conditional probability addresses the ideas quantitatively. A partition Π_i establishes Y_i number of mutually disjoint sets of the amino acids. The sets prescribed by partitions Π_i and Π_j may or may not have family members in common. Let us consider replacing component α with β and inquire about the degree of conservation. This is tantamount to asking about the probability that the two components *have* family ties. The term $prob(\beta|\alpha)$ asserts the probability that *given* that α belongs to a family in the component space, β is *also* a member of the family. We get a handle on the probability from the minimalist definition from Chapter Two:

$$prob(\beta \mid \alpha) = \frac{\text{number of } (\alpha, \beta) \text{ shared families}}{\text{total families}}$$

$$= \frac{\text{number of } (\alpha, \beta) \text{ shared families}}{\sum_i Y_i}$$

The summation in the denominator is over all sets of all partitions. The denominator terms depend on the partitions *we* apply to the component space. This is why conservation is open-ended in sequence writing.

We demonstrate conservation degree with an example. Consider substitutions $\mathbf{A} \to \mathbf{L}$ and $\mathbf{I} \to \mathbf{L}$, and $\mathbf{Q} \to \mathbf{E}$. Which is the more conservative? Which is least? Revisit the partitions from Chapter Three and the number Y of mutually exclusive sets in each:

1. External, Internal, Ambivalent: $Y_1 = 3$.
2. Acidic, Basic, Non-polar, Polar: $Y_2 = 4$.
3. Acidic, Basic, Aliphatic, Amide, Aromatic, Hydroxyl, Imino, Sulfur-Containing: $Y_3 = 8$.
4. Charged, Electrically-Neutral: $Y_4 = 2$.
5. Lightweight, Middle-Weight, Heavyweight: $Y_5 = 3$.

The partition and set properties take us to:

$$prob(L \mid A) = \frac{\text{number of } (L, A) \text{ shared families}}{\text{total families}} = \frac{0+1+1+1+1}{3+4+8+2+3} = 0.20$$

$$prob(L \mid I) = \frac{\text{number of } (L, I) \text{ shared families}}{\text{total families}} = \frac{1+1+1+1+1}{3+4+8+2+3} = 0.25$$

$$prob(E \mid Q) = \frac{\text{number of } (Q, E) \text{ shared families}}{\text{total families}} = \frac{1+0+0+0+1}{3+4+8+2+3} = 0.10$$

We witness the hierarchy of conservation and are able to explore other scenarios.

C   EXPANDING TEMPLATES TO OBTAIN NEW SEQUENCES

Every sequence presents a base structure in family terms, for example, lysozyme $\leftrightarrow a_{11}b_{20}n_{47}p_{52}$. This can be regarded as a template for expanding and writing new sequences. A template stakes out a miniscule fraction of 20^N space. Even so, it grants enormous territory for maneuvering, not to mention originality, for the *possible* abnp-spellings for $a_{11}b_{20}n_{47}p_{52}$ number:

$$\frac{130!}{11!20!47!52!} \approx 3.2 \times 10^{66}$$

Where and how do we start?

The answer is that, just as in Section B, we have to decide how far we wish to start from the *special* expansion of $a_{11}b_{20}n_{47}p_{52}$, i.e., that established by a natural protein. Do we want the apple to lie close to the tree, the next tree over, or somewhere in the next orchard? To see how things work, consider five examples, all with base $a_{11}b_{20}n_{47}p_{52}$:

bnnabpannbpnbbnpnappbpnpnnpnnpnnbnapppppbnppppnpabppappnnpnppbpnpp
apbpnpnnpnpbnpppnnnpapnnannnpnbbnnbanppnbnnnnnbpbpppbanbppnppppn   (9)

bnnabpannbpnbbn**np**appbpnpnnpnnpnnbnapppppnbppppnpabppappnnpnppbpnppa
pbpnpnnpnpbnpppnnnpapnnannnpnbbnnbanppnbnnnnnpbbpppbanbppnppppn   (10)

nnnnnnnnnnnnnnnnnnnnnnnnnnnnnnnnnnnnnnnnnnnnnnnnnppppppppppppppppppppppppppppppppppppppp
pppppppppppppppppppppppaaaaaaaaaaaabbbbbbbbbbbbbbbbbbbbbb   (11)

appnpbppbnnpabnapbnpnapbppppnpbnnpnpnpnnpanbpnapppnnnpbnnpbbnnbpbppppbnn
nnbppbpnnnnnppapappnannbnpppnbnnppppnpnpnpbnpnbbpnnpnppaanpnp   (12)

papnnpnnpppapnpppnpppnpbppbnanpbnppaanppnbppppnanbpnnbppnnnbpnpapbbnnbbnp
pnnnpbppbppnanpbnpnpnnnnnnbpannpbppbbnppbpnbnappppnnnnpann   (13)

Sequence (9) aligns perfectly with lysozyme while Sequence (10) is very close—the two variant sites are highlighted in bold underline. Sequence (11) is obviously miles away, but (12) and (13) are not so clear-cut. Are they near or far from the special at the low information levels?

The question returns us to the organic compounds discussed in Chapter Six. A given base in atomic terms allows multiple isomers. A few consonants with $C_7H_{10}O$ are as follows:

The corresponding base+(2) structures are as follows:

$$(C=C)_1 (C=O)_1 (C-C)_6 (C-H)_{10}$$

$$(C=O)_1 (C-C)_8 (C-H)_{10}$$

$$(C=C)_3 (C-O)_1 (C-C)_3 (C-H)_{10} (O-H)_1$$

Thus if six-member rings were *the* motif, and we wanted close-by variants, we might write:

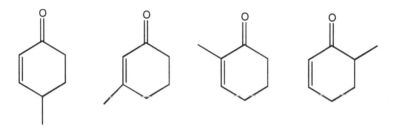

This makes sense because the above structures match the motif in base and base+(2) information. In contrast, if we wished to explore faraway territory, we might write:

The bases matches the motif, but diverge at the base+(2) levels.

The same ideas apply to proteins at the supra-atomic level. The base+(2) representations for the examples are as follows:

$(bb)_2(ba)_2(bp)_9(bn)_7(ab)_2(ap)_5(an)_4(pb)_7(pa)_5(pp)_{21}(pn)_{19}(nb)_8(na)_4(np)_{17}(nn)_{17}$ $\leftrightarrow$ Sequence (9)

$(bb)_2(ba)_2(bp)_9(bn)_7(ab)_2(ap)_5(an)_4(pb)_7(pa)_{5+1}(pp)_{21}(pn)_{19-1}(nb)_8\mathbf{(na)_{4-1}(np)_{17-1+1}(nn)_{17+1}}$

$\leftrightarrow$ Sequence (10)

$(bb)_{19}(ab)_1(aa)_{10}(pa)_1(pp)_{51}(np)_1(nn)_{46}$ $\leftrightarrow$ Sequence (11)

$(bb)_2(bp)_8(bn)_{10}(ab)_1(aa)_1(ap)_6(an)_3(pb)_{11}(pa)_5(pp)_{19}(pn)_{16}(nb)_6(na)_4(np)_{19}(nn)_{18}$ $\leftrightarrow$ Sequence (12)

$(bb)_3(bp)_{10}(bn)_7(aa)_1(ap)_4(an)_6(pb)_{10}(pa)_6(pp)_{20}(pn)_{16}(nb)_7(na)_4(np)_{17}(nn)_{18}$ $\leftrightarrow$ Sequence (13)

We measure distances in the usual way using vectors and angles whence close-lying structures have vectors that point in nearly the same direction. Four of the components for Sequence (9) are as follows:

$$\text{bb}: \sqrt{\frac{2}{130-1}} \quad \text{ba}: \sqrt{\frac{2}{130-1}} \quad \text{bp}: \sqrt{\frac{9}{130-1}} \quad \text{bn}: \sqrt{\frac{7}{130-1}}$$

The base of Sequence (10) demonstrates identical strengths along these axes. In contrast, the base of Sequence (13) presents:

$$\text{bb}: \sqrt{\frac{3}{130-1}} \quad \text{ba}: \sqrt{\frac{0}{130-1}} \quad \text{bp}: \sqrt{\frac{10}{130-1}} \quad \text{bn}: \sqrt{\frac{7}{130-1}}$$

The angle between the vectors arrives from the inverse cosine formula:

$$\theta_{(9),\,(13)} = \cos^{-1}\left\{\sum_{i=1}^{i=16}\left(prob_i^{(9)}\right)^{1/2}\cdot\left(prob_i^{(13)}\right)^{1/2}\right\}$$

The reader should verify that the angle works out to be 0.214 radians or 12.3°. Other calculations are just as straightforward.

It may seem that expansions close to the special are limited. But this is far from the case as seen by experiments: we form random expansions of $a_{11}b_{20}n_{47}p_{52}$ and measure how far their base+(2) information lies from lysozyme. This gives us a sense of the distributions allowed by $a_{11}b_{20}n_{47}p_{52}$. For example, the results for several thousand expansions are illustrated in Figure Eleven. The distribution is normal as can be shown by parametric testing. The base+(2) vector for lysozyme projects at 0.00 radians with respect to itself. In turn, lysozyme places at the far lower left indicated by the large filled circle. In contrast, the placement of Sequence (13) is at 0.214 radians marked by the open circle at the upper right. The mean distance from lysozyme is ca. 0.165 radians while the standard deviation is ca. 0.04266 radians. All observations can be translated to normal Z-scores, for example,

$$Z_{\text{lysozyme}} \approx \frac{0-0.165}{0.04266} \approx -3.87$$

$$Z_{\text{sequence (13)}} \approx \frac{0.214-0.165}{0.04266} \approx +1.15$$

The distribution placements and Z-scores of the base+(2) structures (11) and (12) are left as an exercise.

Then how many $a_{11}b_{20}n_{47}p_{52}$ expansions place close to lysozyme? Because we approximate the distributions using random expansions, we do not observe quite the same results with each experiment. There is some wobble in the procedure that positions lysozyme at slightly different distances from the mean of a distribution. The wobble corresponds to ca. 10% of the standard deviation in Figure Eleven.

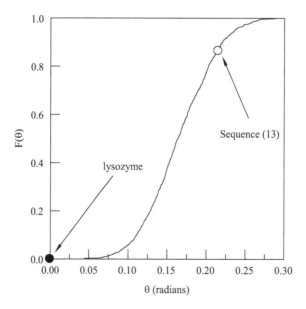

FIGURE ELEVEN Cumulative distribution of distances of random spellings of $n_{47}p_{52}a_{11}b_{20}$ from wild-type lysozyme. The distances are computed from vectors underpinned by base+(2) information in family terms.

Conveniently, this enables approximating the probability that a random expansion places at the same spot as lysozyme within experimental error. The probability is estimated by the normal law density:

$$prob \approx \frac{1}{\sqrt{2\pi}} \exp\left[\frac{-Z^2}{2}\right] \cdot \Delta Z \approx \frac{1}{\sqrt{2\pi}} \exp\left[\frac{-3.87^2}{2}\right] \cdot (0.10) \approx 2.2 \times 10^{-5}$$

The probability corresponds to one chance in 44,800. Point being made, perhaps surprisingly: we have plenty of choices for template expansions that stay close to lysozyme because the sequence space is so generous. The space is of size:

$$prob \times \frac{130!}{11!20!47!52!} \approx (2.2 \times 10^{-5})(3.2 \times 10^{66}) \approx 7.0 \times 10^{61}$$

There is a second point to make which should already be apparent. The sequence for lysozyme is *extraordinarily* selected and rarefied: even its four-letter version is one of 3.2×10^{66} possibilities. The nearest-neighbor information *alone* is also highly selected: the base+(2) structure in abnp-terms is one in 45K.

Then we have multiple ways to proceed with expansions. To access one of the 45K, we trade one or a few components from different families and track the distance traveled via the base+(2) information. For example, we have the following where the trade-sites have been marked in bold underline as follows:

bnnabpannbpnbbnpnappbpnpnnpnnpnnbn**n**pppppbnppppnpabppappnnpnppbpnppap bpnpnnpnpbnpppnnnpapnnannnpnbbnnbanppnbnnn**a**nbpbpppbanbppnppppn

The base+(2) information is then:

$$(bb)_2(ba)_2(bp)_9(bn)_7(ab)_2(\mathbf{ap})_{5\text{-}1}(\mathbf{an})_{4+1}(pb)_7(pa)_5(pp)_{21}(pn)_{19}(nb)_8(\mathbf{na})_{4\text{-}1+1}(np)_{17}(nn)_{17}$$

To take things home, we overlay the sequence using the 20-letter alphabet. To stay extra close to the motif, we maintain the base structure of lysozyme. The composition is automatically a close-lying permutation of the motif, for example,

KVFERCELARTLKRLGMDGYRGISLANWMCLAKW**AS**GYNTRATNYNAGDRS TDYGIFQINSRYWCNDGKTPGAVNACHLSCSALLQDNIADAVACAKRVVRDPQ GIRAWV**E**WRNRCQNRDVRQYVQGCGV

The two changes have been marked in bold underline. We have barely altered the mass distribution, N_D count, and long-distance relations. Moreover, the sequence maintains as an outlier on suprisal state plots as in Figure Four. To stray from the base of lysozyme means otherwise and requires diligence tracking the information changes.

The same guidelines apply when we explore expansions far from the motif. If we preserve the lysozyme base, but travel into uncharted base+(2) territory, we obtain letter strings like:

AMAIAMPPWVVILVIFIIVLWLFAVAALWWWLVAVAVALLAALAAANSGQGN YQCCSYTYSYTQGQSSGNYCSNQTYTNTNCGCGNNNGNCGCGQGCGDDEE DDEDDDDHKRKKKKKRRRRRRRRRRRR   (14)

PIVWPAAIAWWMMLLIVLIIWWFVFLVVLLAAVLLAVVVAAAAAAAASNNNT NQTGGGSNTNNCSYGGCNQTTSGQNNQCSYYGYYQYSQGGCCCCCGGEEEDD DDDDDDKKRHRRKKKRRKRRRRRRRRR   (15)

DQTVYHQTKFIGEKIECKATVEGRQTTYFGKLIQINVIQDLRSADCSGAMLGRPPG
RKLMRQRYYNCRVLVARYSRNLVVAVGSDCDGNWDLWRVNSSWRLWNCC
NWANANRAYARRNAAGAGCDDACAG   (16)

EGCWTHTNKIITEKPETKIYWDTRYSNGFQRLVYFSIASDAKQVDYGQLIWCKW
MYRRMLRSRQCNSRWPAVRGYRQLVALLCCDSDGNVDVARVQCNLRLVCN
GNAVNACRANARRNAAGAGGDDAGAG   (17)

YDNFATWANCYEQVSCYAYGSITKSYRLELCKAYGDDPTSLRTGQIDPRQFM
RTQWIVKCWSESHRLLKKIQQWAANRGNRGNLDLCRANLGIAAAVARCDVVGR
CCRRVGGRNVRADGNNNWVVMGDAA   (18)

QDGVIYPLNYTDGWGTNWSQNLNHQGKPELYRMYCEDLTTLKYNGMEAKY
WVRQTVALRSLQDQKRFLKRFNGIWVGRNGRGNADASRWNVNIIIVAVRGDAVG
RSCRRASCRCVRADSCCCAAAACDAA   (19)

The template expansions using the 20-letter alphabet correspond as follows: Sequences (14) and (15) derive from (11); Sequences (16) and (17) derive from (12); Sequences (18) and (19) derive from (13). Needless to say, we have we have no shortage of variations to explore. But how should we critique efforts such as above and beyond? Surely, the grammar in (14) and (15) is not in the spirit of natural sequences on so many accounts. For openers, the abnp-fractions are radically different across sections. We incur *infinite* Kullback–Leibler penalties if we assume that the fractions in one part of the sequence match another part. Just as glaring, the N_D counts fall miles short for $N = 130$ proteins. The more interesting question is Do Sequences (14) and (15) miss the boat in every respect? The answer is negative as shown by $\Phi(k)$ in Figure Twelve. We see how the cumulative mass distributions for both (14) and (15) scale more or less linearly with the sequence index. Moreover, they block out domains involving tens of amino-acid units, just like in natural proteins! Point being made: even blatantly abiotic structures do not necessarily fail on all fronts. This is important. The most primitive of sequences eons ago may not have been as scrambled as in modern day. Yet they still carried some of the global characteristics of natural proteins.

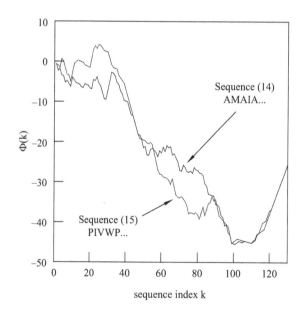

FIGURE TWELVE Cumulative mass distributions for abnp-template expansions corresponding to Sequences (14) and (15).

Sequences (16)–(19) can be similarly critiqued. On the positive side, they present randomly as in natural sequences. But certain features should raise flags. In particular, the Kullback–Leibler penalties are about eight times higher than for motif lysozyme. The long-distance relationships present as exponential distributions. The mass information in $\Phi(k)$ is even more diagnostic as shown in Figures Thirteen and Fourteen. In Figure Thirteen, we are pointed to the more favorable grammar portions of (16) and (17): that which scales more linearly and blocks out large-scale domains. In Figure Fourteen, we see that Sequences (18) and (19) do not offer mass information conducive to natural proteins. These are examples of compositions we should scratch and start anew! With so many sequences in 20^N space, we are bound to encounter many unfavorable ones.

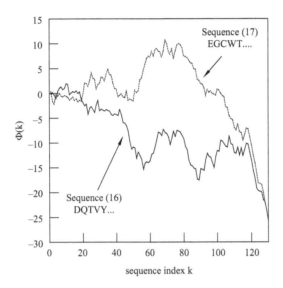

FIGURE THIRTEEN Cumulative mass distributions for abnp-template expansions corresponding to Sequences (16) and (17).

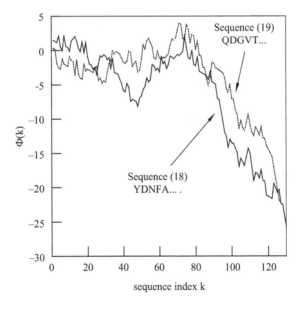

FIGURE FOURTEEN Cumulative mass distributions for abnp-template expansions corresponding to Sequences (18) and (19).

In closing this section, let us not give the impression that only random compositions are appropriate. Recall the keratins of Chapter Three having repeating units, for example,

ITYFAPMCNIITYFAPMCNIITYFAPMCNIITYFAPMCNIITYFAPMCNIITYFAP
MCNIITYFAPMCNIITYFAPMCNIITYFAPMCNIITYFAPMCNIITYFAPMC
NIITYFAPMCNIITYFAPMCNIITYFAPMCNIITYFAPMCNIITYFAPMCL

These are as natural as any proteins and provide templates just inviting as random ones, namely:

nppnnnnppnnppnnnnppnnnppnnnnppnnppnnnnppnnppnnnnppnnppnnnnppnnppnnnn
ppnnppnnnnppnnppnnnnppnnppnnnnppnnppnnnnppnnppnnnnppnnppnnnnppnnppnn
nnppnnppnnnnpn

The periodic sequences are distinctive for their high compressibility:

$$(\mathbf{ITYFAPMCNI})_{16}\text{ITYFAPMCL} \quad \leftrightarrow \quad (\mathbf{nppnnnnppn})_{16}\text{nppnnnnpn}$$

These can serve as motifs for writing new and interesting sequences, for example,

LGSPACNSLGSPACNSLGSPACNSLGSPACNSLGSPACNSLGSPACNSLGSPACNSL
GSPACNSLGSPACNS

The point is that the lack of randomness is not a deal breaker. Indeed BLAST searches with periodic sequences based on keratins generally result in abundant hits. The above example indeed is a close match with uncharacterized proteins from *Xenopus tropicalis*. Periodic sequences are the easiest to compose and follow the spirit of large numbers of proteins. This can be explored endlessly in writing projects.

D WRITING SEQUENCES FROM SCRATCH

The challenges of starting with a blank page or computer screen bring to mind sentiments expressed by the legendary pianist Arthur Rubinstein. He declared that piano playing was either easy or impossible; it was never hard. This is subject to debate among pianists, of course. Yet arguably, an analogous sentiment is spot-on for sequence writing. Substituting motifs and expanding templates are pretty easy whereas biochemically-plausible compositions from scratch would seem impossible. With every unit put to page or screen, for example, **RLEA...**, we are placing a point in a 20^N space, viz.

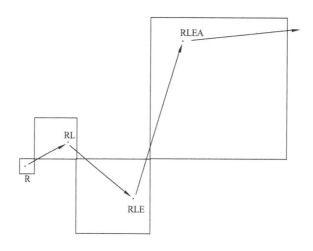

Writing a sequence means hop-scotching through an exponentially-expanding territory. To hit points sampled by evolution or could well be sampled in the future has vanishing probability. But as in piano playing, getting close to phrases or hitting a few right notes is not out of the question. Sometimes we write a sequence containing fragments that overlap entries in databases. The point we place in a 20^N space, for example, **RLEAGSD...** (upper text, open circle), turns out to be neighbor to a protein (lower text, filled circle) encoded by a real genome, viz.

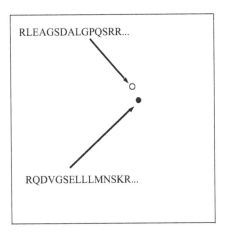

RLEAGSDALGPQSRR...

RQDVGSELLLMNSKR...

How do we compose from scratch? The question takes us back to Chapter Five concerning the base populations of proteomes. Recall that in modeling distributions, we had to construct bases that were skewed with respect to components. Egalitarian distributions—5% A, 5% V, 5% L, etc.—were to be avoided because the end products were too close information-wise. They presented diversity incommensurate with the complexity of a biological system. Besides, sequences with uniform component fractions do not express as outliers on surprisal plots such as Figure Four. Concomitantly, bases needed to reflect canonical selection weights among the fundamental families. When it comes to writing, we have to keep these guidelines in mind along with base+ information, both alphabetical and numerical. We illustrate four examples in this concluding section accompanied by critiques. Further explorations are encouraged as exercises and indeed projects. Writing a sequence is not unlike a board game for which a few strategies are well established. The illustrations here feature $N = 130$, same as for lysozyme.

Our lexicon is the familiar {A, V, L, I, P, F, W, M, G, S, T, C, Y, N, Q, D, E, K, R, H} which we keep in full view during writing—we do not want to overlook components as we are prone to do when writing at whim. To begin the first example, we may recall that leucine **L** is the component most expressed by proteomes. It is then fully reasonable to initiate a composition with **L**. We get the ball rolling by writing:

L

We further appreciate that proteins give the appearance of random component selection. We would subsequently bet that **LL** is a commonly expressed word. We write:

LL

and feel at ease. But non-polars alone do not make a useful molecule. To add a surface component, we write, say,

LLD

But then again, non-polars *are* the major components and the number for any single non-polar needs to be skewed (cf. Chapter Five). Thus we write with justification:

LLDIL

And so it goes. We continue adding letters with one eye on the lexicon and the other on the composition. We try to keep the unit placements as random as possible by not over-thinking. Eventually our efforts yield:

LLDLILLELLEPLLICSAELGCLADLRTTEFELCSLNTNSNFIEEFGAECQRMGM
RRASMSVLAGLNSTNGDTECNNSNKKEPVGQSVTPLRIKLNCVSTLTALICF
VCLAVGSSTTLDTSEPYQVPLEL   (20)

Sequence (20) bears traits of bona fide proteins, as should be clear. It hosts eight cysteines at scattered sites in bold font, thus accommodating four cross links. Further, we have avoided periodicity from start to finish. But before we take any bows, we need to critique the writing along several dimensions. It can only serve as a crude draft and indeed, a few tests will show it to be glaringly ungrammatical and deserving of the wastebasket.

Critiques begin by tabulating the base structure in component and family terms:

$$A_7V_7L_{24}I_5P_5F_4W_0M_3G_7S_{12}T_{11}C_8Y_1N_9Q_3D_4E_{12}K_3R_5H_0 \leftrightarrow a_{16}b_8n_{55}p_{51}$$

Already we see that the a,b-fractions are inverted from what they should be according to canonical weights. The individual components are skewed, but overly so: there is excessive L and absent W and H. Pictures fill out critiques as in Figure Fifteen. The panels show $\Phi(k)$, $F(n)$, and the base and sequence surprisal placements in component and family terms. We observe that $\Phi(k)$ follows the model of canonical proteins given the linearity—we succeeded on this account. However, $F(n)$ deviates from first-order exponential while the surprisal placements are overly remote. Delving deeper, we inspect N_D and the Kullback–Leibler information. We count $N_D=90$—the vocabulary is woefully deficient. The information makes for a significant penalty comparing the abnp-distributions of the $1 \leq k \leq 65$ and $66 \leq k \leq 130$ sectors. The penalty is about ten times that of typical proteins.

Critiques lay the ground for corrections. To improve **LLDL...**, we would enrich the word count by losing L in places and incorporating W and H into the base. This will boost N_D to more reasonable levels. We would further smooth the abnp-distributions through interchanges to lower the Kullback–Leibler penalties. Out of curiosity, we should examine databases to see if **LLDL...** comes close to anything. Not surprisingly, it does as a BLAST search finds overlap with a bacterial GTP-cyclase:

```
Query    20   LGCLADLRTTEFELCSLNTNSNFIEEFGAECQRMGMRRASMSVLAGLNSTNGDTEC
              NNSN    79
              +  L +LR    +  SL TN+ ++E++   E ++ G+    ++S    L+S N DT  N +
Sbjct   110   INTLNELRPLGLQRISLTTNAYYLEKYAVELKQAGLDDLNIS----LDSINPDTFFN
              MTQ    165
Query    80   KK-EPVGQSVTPLR-----IKLNCV    98
              K   EPV + +    +      IKLNCV
Sbjct   166   KPLEPVLKGIHAAKDVNIPIKLNCV    190
```

The closeness should not be overly interpreted. It only demonstrates that attention to a handful of information properties gets from-scratch compositions into the ballpark of natural proteins. Even deficient piano playing gets a few notes and phrases right.

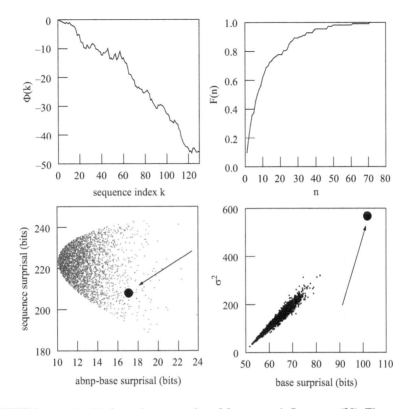

FIGURE FIFTEEN   Low-level information properties of from-scratch Sequence (20). The upper left panel shows the cumulative mass distribution. The upper right panel illustrates $F(n)$. The placements of the abnp-base and sequence surprisals appear in the lower left panel. The state point for the base surprisal and deviation is shown in the lower right panel. The small circles in the lower panels apply to random sequences and bases.

Our second example learns from the experience of the first. In particular, we try to give greater attention to the (ordinarily) minor components. Concomitantly we aim for a richer vocabulary so as to not fall short on N_D. At the same time, we keep the writing as random as possible and avoid periodicities. We begin with, say,

FRAD

This is elaborated upon to obtain:

FRADRDQTDFFWFKDLRAKKARNQKSMQFERQMTPPRRQGMFAAIFATLLQDF
EGDVEQIRGVQPMALKECEGSRKMAPLTCPQRACKMRARESLRAPALRNSCFA
RRLVGFFSNSMWPQVPSATDEMVK   (21)

This seems an improvement over Sequence (20). For openers we count $N_D = 103$ and have kept the cysteines at an even number 4. Tabulating the base structures gives:

$$A_{14}V_5L_8I_2P_8F_{11}W_2M_8G_5S_7T_5C_4Y_0N_3Q_{10}D_7E_7K_8R_{16}H_0 \leftrightarrow a_{14}b_{24}n_{58}p_{34}$$

The individual components appear more reasonably skewed. However, we have wrongly weighted the polars and non-polars. The panels in Figure Sixteen provide more extended critiques. On the positive side, the surprisal placements are satisfactory and $F(n)$ diverges only a little from first-order

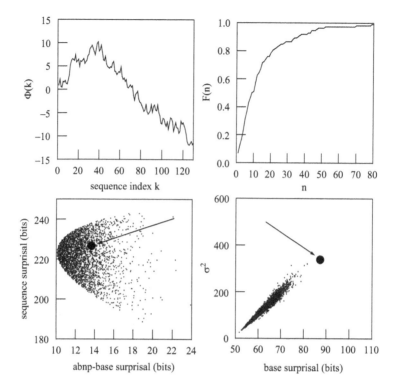

FIGURE SIXTEEN   Low-level information properties of from-scratch Sequence (21). The upper left panel shows the cumulative mass distribution. The upper right panel illustrates $F(n)$. The placements of the abnp-base and sequence surprisals appear in the lower left panel. The state point for the base surprisal and deviation is shown in the lower right panel. The small circles in the lower panels apply to random sequences and bases.

exponential. On the negative side, FRAD… misses the boat in the mass distribution. The sequence is freighted with heavyweight units, especially at the beginning. Critiques point to corrections. In a second draft, we would substitute several F for lighter weight components—non-polar ones at that. The grammatical deficiencies notwithstanding, **FRAD…** overlaps slightly with a natural protein, specifically a glycoprotein encoded by *Rhizoctonia solani*:

```
Query   35   PPRRQGMFAAIFATLLQDFEGDVEQIRGVQPMALKECEGSR   75
             PP + GMF++IF+ + ++FE  VE +RGV P +  + E  R
Sbjct   23   PPAKTGMFSSIFSAVSREFESFVENVRGVDPSSRVQAEEPR   63
```

As is typical, scratch-compositions come close in a few respects to natural proteins.

Our third example takes lessons from the first two. We keep the lexicon A V L I P… in view while writing and give components and family members their due. We practice a rich vocabulary and avoid periodicity. To get the ball rolling, we write:

GLGLQY

In so doing, we are trying to balance the light- and heavyweight amino acids and aim for high word counts. Eventually these efforts take us to:

GLGLQYGVGYDGGPQSSIIPGNGAAAVVTHLTTGSLNNQLAPSNSGSLIKE
AGNDGCAVGLHVGYPFWLLKEGQVHMKNLLTSVYTIVNDMGERLVKRTAQ
NEEGETSRWIHDTNFSKSCDARWNKAVAP   (22)

The base structures are as follows:

$$A_{10}V_{10}L_{12}I_5P_5F_2W_3M_2G_{17}S_{10}T_8C_2Y_4N_{10}Q_5D_5E_6K_6R_4H_4 \leftrightarrow a_{11}b_{14}n_{49}p_{56}$$

This time, all 20 components are featured and look to the frames in Figure Seventeen for additional counsel.

We did better than the first two examples! GLGLQY… is an outlier on the surprisal placements, but not overly so. We have dispersed an even number of cysteine units and $F(n)$ is nearly exponential. On the downside, our attempt to shore up the non-polars meant too heavy a hand and we neglected the basics and polars. Mass-distribution-wise, $\Phi(k)$ defines two linear domains and is biased toward the lightweight components. Not unexpectedly, the composition comes close to database proteins, for example, a ribosomal protein encoded by *Bacillus zhangzhouensis*, viz.

```
Query    3    GLQYGVGYDGGPQSSIIPGNGAAAVVTHLTTGSLNNQLAPSNSGSLIKEAGNDGC
              AVGLH    62
              G+Q G+     GP++ I PGN    +    + T   N +L P   G L++ AG     +G
Sbjct  109    GIQVGIEVTSGPEADIKPGNALPLINIPVGTVVHNIELKPGKGGQLVRSAGTSAQ
              VLGKE   168
```

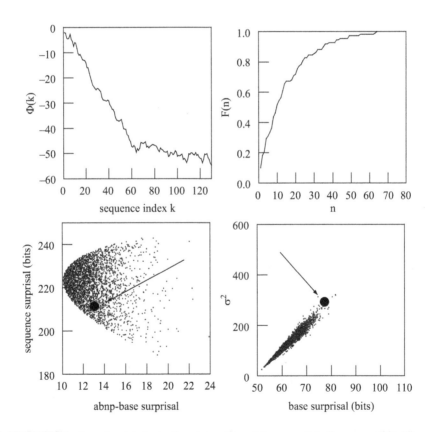

FIGURE SEVENTEEN Low-level information properties of from-scratch Sequence (22). The upper left panel shows the cumulative mass distribution. The upper right panel illustrates *F(n)*. The placements of the abnp-base and sequence surprisals appear in the lower left panel. The state point for the base surprisal and deviation is shown in the lower right panel. The small circles in the lower panels apply to random sequences and bases.

We are writing biological sentences while keeping a few principles in mind. And we are doing considerably better than our Chapter Six attempt where we typed letters off the top of our head. All the same, the results prove mixed as shown by critiques. These examples and more point to another method, one that entails some preparation.

Instead of writing sequences with the lexicon in view, let us use outlines: base structures that conform to information properties we desire for a sequence. Outlines have sections, right? Hence we go further by dividing bases into groups that are balanced in the abnp-distributions. We use the groups for constructing two or more sequences which can then be combined. As always, results can be critiqued to point the way to second drafts. Essays follow deliberate outlines and thermodynamic behavior follows specific components and mole amounts. This is the best model for writing sequences from scratch.

For example, we write for an $N = 130$ base:

$$A_{10}V_8L_{15}I_3P_4F_2W_2M_5G_{12}S_4T_8C_8Y_2N_5Q_8D_6E_7K_6R_{12}H_3 \leftrightarrow a_{13}b_{21}n_{49}p_{47}$$

We divide this into two pieces of similar size:

$$A_5V_4L_8I_1P_2F_1W_1M_3G_6S_2T_4C_4Y_1N_3Q_4D_3E_4K_3R_6H_2$$
$$A_5V_4L_7I_2P_2F_1W_1M_2G_6S_2T_4C_4Y_1N_2Q_4D_3E_4K_3R_6H_1$$

We initiate writing with *this information* in view in place of the lexicon. We learn from databases that sequences frequently begin with M. We use this to get the ball rolling and write:

MLQA

Elaboration takes us to

MLQASRRQVGLYERGTKRLLVHFLTTPLEETDEISANGVNDQACVLACKPCGQRK WGCGLNDHMAMR

The second part need not start with M. We write, say,

APIY

Elaboration takes us to:

APIYFTMCRTCTTEMELLCLRDPLVQGGAIVRDRKEKGDCNALAVGASGQSLN VHGLQQWKRR

Combining the two sections gives:

MLQASRRQVGLYERGTKRLLVHFLTTPLEETDEISANGVNDQACVLACKPCGQRK WGCGLNDHMAMRAPIYFTMCRTCTTEMELLCLRDPLVQGGAIVRDRKEKGDCN ALAVGASGQSLNVHGLQQWKRR   (23)

We are taking our cues from the outlines and not so much from the lexicon. The latter provides insufficient support for our limited internal-processing skills. Figure Eighteen illustrates the critique panels for evaluation. Note how the composition succeeds along several fronts of global information. It is emblematic of what can be obtained by attention to base and base+ structures. What lab

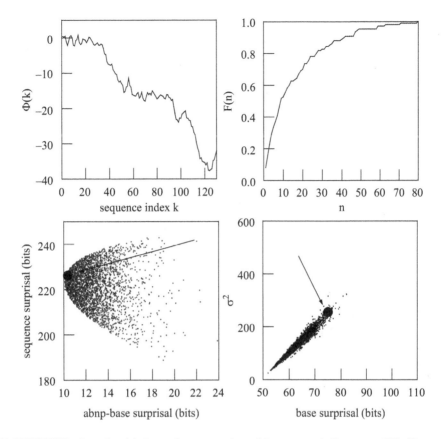

FIGURE EIGHTEEN Low-level information properties of from-scratch Sequence (23). The upper left panel shows the cumulative mass distribution. The upper right panel illustrates *F(n)*. The placements of the abnp-base and sequence surprisals appear in the lower left panel. The state point for the base surprisal and deviation is shown in the lower right panel. The small circles in the lower panels apply to random sequences and bases.

experiments does it motivate? How would such a protein fold? What signaling or catalytic functions would it carry? These are the most intriguing questions.

The major points of this chapter are as follows:

1. Writing sequences means placing points in spaces of size 20^N. Low-level information is, by no stretch, the end-all and be-all. However, attention to the low levels is time well spent as the global information provides critiques and guidance. The more straightforward methods entail substituting motifs and expanding templates with attention to word counts, mass distribution, and interval structures.
2. The degree of conservation for a substitution is fluid. It depends on the partitions established in advance or over the course of writing.
3. Composing biologically-plausible sequences from scratch is the most challenging. It profits from groundwork whereby a canonical base structure is first established and then divided into sectors. This is the outline method of writing protein sequences.
4. It is typical that newly-composed sequences have neighbors in 20^N space visited by evolution. Attention to global information properties gets sequences close to biologically relevant proteins.

EXERCISES

1. Consider the sequence space available to $N=130$ proteins. The primary structure of lysozyme corresponds to one point in the space. If an analog model of the space has area 100 square miles, how many points lie in one square Angstrom?

2. Revisit the first set of base structures of Section B:

$A_{17}V_8L_8I_5P_2F_2W_3M_2G_8S_{14}T_4C_7Y_5N_9Q_5D_8E_3K_5R_{14}H_1$ $\leftrightarrow$ Sequence (1)
$A_8V_9L_9I_7P_2F_4W_4M_4G_{10}S_8T_5C_7Y_9N_7Q_6D_7E_4K_6R_{12}H_2$ $\leftrightarrow$ Sequence (2)
$A_{11}V_6L_{10}I_8P_2F_3W_4M_3G_9S_4T_8C_7Y_5N_9Q_{10}D_7E_4K_7R_{11}H_2$ $\leftrightarrow$ Sequence (3)
$A_{13}V_6L_7I_5P_4F_6W_3M_3G_{10}S_9T_7C_6Y_8N_8Q_4D_8E_3K_7R_{10}H_3$ $\leftrightarrow$ Sequence (4)
$A_{14}V_9L_8I_5P_2F_2W_5M_2G_{11}S_6T_5C_8Y_6N_{10}Q_6D_8E_3K_5R_{14}H_1$ $\leftrightarrow$ lysozyme

   Consider information vectors underpinned by the bases and take lysozyme's to mark an origin. How far from the origin do the other structures place? Sketch a plot of the type $g(\theta)$ discussed in Chapter Seven.

3. Revisit Sequence (1) listed in Section B:

   KVFERCELARTLKRLGMDGYRGISLANWMCLAKWESSSSSRATNYNAGDRSTD
   YGIFQINSRYWCNDGKTPGAVNACHLSCSALLQDNIADAVACAKRVVRDPQ
   GIRAAAAARNRCQNRDVRQYVSSSSV   (1)

   The sequence was shown to be N_D-deficient. Propose modifications which bring N_D up to speed.

4. Sequence (5) in Section B presented a number of awkward substitutions. Can you identify their locations? Why are they awkward?

5. Revisit the second set of base structures in Section B:

$A_{12}V_{12}L_8I_5P_2F_2W_3M_2G_8S_5T_4C_7Y_5N_9Q_5D_8E_8K_5R_{19}H_1$ $\leftrightarrow$ Sequence (5)
$A_{10}V_8L_5I_4P_3F_2W_5M_2G_6S_9T_7C_7Y_4N_8Q_9D_{11}E_4K_7R_{12}H_7$ $\leftrightarrow$ Sequence (6)
$A_{14}V_9L_8I_6P_5F_3W_5M_3G_{10}S_4T_5C_8Y_3N_7Q_8D_9E_4K_5R_{13}H_1$ $\leftrightarrow$ Sequence (7)
$A_{12}V_8L_6I_5P_2F_4W_7M_3G_{11}S_9T_6C_7Y_5N_{10}Q_5D_5E_3K_6R_{15}H_1$ $\leftrightarrow$ Sequence (8)

   Which base lies closest to lysozyme?

6. Sequence (5) in Section B is N_D-deficient. Propose modifications that address this.

7. Consider the sequence and base information of lysozyme. Propose variants with abnp-conservative substitutions to increase the distance incrementally.

8. Repeat Exercise Seven, only now let the substitutions be non-conservative.

9. Consider the following $N = 130$ sequence written from scratch:

   EEKEQVERREVLKGRNLGKKNTPNVRERRFEKSLSDSNGEKGTFRHREEKEED
   KGVLEVSEHLDKVPLVELGDLLDRAVEDMCQPLCGKYHSVLLEGV
   SNLLPLKIVPDDQEEVAFPCYLGEWCQQKECV   (24)

   Critique the composition with the assistance of Figure Nineteen.

10. Revisit Sequence (23) of the from-scratch designs:

    MLQASRRQVGLYERGTKRLLVHFLTTPLEETDEISANGVNDQACVLACKPCGQR
    KWGCGLNDHMAMRAPIYFTMCRTCTTEMELLCLRDPLVQGGAIVRDRK
    EKGDCNALAVGASGQSLNVHGLQQWKRR   (23)

    Use a BLAST search to learn what proteins share properties with the sequence.

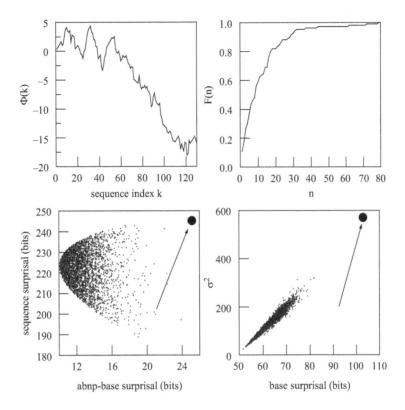

FIGURE NINETEEN Low-level information properties of from-scratch Sequence (24) in Exercise Nine. The upper left panel shows the cumulative mass distribution. The upper right panel illustrates *F(n)*. The placements of the abnp-base and sequence surprisals appear in the lower left panel. The state point for the base surprisal and deviation is shown in the lower right panel. The small circles in the lower panels apply to random sequences and bases.

NOTES, SOURCES, AND FURTHER READING

The design of proteins is an endless frontier. So many properties have to be taken into account: electronic stability, solubility, folding kinetics, surface interactions, and more. The literature is rich with research along these lines, only an ocean drop of which is listed here [5–18]. The approach taken in this chapter has been toward the craft of designing proteins. The focus has been restricted to the base and primary structure levels for their accessibility sans sophisticated computation or directed evolution. Emphasis has been placed on the more rational choices and the means to critique sequences.

1. Copeland, R. A. 2005. *Evaluation of Enzyme Inhibitors in Drug Discovery*, Wiley, Hoboken, NJ.
2. Karlin, S., Buche, P., Brendel, V., Altschul, S. F. 1991. Statistical methods and insights for protein and DNA sequences, *Annu. Rev. Biophys. Biophys. Chem.* 23, 407–439.
3. Jaynes, E. T. 1979. Where Do We Stand on Maximum Entropy? in *The Maximum Entropy Formalism*, M. Tribus, R. D. Levine, eds., MIT Press, Cambridge, MA.
4. Applebaum, D. 1996. *Probability and Information: An Integrated Approach*, Cambridge University Press, Cambridge, UK.
5. Jardetsky, O., Lefèvre, J.-F., Holbrook, R. 1998. *Protein Dynamics, Function, and Design*, Plenum Press, New York.
6. Jackel, C., Kast, P., Hilvert, D. 2008. Protein design by directed evolution, *Ann. Revs. Biophys. Chem.* 37, 153.

7. Watson, J. D., Laskowski, R. A., Thornton, J. M. 2005. Predicting protein function from sequence and structural data, *Curr. Opin. Struct. Biol.* 15(3), 275–284.

8. Broglia, R. A., Shakhnovich, E. I., eds. 2001. *Proceedings of the International School of Physics*, Course CXLV, Protein Folding, Evolution, and Design, IIOS Press, Amsterdam.

9. Pande, V. S., Grosberg, A. Y., Tanaka, T. 2000. Heteropolymer freezing and design: Toward physical models of protein folding, *Rev. Mod. Phys.* 72, 259–314.

10. Dewey, T. G. 1997. The algorithmic complexity of a protein, *Phys. Rev. E.* 56, 4545.

11. Fiser, A. 2009. Comparative Protein Structure Modeling, in *From Protein Structure to Function with Bioinformatics*, D. J. Rigden, ed., Springer, Berlin, p. 57.

12. Bowie, J. U., Luthy, R., Eisenberg, D. 1991. A method to identify protein sequences that fold into a known three-dimensional structure, *Science* 253, 164–170.

13. Wootton, J. C. 1994. Sequences with "Unusual" amino acid compositions, *Curr. Opin. Struct. Biol.* 4, 413–421.

14. Köhler, V., ed. 2014. *Protein Design: Methods and Applications*, 2nd ed., Humana Press, New York.

15. Jensen, K. J. 2009. *Peptide and Protein Design for Biopharmaceutical Applications*, Wiley, Hoboken, NJ.

16. Guerois, R., Lopez de la Paz, M. 2006. *Protein Design: Methods and Applications*, Humana Press, Totowa, NJ.

17. Svendsen, A. 2004. *Enzyme Functionality: Design, Engineering, and Screening*, Marcel Dekker, New York.

18. Angeletti, R. H. 1998. *Proteins: Analysis and Design*, Academic Press, San Diego, CA.

Nine Horizons

We circle back to themes of the text, sources, and further reading: the complexities of protein information at multiple structure levels. A few schematics and equations offer a computational perspective which points to questions to consider for the future.

A protein presents levels of structure information as discussed in Chapter One and beyond. The levels have motivated decades of physical and chemical studies of the molecules as individuals. These are the best of times given the power and economics of gene expression, diffraction, magnetic resonance, and computational techniques—to name a few. At the same time, proteins are the products of evolution. Thus, all their facets are geared toward dynamic, cooperative behavior in systems far from equilibrium. This necessitates the structure information to be thoroughly intertwined, both within and across species. Archetypes such as lysozyme are important to investigate in solid and liquid phases, and in silico. However, their full-bore significance obtains only in living cellular environments. Why is lysozyme's sequence **KVF**ER...**GC**GV and not the close-lying anagram **KFV**ER...**CG**GV? The answer most likely involves all elements of the human genome.

A way of thinking about protein information goes beyond labels—transferases, isomerases, defensins, etc. An enzyme ferries hundreds to thousands of functional groups through Brownian motion. Much the same groups are carried by cohort proteins inside a cell. In turn, the proteins overlap substantively in shape, solubility, and charge properties. We point out that analogous situations feature in modern day computation [1–3]. Hardware engages specific operations to complete a task. The same operations figure in countless other tasks, some yet to be engineered or in the rearview mirror. The software that directs the hardware is labeled according to overall purpose—sorting, factoring, etc. The brevity is out of convenience and does no justice to the details of operations and overlap with others. Labels fall especially short in parallel computations where multiple processors are coordinated. A few schematics and equations reflect on the intricacies of computation. The purpose is to draw analogies and, in turn, highlight questions for the future about protein information.

Parallel computations present events that have to happen *and* be perfectly coordinated: register shifts, twos-complement operations, and so forth. The events can't transpire just in any order or independently. Procedure A must execute in conjunction with B, C, and D to lay the groundwork for E, F, and beyond. We can think of the events as presenting a finite space:

Each point in the space corresponds to a fundamental operation while neighborhoods apply to sets of operations. Parallel computations require the thorough coordination of algorithms and hardware. These can be imagined as covers with jurisdictions: a given cover (circle) holds sway over neighborhoods of the event space.

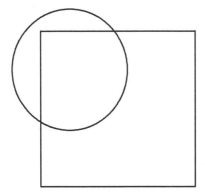

A cover need not be restricted to the space, for it can hold potential for directing operations outside the space. It can feature excess capacity which is useful for applications down the road.

For computations to succeed, a thorough covering of the space is required via multiple algorithms:

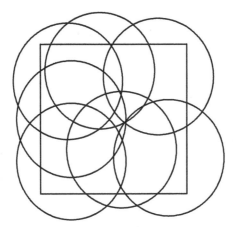

This is not an easy bar to scale or maintain since systems are subject to defects. These can include holes.

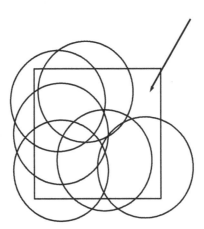

The events in hole-regions need to happen and in a coordinated fashion. However, they lack the requisite algorithms and the system must make do with the remaining ones. Other defects express when algorithms become corrupt (dotted circles) and are unable to do their job. The manifestations include straying from jurisdictions or the event space altogether.

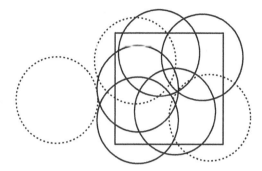

Corrupt algorithms may direct some operations, but not ones vital to the system. Where they are not part of the solution, they are the causes of multiple problems. Another defect is where algorithms not only fall down on the job, but also get in the way of neighbors. They occlude the covering ability of neighbor algorithms as suggested by the filled circles:

Computational systems are nonequilibrium thermodynamic at heart: they require input work and dissipate heat in the processing of information [4]. When they are in cruise mode, everything works fine and the conditions are approximately steady-state. When defects nucleate and propagate, however, deleterious processes become overriding and potentially fatal.

A few equations draw upon the schematics. Let us represent a state of the system by a vector with dimension D equal to the number of covers $\leftrightarrow$ algorithms:

$$\vec{V} = a_1\hat{x}_1 + a_2\hat{x}_2 + a_3\hat{x}_3 + \cdots + a_D\hat{x}_D$$

$$= \sum_{i=1}^{i=D} a_i\hat{x}_i$$

where $\hat{x}_i$ are orthogonal unit vectors, one for each algorithm and the a_i represent weight coefficients. Needless to say, the coordination of events means a finely-tuned state that stands apart from all others. It requires that *all* the algorithms do their job perfectly *and* cooperatively. The state vector reflects this specialness via weight coefficients that are uniformly positive, more or less equal, and span the space of events. No algorithm is more crucial than another, hence the egalitarian weights. The vectors for dysfunctional states are markedly different: they express disparate mixtures of weight coefficients, positive, negative, and zero (or nearly so) in places. They reflect that some of the algorithms are over-expressed, getting in the way of others, or failing outright.

Matrix-eigenvalue and -eigenvector equations offer models for multi-dimensional, cooperative systems. They are well established in studies of molecular orbitals, coupled oscillators, and electric circuits, to name a few. We direct this equation genre by writing:

$$\tilde{M}\vec{V} = \lambda\vec{V}$$

We can think of matrix $\tilde{M}$ as composed of real numbers that weigh how far the covers $\leftrightarrow$ algorithms place from one another in their information. They can't succeed zero distance apart because then computational diversity would be lacking. Yet the algorithms cannot function properly if they are too far apart, for their cooperation is then imperiled. It is as far from algorithm A as it as to B and vice-versa. Thus, $\tilde{M}$ is symmetric with elements:

$$m_{ij} = m_{ji}$$

Further, an algorithm works at zero distance from itself. Thus, for all diagonal elements,

$$m_{ii} = m_{jj} = 0$$

We are describing special case of a matrix with *total* positivity: the values of all $k \times k$ determinants so imbedded are greater than or equal to zero [5,6]. Such matrices are wellsprings for unusual sets of eigenvalues and eigenvectors. In the application at hand, *one* value- and vector pair is guaranteed to stand apart from the others.

We bring proteins to the table by a change in vocabulary. Event spaces translate to sets of processes which transpire in parallel and cooperatively in a biological system: transcription, replication, oxidation-reduction, etc. The hardware obtains by the proteins encoded by the genome. The algorithms or software is the structure information beginning with amino acid composition. One protein constitutes an *integrated* hardware-software component that impacts events in conjunction with neighbor proteins.

For a simple application, consider the sets of proteins encoded by viruses. The elements of sets are relatively few in number and task-specific. Further, their information and functions are highly evolved and cooperative and, in most cases, subject to regular updates: some viral proteomes evince hardware and software changes on a yearly basis.

For an example, consider the 12 proteins encoded by influenza A viruses. The sequences for a particular strain (Hong Kong, 1977) are listed in part as follows along with labels in bold font. The complete sequences are readily available from the Influenza Research Database (IRD).

PB2: MERIKELRNLMSQSRTREILTKTTVDHMAIIKKYTSGRQEKNPSLRM....
PB1: MDVNPTLLFLKVPAQNAISTTFPYTGDPPYSHGTGTGYTMDTVNRT....
PB1-F2: MGQEQGTPWIQSTGHISTQKGEDGQKTPKLEHRNSTRLMGHYQKT....
NS2: MDPNTVSSFQDILMRMSKMQLGSSSEDLNGMITQFESLKLYRDSLGE....
NS1: MDPNTVSSFQVDCFLWHVRKQVADQELGDAPFLDRLRRDQKSLRGRG....
MP2: MSLLTEVETPIRNEWGCRCNDSSDPLVVAASIIGILHLILWILDRLFFKCIYR....
MP1: MSLLTEVETYVLSIVPSGPLKAEIAQRLEDVFAGKNTDLEALMEWLKTRPIL....
PA: MEDFVRQCFNPMIVELAEKAMKEYGEDLKIETNKFAAICTHLEVCFMYSDF....

PAX:MEDFVRQCFNPMIVELAEKAMKEYGEDLKIETNKFAAICTHLEVCFMYSDF....
NA: MNPNQKIITIGSICMAIGIISLILQIGNIISIWVSHSIQTGSQNHTGICNQRIITYE....
HA:MKAKLLVLLCALSATDADTICIGYHANNSTDTVDTVLEKNVTVTHSVN....
NP: MASQGTKRSYEQMETDGERPNATEIRASVGKMIDGIGRFYIQMCTELKLSDYE....

The proteins are close in their low-level information. If we apply the base+(2) abnp-vector methods of Chapters Six and Seven, we quantify all the distances in radians. These can serve as the matrix elements m_{ij} and matrix symmetry and positivity with zeroes along the diagonal. The off-diagonal elements are nonzero and not too far apart in magnitude. The set of influenza proteins underpins the following matrix equation; the row/column numbering corresponds to the above order of sequences: **PB2** $\leftrightarrow$ 1, **PB1** $\leftrightarrow$ 2, etc.

$$
\begin{bmatrix}
0 & 0.121 & 0.371 & 0.227 & 0.173 & 0.208 & 0.125 & 0.157 & 0.162 & 0.208 & 0.167 & 0.0987 \\
0.121 & 0 & 0.326 & 0.224 & 0.178 & 0.205 & 0.107 & 0.165 & 0.182 & 0.165 & 0.111 & 0.0921 \\
0.371 & 0.326 & 0 & 0.481 & 0.420 & 0.450 & 0.390 & 0.433 & 0.430 & 0.319 & 0.305 & 0.315 \\
0.227 & 0.224 & 0.481 & 0 & 0.231 & 0.248 & 0.216 & 0.185 & 0.207 & 0.293 & 0.271 & 0.229 \\
0.173 & 0.178 & 0.420 & 0.231 & 0 & 0.139 & 0.189 & 0.106 & 0.144 & 0.279 & 0.234 & 0.148 \\
0.208 & 0.205 & 0.450 & 0.248 & 0.139 & 0 & 0.226 & 0.142 & 0.159 & 0.293 & 0.257 & 0.185 \\
0.125 & 0.107 & 0.390 & 0.216 & 0.189 & 0.226 & 0 & 0.193 & 0.214 & 0.195 & 0.178 & 0.150 \\
0.157 & 0.165 & 0.433 & 0.185 & 0.106 & 0.142 & 0.193 & 0 & 0.0877 & 0.257 & 0.212 & 0.157 \\
0.162 & 0.182 & 0.430 & 0.207 & 0.144 & 0.159 & 0.214 & 0.0877 & 0 & 0.298 & 0.242 & 0.157 \\
0.208 & 0.165 & 0.319 & 0.293 & 0.279 & 0.293 & 0.195 & 0.257 & 0.298 & 0 & 0.102 & 0.176 \\
0.167 & 0.111 & 0.305 & 0.271 & 0.234 & 0.257 & 0.178 & 0.212 & 0.242 & 0.102 & 0 & 0.127 \\
0.0987 & 0.0921 & 0.315 & 0.229 & 0.148 & 0.185 & 0.150 & 0.157 & 0.157 & 0.176 & 0.127 & 0
\end{bmatrix}
$$

$$
\cdot
\begin{bmatrix}
a_1\hat{x}_1 \\
a_2\hat{x}_2 \\
a_3\hat{x}_3 \\
a_4\hat{x}_4 \\
a_5\hat{x}_5 \\
a_6\hat{x}_6 \\
a_7\hat{x}_7 \\
a_8\hat{x}_8 \\
a_9\hat{x}_9 \\
a_{10}\hat{x}_{10} \\
a_{11}\hat{x}_{11} \\
a_{12}\hat{x}_{12}
\end{bmatrix}
= \lambda \cdot
\begin{bmatrix}
a_1\hat{x}_1 \\
a_2\hat{x}_2 \\
a_3\hat{x}_3 \\
a_4\hat{x}_4 \\
a_5\hat{x}_5 \\
a_6\hat{x}_6 \\
a_7\hat{x}_7 \\
a_8\hat{x}_8 \\
a_9\hat{x}_9 \\
a_{10}\hat{x}_{10} \\
a_{11}\hat{x}_{11} \\
a_{12}\hat{x}_{12}
\end{bmatrix}
$$

The equation is readily solved to yield the spectrum of eigenvalues $\{\lambda_i\}$:

$$\{-0.7690, -0.4443, -0.2962, -0.1871, -0.1261, -0.1197, -0.08498,$$

$$-0.07437, -0.06625, +2.532\}$$

The values are listed in ascending order. Note how the right-most value stands alone by its positive sign and magnitude. The allied eigenvector is just as distinctive by having uniformly-positive weight factors of comparable magnitude:

$$\vec{V} = 0.246\hat{x}_1 + 0.228\hat{x}_2 + 0.452\hat{x}_3 + 0.327\hat{x}_4 + 0.269\hat{x}_5 + 0.298\hat{x}_6 + 0.263\hat{x}_7 + 0.253\hat{x}_8$$
$$+ 0.274\hat{x}_9 + 0.298\hat{x}_{10} + 0.260\hat{x}_{11} + 0.212\hat{x}_{12}$$

In contrast, for example, the lowest eigenvalue $\lambda = -0.7690$ pairs with eigenvector:

$$\vec{V} = -0.109\hat{x}_1 - 0.0202\hat{x}_2 + 0.732\hat{x}_3 - 0.206\hat{x}_4 - 0.265\hat{x}_5 - 0.266\hat{x}_6 - 0.0743\hat{x}_7 - 0.308\hat{x}_8$$
$$- 0.292\hat{x}_9 + 0.234\hat{x}_{10} + 0.151\hat{x}_{11} - 0.0575\hat{x}_{12}$$

Note the mix of positive and negative weights dispersed over a range of magnitudes.

Rescaling and plotting the spectrum best conveys the story. If we compute the average and standard deviation of the eigenvalues, we obtain Z-factors as follows:

$$Z_i = \frac{\lambda_i - \langle \lambda \rangle}{\sigma_\lambda}$$

The distribution then appears in Figure One. Note how the maximum Z is special indeed: it places more than three standard deviations above the second in line. Concomitantly, the trailing eigenvalues clump together with little to distinguish them.

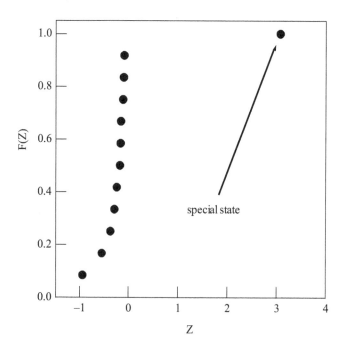

FIGURE ONE   Distribution of eigenvalues allied with set of H1N1 influenza proteins (Hong Kong, 1977). The values have been re-scaled in units of the standard deviation. The matrix elements follow from computing inter-protein distances based on base+(2) information.

How do defects alter the picture? Computational holes arise from inactive or absent algorithms. Then if, say, two of the viral proteins failed to be encoded or synthesized, the effect would be zeroes along particular rows and columns of the matrix:

$$
\begin{bmatrix}
0 & 0.121 & 0.371 & 0.227 & 0.173 & 0.208 & 0.125 & 0.157 & 0 & 0 & 0.167 & 0.0987 \\
0.121 & 0 & 0.326 & 0.224 & 0.178 & 0.205 & 0.107 & 0.165 & 0 & 0 & 0.111 & 0.0921 \\
0.371 & 0.326 & 0 & 0.481 & 0.420 & 0.450 & 0.390 & 0.433 & 0 & 0 & 0.305 & 0.315 \\
0.227 & 0.224 & 0.481 & 0 & 0.231 & 0.248 & 0.216 & 0.185 & 0 & 0 & 0.271 & 0.229 \\
0.173 & 0.178 & 0.420 & 0.231 & 0 & 0.139 & 0.189 & 0.106 & 0 & 0 & 0.234 & 0.148 \\
0.208 & 0.205 & 0.450 & 0.248 & 0.139 & 0 & 0.226 & 0.142 & 0 & 0 & 0.257 & 0.185 \\
0.125 & 0.107 & 0.390 & 0.216 & 0.189 & 0.226 & 0 & 0.193 & 0 & 0 & 0.178 & 0.150 \\
0.157 & 0.165 & 0.433 & 0.185 & 0.106 & 0.142 & 0.193 & 0 & 0 & 0 & 0.212 & 0.157 \\
0 & 0 & 0 & 0 & 0 & 0 & 0 & 0 & 0 & 0 & 0 & 0 \\
0 & 0 & 0 & 0 & 0 & 0 & 0 & 0 & 0 & 0 & 0 & 0 \\
0.167 & 0.111 & 0.305 & 0.271 & 0.234 & 0.257 & 0.178 & 0.212 & 0 & 0 & 0 & 0.127 \\
0.0987 & 0.0921 & 0.315 & 0.229 & 0.148 & 0.185 & 0.150 & 0.157 & 0 & 0 & 0.127 & 0
\end{bmatrix}
$$

$$
\begin{bmatrix}
a_1\hat{x}_1 \\
a_2\hat{x}_2 \\
a_3\hat{x}_3 \\
a_4\hat{x}_4 \\
a_5\hat{x}_5 \\
a_6\hat{x}_6 \\
a_7\hat{x}_7 \\
a_8\hat{x}_8 \\
a_9\hat{x}_9 \\
a_{10}\hat{x}_{10} \\
a_{11}\hat{x}_{11} \\
a_{12}\hat{x}_{12}
\end{bmatrix}
= \lambda \cdot
\begin{bmatrix}
a_1\hat{x}_1 \\
a_2\hat{x}_2 \\
a_3\hat{x}_3 \\
a_4\hat{x}_4 \\
a_5\hat{x}_5 \\
a_6\hat{x}_6 \\
a_7\hat{x}_7 \\
a_8\hat{x}_8 \\
a_9\hat{x}_9 \\
a_{10}\hat{x}_{10} \\
a_{11}\hat{x}_{11} \\
a_{12}\hat{x}_{12}
\end{bmatrix}
$$

The defects do not alter the matrix elements in place, but their action severely impacts the spectrum and eigenvectors. The spectral distribution for the above is shown in Figure Two and allows comparison with that for hole-free conditions. We witness that the largest eigenvalue for hole-defects (open circles) is diminished in magnitude and stands less apart from the crowd near $\lambda \approx 0$. There is more to take home from the eigenvector: the component weights lose their egalitarian character by the near-zero node along a dimension indicated by the bold vertical arrow:

$$
\vec{V}_{\text{hole}} = 0.256\hat{x}_1 + 0.241\hat{x}_2 + 0.478\hat{x}_3 + 0.340\hat{x}_4 + 0.276\hat{x}_5 + 0.307\hat{x}_6 + 0.277\hat{x}_7 + 0.259\hat{x}_8
$$

$$
\downarrow
$$
$$
+ 0.279\hat{x}_9 + 0.0600\hat{x}_{10} + 0.285\hat{x}_{11} + 0.232\hat{x}_{12}
$$

The schematics portrayed stray algorithms and occlusions. For proteins, these defects would apply when one or more components (e.g., enzymes) operate too far apart and lack the capacity for cooperation. Matrix-wise, the defects impose maximum distances along some of the rows and columns. We recall from Chapter Two that for information vectors, the maximum distance

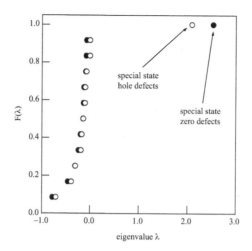

FIGURE TWO Distribution of eigenvalues allied with influenza proteins subject to hole-defects. The matrix elements follow from computing inter-protein distances based on base+(2) information. Zeroes have been applied to two rows and columns of the distance matrix.

approaches $\pi/2$. In place of the hole-defects in the previous matrix, we inject near-maximum radian values to obtain:

$$
\begin{bmatrix}
0 & 0.121 & 0.371 & 0.227 & 0.173 & 0.208 & 0.125 & 0.157 & 1.50 & 1.50 & 0.167 & 0.0987 \\
0.121 & 0 & 0.326 & 0.224 & 0.178 & 0.205 & 0.107 & 0.165 & 1.50 & 1.50 & 0.111 & 0.0921 \\
0.371 & 0.326 & 0 & 0.481 & 0.420 & 0.450 & 0.390 & 0.433 & 1.50 & 1.50 & 0.305 & 0.315 \\
0.227 & 0.224 & 0.481 & 0 & 0.231 & 0.248 & 0.216 & 0.185 & 1.50 & 1.50 & 0.271 & 0.229 \\
0.173 & 0.178 & 0.420 & 0.231 & 0 & 0.139 & 0.189 & 0.106 & 1.50 & 1.50 & 0.234 & 0.148 \\
0.208 & 0.205 & 0.450 & 0.248 & 0.139 & 0 & 0.226 & 0.142 & 1.50 & 1.50 & 0.257 & 0.185 \\
0.125 & 0.107 & 0.390 & 0.216 & 0.189 & 0.226 & 0 & 0.193 & 1.50 & 1.50 & 0.178 & 0.150 \\
0.157 & 0.165 & 0.433 & 0.185 & 0.106 & 0.142 & 0.193 & 0 & 1.50 & 1.50 & 0.212 & 0.157 \\
1.50 & 1.50 & 1.50 & 1.50 & 1.50 & 1.50 & 1.50 & 1.50 & 0 & 1.50 & 1.50 & 1.50 \\
1.50 & 1.50 & 1.50 & 1.50 & 1.50 & 1.50 & 1.50 & 1.50 & 1.50 & 0 & 1.50 & 1.50 \\
0.167 & 0.111 & 0.305 & 0.271 & 0.234 & 0.257 & 0.178 & 0.212 & 1.50 & 1.50 & 0 & 0.127 \\
0.0987 & 0.0921 & 0.315 & 0.229 & 0.148 & 0.185 & 0.150 & 0.157 & 1.50 & 1.50 & 0.127 & 0
\end{bmatrix}
$$

$$
\cdot
\begin{bmatrix}
a_1\hat{x}_1 \\
a_2\hat{x}_2 \\
a_3\hat{x}_3 \\
a_4\hat{x}_4 \\
a_5\hat{x}_5 \\
a_6\hat{x}_6 \\
a_7\hat{x}_7 \\
a_8\hat{x}_8 \\
a_9\hat{x}_9 \\
a_{10}\hat{x}_{10} \\
a_{11}\hat{x}_{11} \\
a_{12}\hat{x}_{12}
\end{bmatrix}
= \lambda \cdot
\begin{bmatrix}
a_1\hat{x}_1 \\
a_2\hat{x}_2 \\
a_3\hat{x}_3 \\
a_4\hat{x}_4 \\
a_5\hat{x}_5 \\
a_6\hat{x}_6 \\
a_7\hat{x}_7 \\
a_8\hat{x}_8 \\
a_9\hat{x}_9 \\
a_{10}\hat{x}_{10} \\
a_{11}\hat{x}_{11} \\
a_{12}\hat{x}_{12}
\end{bmatrix}
$$

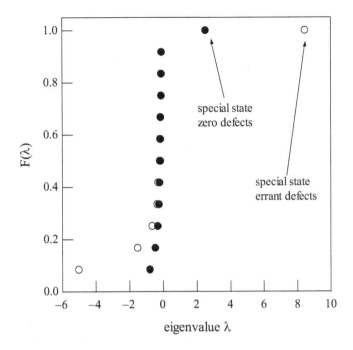

FIGURE THREE Distribution of eigenvalues allied with set of influenza proteins. The matrix elements follow from computing inter-protein distances based on base+(2) information. Near-maximum distance values have been inserted in two rows and columns of the distance matrix.

The spectral distribution (open circles) is shown in Figure Three along with that of defect-free conditions (filled circles). We observe how the stand-alone eigenvalue is significantly enhanced in magnitude. It is as if $\lambda_{\max}$ runs away from the pack. At the same time, the specialness is misleading. Note how the spectrum is more dispersed compared with Figure One. This positions the maximum eigenvalue only about two standard deviations above the crowd. With zero defects, the maximum eigenvalue places more than three standard deviations above the crowd. And just as with hole-defects, the component weights in the eigenvector lose their egalitarian character, viz.

$$\vec{V}_{\text{errant}} = 0.219\hat{x}_1 + 0.216\hat{x}_2 + 0.266\hat{x}_3 + 0.236\hat{x}_4 + 0.223\hat{x}_5 + 0.230\hat{x}_6 + 0.222\hat{x}_7 + 0.221\hat{x}_8$$

$$+ 0.490\hat{x}_9 + 0.490\hat{x}_{10} + 0.225\hat{x}_{11} + 0.215\hat{x}_{12}$$

Cover schematics and matrix models are valuable for the questions they raise and this is the case for considering protein information at the horizons. The models point to the vitality of integrated, parallel systems directed at interdependent tasks. In cruise mode, the systems express robust states that stand far apart from all others. Their state vectors and topmost eigenvalues express as outliers, and not by a little. Yet all systems are susceptible to dysfunction such as when processors and algorithms become defective or fail outright. Such ideas motivate viewing archetypes like lysozyme (KVFER...GCGV) less as individuals and more as components of highly-integrated systems. Hopefully progress at the horizons includes ways and means to elucidate the collective behavior of functional and dysfunctional states. It is reasonable to expect that systems offer combinations of states due to intricate coupling mechanisms. The horizon for investigating these properties looks inviting.

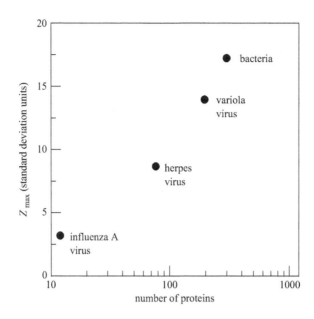

FIGURE FOUR Z_{max} for diverse protein sets. Complete sets were studied for the viral systems. The bacteria set was assembled from randomly selected sequences of the Yersinia Pestis proteome. The plot shows the approximate scaling of the special state eigenvalue with proteome size.

EXERCISES

1. Write an essay discussing a significant unsolved problem regarding proteins.
2. Totally positive matrices can be assembled for protein systems in general: elements are the distances between vectors assembled via base+(2) properties. Figure Four shows how Z_{max} trends as a function of matrix dimension ↔ number of proteins in a system. (a) Use the data to predict Z_{max} for a natural system composed of four thousand proteins. (b) Predict Z_{max} for systems encoding twenty thousand proteins. (c) How would Z_{max} look for a set of independently random proteins?

SOURCES AND FURTHER READING

1. Gibbons, A. M., Spirakis, P. G. 1993. *Lectures on Parallel Computation*, Cambridge University Press, Cambridge, UK.
2. Kanellakis, P. C., Shvartsman, A. A. 1997. *Fault-tolerant Parallel Computation*, Kluwer Academic, Boston, MA.
3. Roosta, S. H. 2000. *Parallel Processing and Parallel Algorithms: Theory and Computation*, Springer, New York.
4. Bennett, C. H. 1982. Thermodynamics of computation—A review, *Intl. J. Theo. Phys.* 21, 905.
5. Karlin, S. 1968. *Total Positivity*, Stanford University Press, Stanford, CA.
6. Gasca, M., Micchelli, C. A., eds. 1996. *Total Positivity and its Applications*, Kluwer Academic, Boston, MA.

Appendix One
Select Properties of the Amino Acids

Properties include molecular structure diagrams, abbreviations, molecular formula, and molar mass. The pK_a values apply to the carboxylic acid group(s) and the amino group(s) when protonated. The third number applies to side-chain properties if applicable. The amino acids are listed in the same order featured in base structures throughout this text: AVLIPFWMGSTCYNQDEKRH. The order and properties are elaborated upon by Lehninger [1].

Alanine ↔ Ala ↔ A

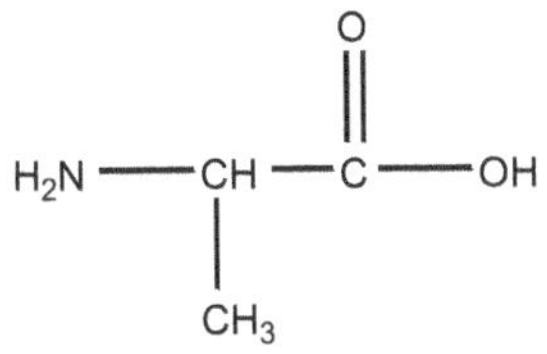

$C_3H_7NO_2$ 89.09 g/mol $pK_a = 2.35, 9.69$

Valine ↔ Val ↔ V

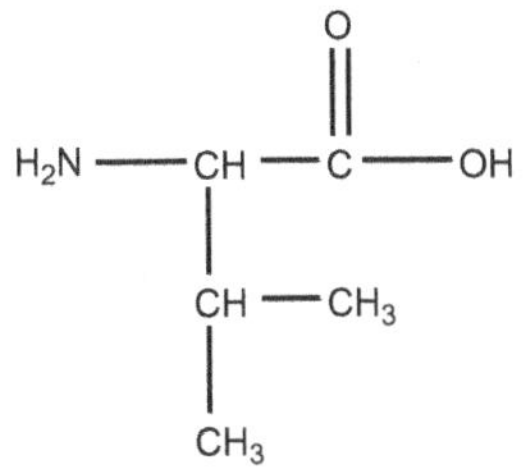

$C_5H_{11}NO_2$ 117.15 g/mol $pK_a = 2.32, 9.62$

Leucine ↔ Leu ↔ L

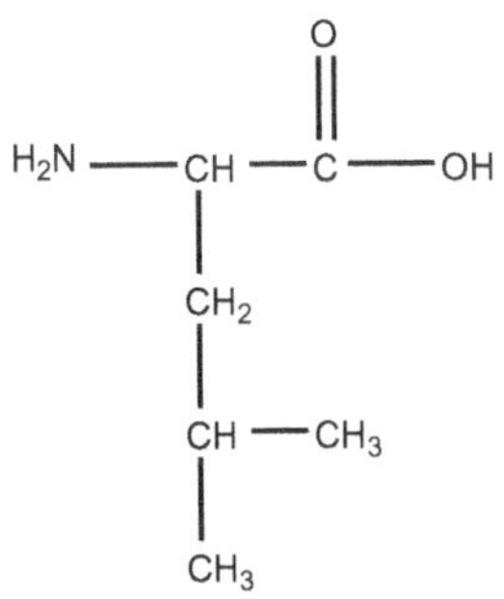

$C_6H_{13}NO_2$ 131.2 g/mol $pK_a = 2.36, 9.60$

Isoleucine ↔ Ile ↔ I

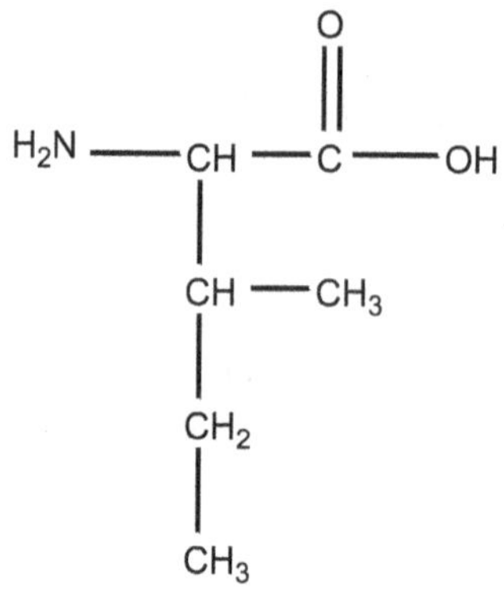

$$H_2N-CH-C(=O)-OH$$

(structure of isoleucine)

$C_6H_{13}NO_2$ 131.2 g/mol $pK_a = 2.32, 9.76$

Proline ↔ Pro ↔ P

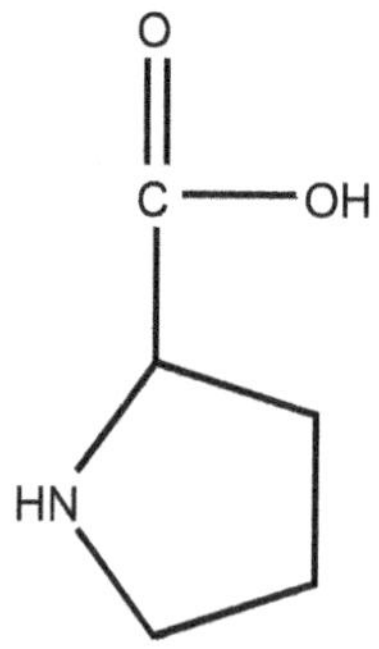

(structure of proline)

$C_5H_9NO_2$ 115.13 g/mol $pK_a = 1.99, 10.96$

Phenylalanine ↔ Phe ↔ F

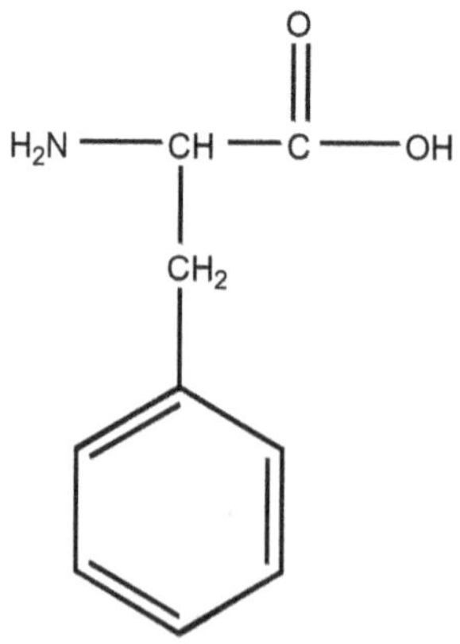

(structure of phenylalanine)

$C_9H_{11}NO_2$ 165.2 g/mol $pK_a = 2.20, 9.31$

Tryptophan ↔ Trp ↔ W

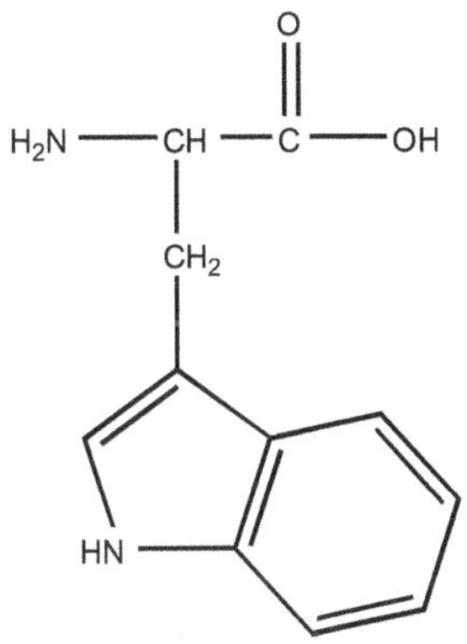

$C_{11}H_{12}N_2O_2$ 204.2 g/mol $pK_a = 2.38, 9.39$

Methionine ↔ Met ↔ M

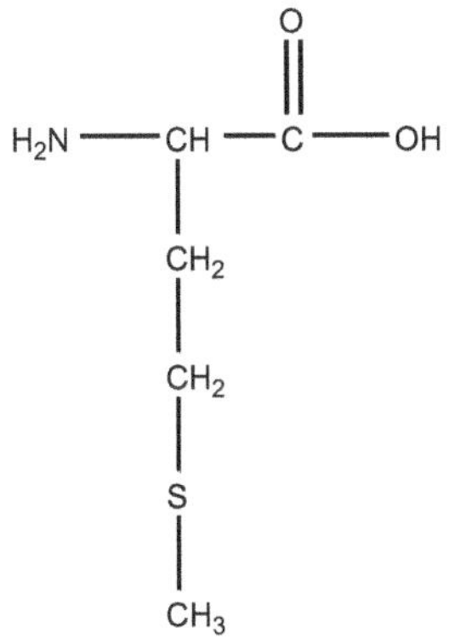

$C_5H_{11}NSO_2$ 149.2 g/mol $pK_a = 2.28, 9.21$

Glycine ↔ Gly ↔ G

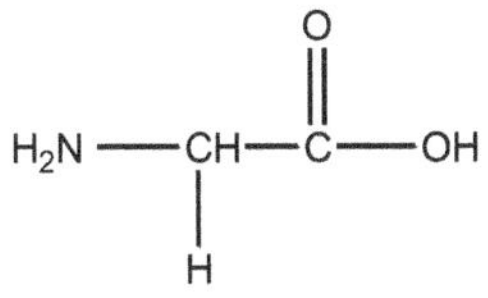

$C_2H_5NO_2$ 75.07 g/mol $pK_a = 2.34, 9.6$

Serine ↔ Ser ↔ S

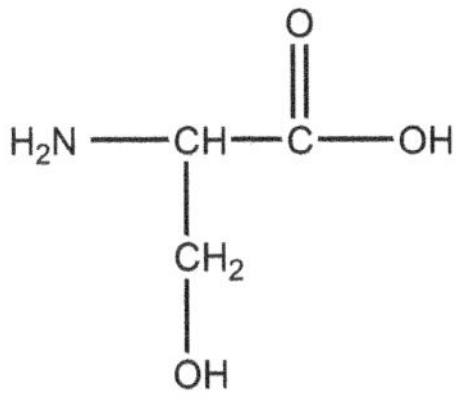

$C_3H_7NO_3$ 105.1 g/mol $pK_a = 2.21, 9.15$

Threonine ↔ Thr ↔ T

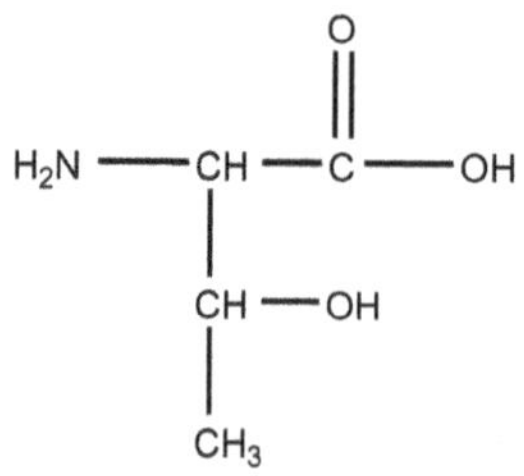

C$_4$H$_9$NO$_3$ 119.1 g/mol   pK$_a$ = 2.63, 10.43

Cysteine ↔ Cys ↔ C

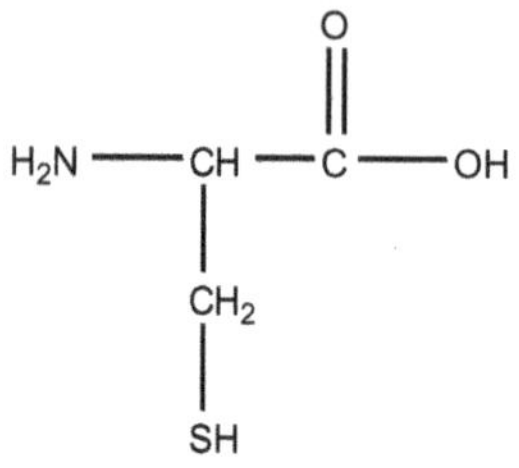

C$_3$H$_7$NO$_2$S   121.2 g/mol   pK$_a$ = 2.35, 9.69

Tyrosine ↔ Tyr ↔ Y

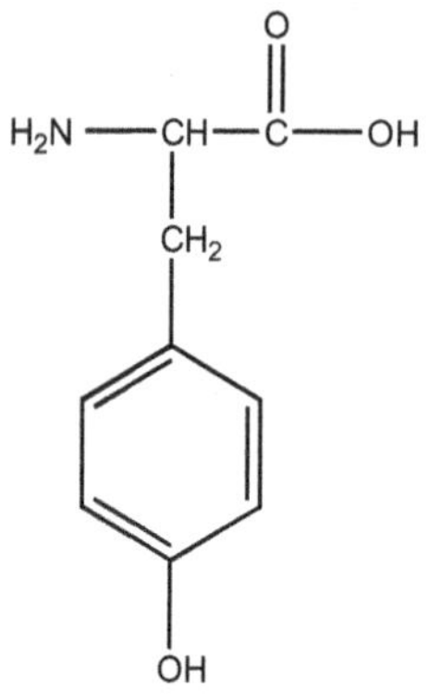

C$_9$H$_{11}$NO$_3$ 181.2 g/mol   pK$_a$ = 2.20, 9.21, 10.46

Asparagine ↔ Asn ↔ N

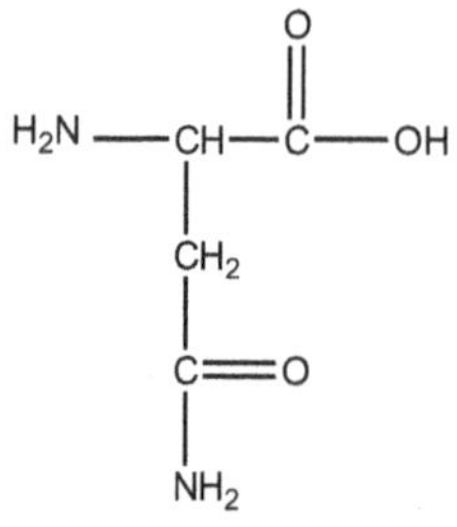

C$_4$H$_8$N$_2$O$_3$ 132.1 g/mol   pK$_a$ = 2.02, 8.80

Glutamine ↔ Gln ↔ Q

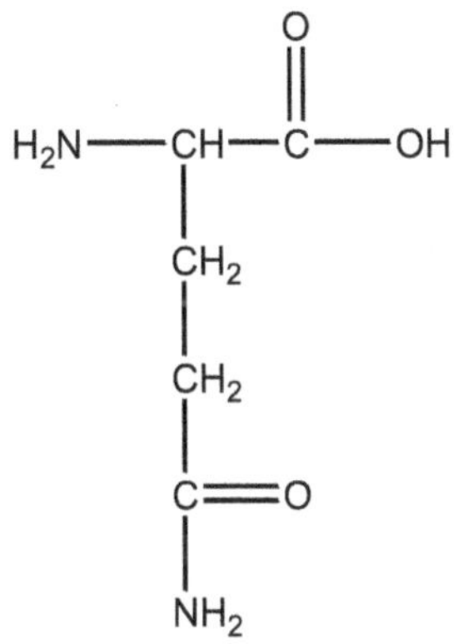

$$H_2N-CH-C-OH$$

(structure: glutamine)

$C_5H_{10}N_2O_3$ 146.2 g/mol $pK_a = 2.2, 9.1$

Aspartic Acid ↔ Asp ↔ D

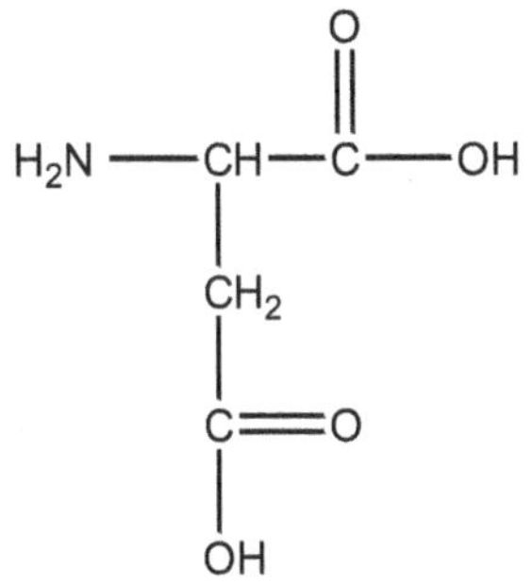

(structure: aspartic acid)

$C_4H_7NO_4$ 133.1 g/mol $pK_a = 1.99, 9.90, 3.90$

Glutamic Acid ↔ Glu ↔ E

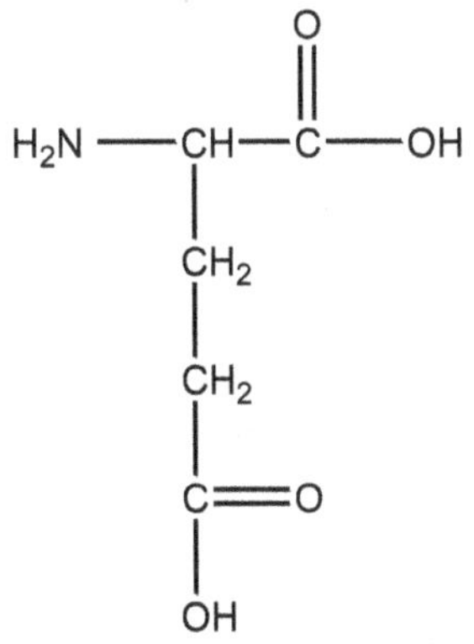

(structure: glutamic acid)

$C_5H_9NO_4$ 147.1 g/mol $pK_a = 2.10, 4.07, 9.47$

Lysine ↔ Lys ↔ K

$C_6H_{14}N_2O_2$ 146.2 g/mol pK_a = 2.16, 9.06, 10.54

Arginine ↔ Arg ↔ R

$C_6H_{14}N_4O_2$ 174.2 g/mol pK_a = 1.82, 8.99, 12.48

Histidine ↔ His ↔ H

$C_6H_9N_3O_2$ 155.2 g/mol pK_a = 1.80, 9.33, 6.04

REFERENCE

1. Lehninger, A. L. 1970. *Biochemistry*, Worth Publishers, New York.

Appendix Two
Essential Properties of Steady-State Systems

We are used to viewing the graphics for protein structures that are established under (more or less) equilibrium conditions. See Chapter One for representations of archetypes and the Protein Data Bank (PDB) for 10^5 more examples. Equilibrium conditions, however, are thermodynamically rarefied *and* antithetical to active proteins. Equilibrium conditions apply to closed systems, admitting neither increases nor decreases of energy and material. Further, they express maximum entropy subject to constraints—walls, components, and so forth. The constraints force the time averages of macroscopic variables to hold constant. Among the consequences, equilibrium systems are devoid of persistent gradients in the intensive properties: temperature, density, and electrochemical potentials. They are further impervious to the effects of catalysts. To be sure, fluctuations push and pull systems away from maximum entropy. But equilibrium states are restorative and accidental gradients are damped out sooner than quicker. In effect, the equilibration process aims at maximizing the entropy and zeroing out the gradients. When equilibrium is established, everything that can happen has already happened. History has run its course with nothing on the horizon.

Proteins are the principal agents of biological, near-steady-state systems—where much is happening and has yet to happen. Steady-state systems share some of the rarefied properties of equilibrium systems, but diverge sharply in others [1–3]. Steady-state systems of the biological type are mandatorily open and require the flow-through of energy and material. They are *not* at maximum entropy as gradients are maintained involving charges, reactants, and products. The gradients are critical over space and time. With each, there obtains a thermodynamic force $\vec{F}_i$ which drives material or energy flux $\vec{J}_i$. Forces and fluxes power the system and are subject to cross-coupling. Thus, a perturbation directed at $\vec{F}_i$ generally impacts $\vec{F}_j$ and the attendant fluxes. Cross-couplings are not unique to steady-state systems as they also impact equilibrium systems. Their effects are acutely pronounced in biological systems, however, given the sensitivity to gradients and flow kinetics.

Gradient maintenance in steady-state systems is selective on account of entropy. With each force and flux pair, there is a *net production* of entropy:

$$g_S = \sum_i \vec{F}_i \cdot \vec{J}_i$$

The sum total g_S represents the entropy generated per unit volume in unit time. The value is zero for equilibrium systems, and necessarily positive in steady states. Clearly biological systems survive only if they can expel entropy almost as it is produced. They hold g_S to near minimum by maintaining only the necessary gradients. The superfluous ones are smoothed to mitigate the entropy production. Most visibly, biological systems maintain constant internal temperature and pressure. To function otherwise would mean additional terms in the g_S summation.

Equilibrium systems can be combined, for example, by mixing solution A with B, whereby a new maximum entropy system obtains with time. Steady-state systems are just as prone to mixing. A biological system reflects multiple systems, for example, higher organisms and resident bacterial and viral populations. Proteins are accordingly central to steady state, mixed systems and are responsible for maintaining the critical gradients. It should not surprise that proteins demonstrate a multitude of correspondences across molecules and sets. The correspondences are biological given that systems operate by the genetic code and share functions and evolutionary tracks. At the same time, the correspondences are thermodynamic given the nature of steady-state systems. Correspondences figure prominently in Chapters Three–Eight, not to mention the protein and genomic literature. The listed references provide both an overview and finer details of thermodynamic steady-state systems.

REFERENCES

1. Yourgrau, W., van der Merwe, A., Raw, G. 1982. *Treatise on Irreversible and Statistical Thermophysics*, Dover, New York.
2. Prigogine, I. 1980. *From Being to Becoming: Time and Complexity in the Physical Sciences*, W. H. Freeman, San Francisco, CA.
3. Prigogine, I. 1967. *Introduction to Thermodynamics of Irreversible Processes*, Interscience Publishers, New York.

Appendix Three
The Stirling Formula for Large Factorials

The Stirling formula arises in several places of combinatorics, mathematical probability, and functional analysis [1,2]. A student encounters Stirling applications typically in courses in quantum mechanics and statistical thermodynamics. For example, partition functions almost always require Stirling-related approximations in calculations. We sketch briefly how Stirling formulae present multiple layers and even the superficial ones take us pretty far.

Consider the quantity $N!$ Clearly it increases explosively with N:

$$N! = N \cdot (N-1) \cdot (N-2) \cdots 2 \cdot 1$$

Examples: $5! = 120$; $10! = 3{,}628{,}000$; $15! \approx 1.31 \times 10^{12}$; $20! \approx 2.43 \times 10^{18}$.

The logarithms of quantities are tame by comparison:

$$\ln[N!] = \ln\left[N \cdot (N-1) \cdot (N-2) \cdots 2 \cdot 1 \right]$$

Examples: $\ln(5!) \approx 4.79$; $\ln(10!) \approx 15.10$; $\ln(15!) \approx 27.90$; $\ln(20!) \approx 42.34$.

Critically, the logarithm of $N!$ can be subdivided to obtain:

$$\ln(N!) = \ln[N] + \ln[(N-1)] + \ln[(N-2)] + \cdots + \ln[2] + \ln[1]$$

$$= \sum_{i=1}^{i=N} \ln[i]$$

The formula is exact for any integer N. However, we are taken to Stirling by approximating the summation as an integral. For large N and taking the limit, we have:

$$\sum_{i=1}^{i=N} \ln[i] \approx \int_{N'=1}^{N'=N} \ln[N'] \, dN'$$

To prepare for parts integration, we have the substitutions

$$u = \ln[N'], \, dv = dN'$$

$$du = \frac{dN'}{N'}, \, v = N'$$

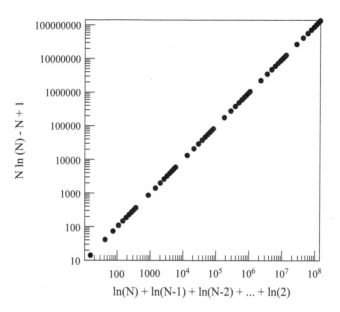

APPENDIX THREE FIGURE   Stirling formula comparison across eight orders of magnitude. The horizontal axis features the "exact" value of the logarithm sum. The vertical axis applies to the lowest order of the Stirling Approximation.

These take us to:

$$uv - \int v\,du$$

$$= N' \cdot \ln\left[N'\right] \mid_1^N - \int_1^N \frac{N'}{N'}dN'$$

$$= N \cdot \ln\left[N\right] - N + 1$$

Check! For $N = 20$, $N \cdot \ln[N] - N + 1 \approx 40.92$ which is close to $\ln(20!) \approx 42.34$. The approximation converges rapidly and is typically simplified further as:

$$\ln(N!) \approx N \cdot \ln\left[N\right] - N$$

The proof is in the pudding. The figure offers visual comparison of Stirling with "exact". On the horizontal axis we plot $\ln(N!) = \ln(N) + \ln(N-1) + \ln(N-2) + \ldots$ while on the vertical axis we plot $\ln[N] - N + 1$. The correspondence is remarkable given the lowest order of approximation.

   More formally, Stirling can be expressed as an asymptotic expansion. This leads to something a more precise, viz.

$$\ln(N!) \approx \left(N + \frac{1}{2}\right) \cdot \ln(N) - N + \ln(2\pi) + o(1)$$

The last term on the right refers to a correction of order unity. The reader should check that the higher order version does not offer significant improvement over that applied in the figure. The references offer further insights into Stirling.

REFERENCES

1. Lawler, G. F., Coyle, L. N. 1999. *Lectures on Contemporary Probability*, American Mathematical Society, Providence, RI, Lecture One.
2. Wilf, H. S. 1978. Asymptotic expansions, in *Mathematics for the Physical Sciences*, Dover Publications, New York.

Appendix Four
The Binomial and Multinomial Distributions

There are countless venues in which probability, draw-pools, and collections intersect. Absent from the intersections in many cases are considerations of draw order. The venues are best modeled by the mathematics of n-choose-m combinations [1]. In binary cases, we are led to the binomial distribution: the probability of k number of successes in n draws from an infinite population:

$$prob(k,n) = \frac{n!}{k! \cdot (n-k)!} \cdot p^k \cdot (1-p)^{n-k}$$

Symbol p refers to the probability of success with any particular draw, while $(1-p)$ is the probability of failure. The key idea is that both terms are taken to be wholly independent of past and future. Note further how they are weighted by the factorial expression. The ratio quantifies the number of distinguishable permutations of success and failure draws. It is enhanced when k and $(n-k)$ are close in value and diminished otherwise. The word choices "success" and "failure" are unfortunate perhaps. It should nonetheless be clear how strings of two unambiguous states obtained by random processes apply far and wide.

The idea of random draws apart from order readily extends to three states:

$$prob(k,l,n) = \frac{n!}{k! \cdot l! \cdot (n-k-l)!} \cdot p^k \cdot q^l (1-p-q)^{n-k-l}$$

If the binomial probability can extend to three states, it should not surprise that multiple states can be accommodated [2]. This is via the multinomial distribution, use of which is illustrated in Chapter Three. The malleability stems from the permutation counting formula and the (assumed) independence of draws. The references provide depth in all matters n-choose-m, l, etc. Often the Stirling Approximation is required to compute the factorials (cf. Appendix Three) [3].

REFERENCES

1. Hamming, R. W. 1991. *The Art of Probability for Scientists and Engineers*, Addison-Wesley, Redwood City, CA.
2. Chung, K. L. 1968. *A Course in Probability Theory*, Harcourt, Brace, and World, Inc., New York.
3. Wilf, H. S. 1978. Asymptotic expansions, in *Mathematics for the Physical Sciences*, Dover Publications, New York.

Appendix Five
Amino Acids and the Genetic Code

The components of a protein are underpinned by triplets of nucleotides in messenger RNA (mRNA): one triplet equals one codon. The transcription code is summarized as follows with further details in Lehninger [1]. U, C, A, and G are the respective abbreviations for uracil, cytosine, adenine, and guanine. The amino acids are specified using the three-letter system (cf. Chapter One) to avoid confusion. CT marks protein chain termination codons.

First Position	U	C	A	G	Third Position
U	Phe	Ser	Tyr	Cys	U
	Phe	Ser	Tyr	Cys	C
	Leu	Ser	CT	CT	A
	Leu	Ser	CT	Trp	G
C	Leu	Pro	His	Arg	U
	Leu	Pro	His	Arg	C
	Leu	Pro	Gln	Arg	A
	Leu	Pro	Gln	Arg	G
A	Ile	Thr	Asn	Ser	U
	Ile	Thr	Asn	Ser	C
	Ile	Thr	Lys	Arg	A
	Met	Thr	Lys	Arg	G
G	Val	Ala	Asp	Gly	U
	Val	Ala	Asp	Gly	C
	Val	Ala	Glu	Gly	A
	Val	Ala	Glu	Gly	G

REFERENCE

1. Lehninger, A. L. 1970. *Biochemistry*, Worth Publishers, New York.

Appendix Six
Accession Details for Relevant Sequences and Sets

INTRODUCTION

Sequence fragments from H1N1 Influenza (Puerto Rico, 1934)
Lysozyme PDB 1REX

CHAPTER ONE

Lysozyme PDB 1REX
Ubiquitin Q81WF7
Human Myoglobin: A0A1K0FU49
Bovine Ribonuclease A PDB 1FS3
Alpha Synuclein P37840
Phopholipase D0VX11
Phospholipase PDB 1POB

CHAPTER TWO

Lysozyme PDB 1REX
Bovine Ribonuclease A PDB 1FS3

CHAPTER THREE

Lysozyme PDB 1REX
Bovine Ribonuclease A PDB 1FS3
Human Myoglobin: A0A1K0FU49
Keratin E2A5K8
Prion Q9UKY0
Alpha Synuclein P37840
Acidic Phospholipase A2 P00597
Polymerase Basic Protein D3J174
Keratin KRTAP5-9
Phosphofructokinase K7DD35
Fibroin P21828
Uncharacterized Protein Q8N1F1
Odontogenesis-associated phosphoprotein Q17RF5
Uncharacterized Protein Q2M2E5
Soft-shell turtle lysozyme P85345
Sheep Lysozyme P80190
Canine Lysozyme P81709
Heliobacter pylori lysozyme E8QNH2
E. coli lysozyme A0A028C8M5

Prion Q9UKY0
Heat Shock V9HW43
Cytochrome P00125
Lymphocyte Antigen P11836-2

CHAPTER FOUR

Lysozyme PDB 1REX
Phospholipase B3RFI8
Human Myoglobin: A0A1K0FU49
Bovine Ribonuclease A PDB 1FS3
Alpha Synuclein P37840
Chicken Lysozyme P00700

CHAPTER FIVE

Epididymis Secretory Protein V9HWD6
Theta Protein P27348
Antigen Q53Z42

CHAPTER FIVE PROTEOMES

Human
Yersinia Pestis UP000000815
Drosophila UP000000803
Bovine UP000009136
Grape UP000009183
Rice
Apis Mellifera UP000005203
Orcinus orca
Physeter catodon
Ciona intestinalis
Echinococcus Granulosus
Strongylocentrotus purpuratus
Parasteatoda tepidariorum
Chelonia mydas
Aotus nancymaae
Halymorpha halys
Gavia stellata
Gekko japonicus
Ukomys damarensis
Colobus angolensis pallatus
Clupea harengus
Abalone Shriv Syndrome NC_011646
Acheta domestica densovirus NC_004290
Acinetobacter phage virus NC_023570
Bacillus phage virus 6A-1 NC_002649
Barley yellow dwarf virus NC_004750
Bat Adenovirus 2 NC_015932
Bovine Leukemia virus NC_001414
Human enteric corona virus strain 4408 NC_012950

Human herpes virus 1 NC_001806
Listeria Phage virus NC_009814
Ebola virus NC_002549
Fowl adenovirus NC_021221
Measles NC_001498
SARS corona NC_004718
Salmonella phage virus NC_021779
Simian T lympho NC_011546
Simian immunodeficiency NC_004455
Strep phage NC_023503
Sudan Ebola NC_006432
Variola NC_001611
Vesicular Stomata NC_001560
Human pappilo virus NC_019023
Human parvo NC_000883
Rubella NC_001545
Hepatitis C NC_004102
Hepatitis E NC_001434
H1N1 Influenza (Puerto Rico, 1934)
H5N1 Influenza (Guangdong, 1996)
H3N2 Influenza (New York, 2004)
H2N2 Influenza (Korea, 1968)
H9N2 Influenza (Hong Kong, 1999)

CHAPTER SIX

Lysozyme PDB 1REX
Human Myoglobin: A0A1K0FU49
Bovine Ribonuclease A PDB 1FS3
Alpha Synuclein P37840

CHAPTER SEVEN

Lysozyme PDB 1REX
Bovine Ribonuclease A PDB 1FS3
Phospholipase PDB 1POB
Insulin (smaller chain) F6MZK5
Insulin (larger chain) C9JNR5
Brazzein PDB 1BRZ
Plant toxin PDB 1GPS
Scorpion neurotoxin PDB 1SNB
Cathepsin PDB 4CIA
Chymotrypsinogen PDB 1EX3
Anticoagulent protein PDB 1TAP

CHAPTER EIGHT

Lysozyme PDB 1REX

CHAPTER NINE

H1N1 Influenza (Hong Kong, 1977)

Appendix Seven
Answers to Selected Exercises

CHAPTER ONE EXERCISES

Exercise One: Use the distance formula and PDB data to find 22.7 Å.

Exercise Three: RLGMDGYRGISLA; use ChemDraw or equivalent to expand. This is a fragment of lysozyme with base structure is $R_2L_2G_3MDYISA$. The number of distinguishable permutations—structural isomers—is

$$\frac{13!}{2!2!3!} \approx 2.59 \times 10^8$$

Exercise Six: The base structure is $A_4V_{10}L_5I_6P_4F_1W_1M_1G_3S_1T_3C_0Y_2N_1Q_3D_4E_5K_5R_5H_0$. Use the permutation formula to count 3.92×10^{64} isomers. Each isomer hosts 1,062 atoms. 10^{-15} moles translate to atoms and grams in respective amounts:

$$2.50 \times 10^{76} \qquad 2.86 \times 10^{53}$$

Exercise Ten: The standard deviation of the internal energy is estimated in joules as:

$$2.1 \times 10^{-19}$$

This is ten or more hydrogen bonds.

Exercise Twenty-One: A student enrolled in a course using this book in draft form submitted an entire poem written using the amino acid alphabet. The poem was entitled HISANDHERWATCHES. Wonderful!

CHAPTER TWO EXERCISES

Exercise One: Information in scenario one equates with $-\log_2(1/52) - \log_2(1/51) - \log_2(1/50) \approx 17$ bits. Information in scenario two equals $-\log_2(1/52) \approx 5.70$ bits.

Exercise Four: For azeotropes, the liquid and vapor compositions are identical. The angle between information vectors is zero radians or degrees. Water/Ethanol azeotropes have mole fractions approximately 0.05/0.95.

Exercise Seven: Look ahead to Chapter Seven for the counting procedure. The information gain is:

$$\log_2(105) \approx 6.7 \text{ bits.}$$

The assumption is that all configurations are regarded as equally plausible. This pushes the information to the maximum possible.

Exercise Eleven: The angle between the vectors is ~22.3°.

CHAPTER THREE EXERCISES

Exercise One: Proceed methodically by writing MGC2GCSEGCGSGCG2C... The base structure is $A_1V_6I_1P_8M_1G_{37}S_{35}C_{59}Y_1N_1Q_6E_1K_8$.

Exercise Two: The answer is definitively yes. The base underpins truly multiple isomers.

Exercise Three: The base structure is given by

$A_{17}V_{20}L_{31}I_{14}P_{14}F_{10}W_5M_7G_{19}S_{20}T_{21}C_2Y_{19}N_{19}Q_{18}D_{13}E_{16}K_{10}R_{15}H_{18}$

Use the binomial formula and the Stirling approximation to estimate the surprisal for the alanine/not alanine content at ~3.5 bits. The most selected component is the most rarefied: cysteine. The least selected is the most numerous: L. The total surprisal is ~101 bits. Note how the surprisal increases with the size of the protein.

Exercise Seven: For 20-component information vectors, the angle is ~17°. For family-term vectors, the angle is ~3.3°.

Exercise Nine: Consider the overlap of families. The respective probabilities are 3/8 and 4/8. The answers change if we have knowledge of the protein sequence or base structure. Probabilities always depend on our state of knowledge prior to an experiment.

Exercise Twelve: The literature reports over one hundred amino acids from post-genomic processing. The number of families exceeds 2^{20+100}. Lots of families!

Exercise Twenty: The cytochrome steers closest, just ahead of the prion. Heat shock brings up the rear. That said, all three proteins hew closely to canonical selection weights.

CHAPTER FOUR EXERCISES

Exercise One: The Burrows–Wheeler transform is RALECREFVK. Completing this exercise need not entail computer programming. The transform offers no savings in code as should be clear.

Exercise Eight: The possible keys number 20! At the rate mentioned, it will take close to a billion centuries to sample the possibilities. This will not happen.

CHAPTER FIVE EXERCISES

Exercise One: Use the multinomial probability to find the surprisal to be 534 bits.

Exercise Two: The information computes as 1.56 bits.

Exercise Three: The geometric mean for a set of real numbers $\{X_1, X_2, ..., X_N\}$ is given by

$$\left[X_1 \cdot X_2 \cdots X_N \right]^{1/N}$$

The volume of the set is closely related as it omits the Nth root operation.

Exercise Five: It is straightforward to construct protein sets of maximum volume. For example, polyalanine, polyglycine, and polyproline express orthogonal information vectors. Obviously, such systems are not realistic for proteomes, organism or viral.

CHAPTER SIX EXERCISES

Exercise One: The base+ structures in ABA terms are $(C–C)_{10}(C–H)_8$, $(C=C)_4(C–C)_4(C–H)_8$, and $(C=C)_2(C–C)_4(C–O)_2(C–H)_{10}$.

Exercise Two: The angle between base+(2) information vectors is ~72.9°. The angle for base+(3) vectors is close to 90°. We don't have to go to very high base+ orders to obtain orthogonal information vectors for two proteins.

Exercise Three: This concerns the base+ interval information for myoglobin. Cysteine is conspicuous by its absence. Histidine is prominent by its abundance.

CHAPTER SEVEN EXERCISES

Exercise One: There are eight paths from one site to the other.
  Exercise Two: The northern route is shorter by one step.

CHAPTER EIGHT EXERCISES

Exercise One: The points per square Angstrom exceed 10^{140}. The sequence space for proteins is imponderably large.
  Exercise Three: For openers, diversify the run of serine units and ditto for the A units.

CHAPTER NINE EXERCISES

Exercise Two: $Z_{max} \approx 74$ for 4,000 proteins. It would exceed 100 for 20,000 proteins. Z_{max} is severely diminished for random proteins as their statistical structures are too close together.

Index